T0270999

Modelling and Optimization of Wave Energy Converters

Wave energy offers a promising renewable energy source, however, technologies converting wave energy into useful electricity face many design challenges. This guide presents numerical modelling and optimization methods for the development of wave energy converter technologies, from principles to applications. It covers the development status and perspectives of wave energy converter systems; the fundamental theories on wave power absorption; the modern wave energy converter concepts including oscillating bodies in single and multiple degree of freedom and oscillating water column technologies; and the relatively hitherto unexplored topic of wave energy harvesting farms. It can be used as a specialist student textbook as well as a reference book for the design of wave energy harvesting systems, across a broad range of disciplines, including renewable energy, marine engineering, infrastructure engineering, hydrodynamics, ocean science, and mechatronics engineering

Modelling and Optimization of Wave Energy Converters

Edited by
Dezhi Ning
Boyin Ding

CRC Press
Taylor & Francis Group
Boca Raton London New York

CRC Press is an imprint of the
Taylor & Francis Group, an **informa** business

First edition published 2023
by CRC Press
6000 Broken Sound Parkway NW, Suite 300, Boca Raton, FL 33487-2742

and by CRC Press
4 Park Square, Milton Park, Abingdon, Oxon, OX14 4RN

CRC Press is an imprint of Taylor & Francis Group, LLC

ISBN: 978-1-032-05739-2 (hbk)
ISBN: 978-1-032-05740-8 (pbk)
ISBN: 978-1-003-19895-6 (ebk)

DOI: 10.1201/9781003198956

Typeset in Latin Modern font
by KnowledgeWorks Global Ltd.

Publisher's note: This book has been prepared from camera-ready copy provided by the authors.

Contents

Preface

Wave energy harvesting is an emerging technology that has been generating serious interest and research development as an alternative renewable energy source. There are currently more than 200 wave energy conversion devices in various stages of testing and demonstration, resulting in numerous industrial and academic endeavours, knowledgebases, and academic literature since the oil crisis. Whilst there already exist a handful of textbooks on the topic of wave energy harvesting, a textbook that systematically guides researchers and engineers into this interesting area is lacking, to the best understanding of the editors of this book.

This book, *Modelling and Optimization of Wave Energy Converters*, presents the fundamental and contemporary state of the art modelling and optimisation approaches that are essential for the design of wave energy converter technologies, drawing on the insights and experience of globally-recognised experts in their various relevant fields. The book content is carefully organised to lead the readership gradually into the story of wave energy harvesting in terms of technology evolutions, economics, development trends, fundamental theories, modern concepts, and cutting-edge methodologies, as supported by over 800 of the most relevant articles in the literature and 6 typical case studies. Not only the breadth and depth of the content but also the applied pyramid structure in presentation make this book unique and a must-read for newcomers to the field of wave energy harvesting.

This book consists of four parts. Section I reviews the history and current status of wave energy converter development, followed by discussions of the future perspectives of the field. Section II introduces the important fundamental theories applicable to wave energy harvesting, such as fluid dynamics, wave-structure interaction solvers, wave energy converter modelling techniques, principles of wave power absorption, and control system and power take-off design. Section III investigates the modelling and performance optimisation of various representative modern wave energy converter technologies, with their theoretical basis thoroughly analysed and their applications examined using case studies that readers are able to replicate. Section IV expands the topics to the modelling and optimization of wave energy converter farms, as a commonly-agreed trend towards the commercialisation of wave energy harvesting.

This book was built upon thousands of hours of effort on the part of the authorship team. Therefore, acknowledgement firstly goes to all the authors of the book, most of whom are in the transition periods of their careers and personal lives but still prioritised this book project towards its successful accomplishment. The authors' endeavours were accompanied by endless support from their families, who are equally acknowledged. Sincere acknowledgement also goes to the editors' affiliations: Dalian University of Technology and the University of Adelaide, for their joint financial support to make this book open-access, benefiting the whole wave energy development community. Alison-Jane Hunter is acknowledged for thoroughly proof reading the book towards its final appearance. Last but not least, huge acknowledgement goes to Taylor & Francis Group for reviewing and publishing the book to expose it to the world's spotlight.

Dezhi Ning   Boyin Ding
Dalian, China   Adelaide, Australia

September 2021

Contributors

In alphabetical order

Bingyong Guo
Northwestern Polytechnical University
Xi'an, China
b.guo@nwpu.edu.cn

Boyin Ding
University of Adelaide
Adelaide, Australia
boyin.ding@adelaide.edu.au

Dezhi Ning
Dalian University of Technology
Dalian, China
dzning@dlut.edu.cn

Leandro Souza Pinheiro da Silva
University of Adelaide
Adelaide, Australia
leandro.dasilva@adelaide.edu.au

Malin Göteman
Uppsala University
Uppsala, Sweden
malin.goteman@angstrom.uu.se

Nataliia Y. Sergiienko
University of Adelaide
Adelaide, Australia
nataliia.sergiienko@adelaide.edu.au

Robert Mayon
Dalian University of Technology
Dalian, China
rmayon@dlut.edu.cn

Rongquan Wang
Dalian University of Technology
Dalian, China
rqwang@dlut.edu.cn

Siming Zheng
University of Plymouth
Plymouth, UK
siming.zheng@plymouth.ac.uk

Yingyi Liu
Kyushu University
Fukuoka, Japan
liuyingyi@riam.kyushu-u.ac.jp

Acronyms

AEP	annual energy production
ALE	arbitrary Lagrangian-Eulerian
AMI	arbitrary mesh interface
ACE	average climate capture width per characteristic capital expenditure
BEM	boundary element method
CAPEX	capital expenditure
CW	capture width
CFD	computational fluid dynamics
CMA	covariant matrix adaptation
DEL	damage equivalent loading
DoF	degree of freedom
DE	differential evolution
EMEC	European Marine Energy Center
F2M	force-to-motion
FD	frequency domain
GA	genetic algorithm
GUI	graphical user interface
HPC	high performance computing
LCOE	levelised cost of energy
MPI	message passing interface
NPV	net present value
OPEX	operational expenditure
OBS	oscillating body system
OWSC	oscillating wave surge converter
OWC	oscillating water column
PA	point absorber
PBP	payback period
PMLG	permanent magnate linear generator
PMSM	permanent magnate synchronous machine
PTO	power take-off

RANS	Reynolds-averaged Navier-Stokes
RCW	relative capture width
SL	statistical linearisation
TRL	technology readiness level
TD	time domain
USD	United States dollar
W2M	wave-to-motion
W2W	wave-to-wire
VOF	volume of fluid
WEC	wave energy converter
WG	wave gauge
WSI	wave-structure interaction

Nomenclature

$\dot{x}$ Time derivative $\dot{x} = \frac{\mathrm{d}x}{\mathrm{d}t}$

ϵ Wave steepness

$\eta(x,y,t)$ Surface elevation

η_{eff} Hydrodynamic efficiency (wave energy capture width ratio/capture factor)

$\eta_{\mathrm{eff,H \to E}}$ Power conversion efficiency (wave-to-electrical)

$\eta_{\mathrm{eff,M \to E}}$ Power conversion efficiency (mechanical-to-electrical)

λ Wavelength

$\mathbf{B_{eq}}$ Equivalent damping matrix

$\mathbf{B}$ Damping matrix

$\mathbf{F}_c$ Control force

$\mathbf{F}_e$ Wave excitation force and moment

$\mathbf{F}_f$ Friction force

$\mathbf{F}_H$ Horizontal wave force

$\mathbf{F}_m$ Mooring force

$\mathbf{F}_r$ Wave radiation force and moment

$\mathbf{F}_s$ Hydrostatic buoyancy force

$\mathbf{F}_V$ Vertical wave force

$\mathbf{F}_v$ Viscous force

$\mathbf{F}_{ext}$ External net force acting on the fluid

$\mathbf{F}_{pto}$ Power take-off force

$\mathbf{K_c}$ Keulegan-Carpenter number

$\mathbf{K_s}$ Hydrostatic stiffness matrix

$\mathbf{K_{eq}}$ Equivalent stiffness matrix

K   Stiffness matrix

$\mathbf{M_{eq}}$ Equivalent inertia matrix

M   Inertia matrix

$\mathbf{S}_\eta$ Wave spectrum

$\mathbf{S_f}$ Force/moment spectrum

$\mathbf{S_x}$ Response spectrum

∇ Nabla operator, in Cartesian coordinates $\nabla = \left(\frac{\partial}{\partial x}, \frac{\partial}{\partial y}, \frac{\partial}{\partial z} \right)$

ω Angular frequency $\omega = 2\pi/T$

$\overline{P}$ Average power

$\phi(\mathbf{x},\omega)$ Fluid velocity potential in the frequency domain

$\Phi(\mathbf{x},t)$ Fluid velocity potential $\mathbf{u} = \nabla\Phi$

Φ_I Velocity potential of the incident waves

Φ_P Velocity potential of the perturbed waves

Φ_R Velocity potential of the radiated waves

Φ_S Velocity potential of the scattered waves

ρ Density

$\mathbf{u}$ Fluid velocity

$\mathbf{x} = (x,y,z)$ Cartesian coordinates

A_i Incident wave amplitude

A_m Frequency-dependent added mass

B_{pto} Power take-off damping coefficient

B_{rad} Frequency-dependent radiation damping

C_g Wave group velocity

C_l Geometrical similar scale factor

E Wave energy available

g Gravitational acceleration constant

H Wave height

h Water depth

H_s Significant wave height

J Time-average incoming wave power per unit width of the wave front

k Wave number $k = 2\pi/\lambda$

K_{pto} Power take-off spring coefficient

m Mass of buoy

m_n n^{th} spectral moment

p Pressure

p_{air} Air pressure

p_{atm} Atmospheric pressure

p_{dyn} Hydrodynamic pressure

T Wave period

T_e Wave energy period

V Volume of buoy

Z_i Frequency-dependent system intrinsic impedance

Z_{pto} Frequency-dependent power take-off impedance

U Complex amplitude of the buoy heave velocity

I

Introduction

CHAPTER 1

Wave energy converter systems – status and perspectives

Robert Mayon[1], Dezhi Ning[1], Boyin Ding[2],
Nataliia Y. Sergiienko[2]
[1]Dalian University of Technology, rmayon@dlut.edu.cn,
[2]University of Adelaide, boyin.ding@adelaide.edu.au

1.1 INTRODUCTION TO WAVE ENERGY CONVERTER SYSTEMS

This chapter presents an overview for the material contained within the subsequent chapters. For the reader who is unfamiliar with the topic of wave energy harvesting or wave energy converter (WEC) systems, this section can serve as a standalone introduction to the field. A background to wave energy is first presented, its importance and the motivations for its continued and future development are discussed. This is followed by a brief history of the progressive evolution of WEC technology. As with all areas of research, it is important to have a familiarity with the timeline of the development of the technology in the respective field; therefore, a brief synopsis of the historical advancements in wave energy converter technology development from the first inception of such devices up to the present time is introduced. We classify the various WEC technologies according to their operating principles, followed by sub-classification based on their deployment and mode of operation. A summary of the most prevalent contemporary WECs and their operating principals is then presented. It is impossible to describe all of the various WEC technologies that have been developed due to the vast number of different designs; however, some of more notable devices will be discussed. These technologies include the various oscillating body systems (OBSs), oscillating water columns (OWCs), wave overtopping converters, and pressure differential devices [38]. The introduction goes on to present some of the economic characteristics and technological factors that are driving the developments in the wave energy sector. The latest advancements in wave energy harvesting

DOI: 10.1201/9781003198956-1

research, for example the integration of such WEC devices with shoreline defence systems and the current research into techniques for the design of more economically viable installations, are discussed. The methods to improve the efficiency of such devices, including the integration of these technologies with complimentary offshore structures such as floating wind turbine platforms, are examined. The introduction section reviews the future directions and technological progressions of wave energy converter systems, including the concept of WEC array fields, which have the potential to be a major turning point in the provision of commercialised, grid-connected WEC technologies in the future.

1.1.1 The origins of wave energy technology development

The concept of using ocean waves as a resource to benefit society is not a particularly novel idea. As far back as 1799, the first patent for a wave-powered device to drive sawmill machinery was registered by Pierre-Simon Girard in France [63] (see Figure 1.1). It consisted of a levered mechanism that was attached to ships moored in a harbour. As the ships rose and fell with wave action, a lever turned on a fulcrum to power onshore machinery. In the 1880s, a patent was awarded to J.M. Courtney of New York for his design of a whistling buoy [344]. This device consisted of a hollow cylindrical column that was partially vertically submerged, such that it trapped a pocket of air at the top. As the waves interacted with the device, air was expelled and inhaled through a hole in the end cap to power a power take-off (PTO) system, which created a whistling noise. This device was used as a navigational aid to ships.

The first documented use of wave energy being transformed into electrical energy occurred in 1909 when a wave-powered system was used to generate electricity for harbour lighting in California [785]. At this time, California was the hub of wave energy experimentation. There were a number of companies attempting to commercialise wave energy through the construction of *wave motors*. The California Wave Power Company [554] was one such company that used waves to generate electricity, which was then used to power small equipment along the pier. The Starr Wave Motor was an ambitious project that commenced construction in California during 1907, with the intent to supply electricity for six surrounding counties [476]. However, the project came to a premature end when the pier upon which the equipment was constructed collapsed.

The first recognised OWC design to power onshore equipment was developed and built around 1910 by M. Bochaux-Praceique near Bordeaux in France [390] (see Figure 1.2). It consisted of a vertical borehole shaft tunnelled into a cliff top, with a horizontal shaft emanating from the vertical shaft through the cliff face, beneath the low tide water mark. The vertical shaft contained a column of water that oscillated vertically as waves surged against the cliff face, thus driving a rudimentary turbine.

The first successful attempt at using wave power to produce electricity on a larger scale took place in the 1940s. In 1947, Yoshio Masuda, a commander in the Japanese Navy, developed an OWC navigational buoy that powered a turbine to generate electricity [503]. Later on, in the 1970s, Masuda was involved in the development of what is now regarded as the first commercial wave energy device, the Kaimei, a

349.

12 juillet 1799.

BREVET D'INVENTION DE QÙINZE ANS,

Pour divers moyens d'employer les vagues de la mer, comme moteurs,

Aux sieurs GIRARD père et fils, de Paris.

————————

LA mobilité et l'inégalité successive des vagues, après s'être élevées comme des montagnes, s'affaissent l'instant après, entraînant dans leurs mouvemens tous les corps qui surnagent, quels que soient leur poids et leur volume. La masse énorme d'un vaisseau de ligne, qu'aucune puissance connue ne serait capable de soulever, obéit cependant au moindre mouvement de l'onde. Qu'on suppose un instant, par la pensée, ce vaisseau suspendu à l'extrémité d'un levier, et l'on concevra l'idée de la plus puissante machine qui ait jamais existé.

C'est principalement sur ce mouvement d'ascension et d'abaissement des vagues, qu'est fondée la théorie des nouvelles machines que nous proposons.

L'application en est aussi simple que l'idée première. Nous avons imaginé plusieurs moyens d'utiliser cette force; mais le moins compliqué de tous consiste à adapter ou à suspendre à l'extrémité

13 *

Figure 1.1: The patent awarded to Pierre-Simon Girard in 1799 for his wave energy converter device.

floating barge which incorporated a number of OWC chambers [545]. Since the 1970s, interest in wave energy as a viable commercial resource has been increasing [644], and the sector has really gathered pace since the last decade of the 20th century. Notwithstanding the fact that the development of wave energy lags behind other forms of renewable energy such as wind and solar power, research and development work in this sector continues apace and progress on new devices with augmented efficiencies is continuously being realised. Whilst wave energy is still regarded as being in the developmental phase, it is moving fast along the technology readiness level (TRL) scale towards being a commercially viable technology [477] (see Section 1.2.2.3).

1.1.2 Classifications of wave energy converter technologies

There have been a number of different technologies developed for harvesting wave energy and WEC devices can be classified according to many different metrics. For example, WECs can be categorised depending on their operating principle, their

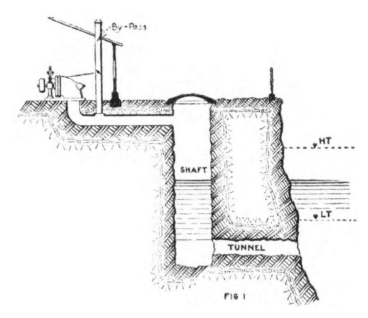

Figure 1.2: An early OWC type device constructed by M. Bochaux-Praceique in 1910, From the magazine *Power*, November 1920.

orientation, their power take off system, their application, etc. [698]. The topic of classifying WECs is complex and this further compounded by the fact that many devices can be assigned to two or more different categories within each of the different classification approaches. Wave energy harvesting is a rapidly evolving field, and great progress is being made in the development of the various technologies to convert wave energy into electrical energy. Increasingly, there are new technology concepts being developed and novel approaches to harvesting wave energy are emerging. Since 2015, it is surmised that there are in excess 1000 wave energy concepts patented each year globally, and many thousands of existing patents have been already registered [368]. There are numerous devices in various stages of development and several full scale plants are either already operational or in the testing phase. As a consequence of this ever-expanding technology development, there are new classification techniques being continuously established. In this section, some of these earlier classification taxonomies are explored in chronological order.

Classification by Orientation, Budal and Falnes, 1975. A well known system to classify WEC technologies depends upon the devices' relative dimensions and orientation to the propagating waves. This method is known as classification by orientation. There are four main categories in this system; terminators, attenuators, point absorbers, and quasi-point absorbers, (see Figure 1.3) [100, 229, 698]. The WECs in each of these categories can be either floating or submerged types of devices. Each of the four types of WEC according to classification by orientation are listed below with their commonly accepted definitions.

- a *terminator* device operates perpendicular to the wave propagation direction (e.g., overtopping devices or oscillating wave surge converter);

- an *attenuator* is oriented parallel to the wave direction and its length is greater than the length of a dominant wave;

- a *point absorber* has dimensions significantly smaller than a wavelength and can absorb power regardless of the direction of wave propagation [100];

- the term *quasi-point absorber* was introduced by [229] in order to describe axisymmetric WECs that are insensitive to the wave direction (similar to point absorbers), but have relatively large dimensions compared with the wavelength (similar to terminators).

The origins of this classification system, and indeed one of the first mention of a point absorber with reference to a WEC, can be traced back to an article written by Budal and Falnes in 1975, [100]. In that paper, the authors described a point absorber as a *system in which the horizontal extent is much smaller than one wavelength*. The authors, Budal and Falnes, also ambiguously alluded to a terminator, or an attenuator type device which they termed a *linear absorber*. They described it as a system that is, *made as a straight construction, at least a few wavelengths long*. In describing the *linear absorber*, the authors reference a study by Salter, published a year earlier, in which he first describes his Edinburgh Duck device, and mentions an array of such WECs being connected by a *common back bone for about 40 vanes* [686]. Since Salter's WEC is generally agreed, nowadays, to be a terminator type device, it can therefore be inferred that Budal and Falnes, in their paper referencing the Edinburgh Duck, were implying that such a *linear absorber* would be a terminator type WEC, without explicitly using that term. Figure 1.3 presents a schematic for the orientation of each of these WEC devices in relation to the incident waves according to this taxonomy of classification by orientation.

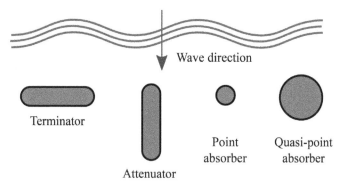

Figure 1.3: WEC classification by orientation [698].

First classification by WEC concepts, Hagerman and Heller, 1988. In the 1970s and 1980s a number of novel design concepts for wave energy converters and technology improvements were developed. Indeed, up to the the early 1990s there were in excess of 1000 patents for a diverse range of WEC devices. Recognising this rapid growth in various WEC concept designs, Hagerman and Heller [324] presented a WEC classification system in 1988 (see Figure 1.4). This was one of the earliest multi-device

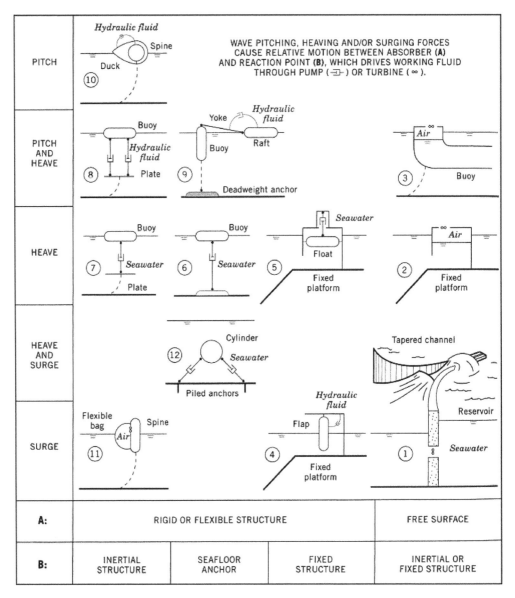

Figure 1.4: WEC classification according to Hagerman, 1988, also appearing in [323].

classification systems and the authors presented specific schematic representations for devices in each category. In their classification approach, the authors categorised the devices according to the mode of energy absorption, i.e., the displacement motion, the type of absorber, and the type of reaction point for the anchoring system. They also made a distinction between the type of working fluid which powered the device's PTO system. The classification matrix did not incorporate a category for those fixed structures such as OWCs, instead, Hagerman and Heller proposed that they belong to heaving type devices. Additionally, the classification system did not incorporate a specific category for pressure differential devices as there had not been a prototype device constructed at that time.

It can be argued that there is an inconsistency in the classification according to mode of energy absorption which Hagerman and Heller adopted. In their system it is not explicitly stated whether the mode of energy absorption, i.e., the degree-of-freedom motion applies to the fluid or to the device. For example, the Edinburgh Duck developed by Salter [686] is included in the pitching category. It is well established that this device displaces with a pitching motion due mainly to the wave surging action against the device, indeed, Salter, in his 1974 paper [686] states *The first step is to get away from the idea of an object bobbing up and down, although, of course, it is this aspect of wave motion which is most apparent. Use of the to and fro movement would be much more rewarding.* Therefore, in this case, the pitching classification seems to apply to the device motion rather than the wave action. The OWC is classed as a heaving type WEC; however, in this case the heaving action is associated with the piston-like motion of the fluid within the chamber. One may claim that this argument over whether the device is heaving, surging or pitching or if the motion designation is applied to the fluid displacement is merely semantics, however it highlights the difficulty in establishing a commonly accepted classification system.

Hagerman and Heller [324] also categorise the devices according to the type of absorber; fabricated structure (rigid or flexible), or free surface of the water. According to their classification, the free surface of the water category comprises two types of device, the overtopping type WEC and the OWC. The rigid or flexible structure type of absorber constitutes all of the other oscillating body type devices. Again, this category distinction may be too simplistic as it can be claimed that the free-surface motion of the sea-water causes the floating buoy to displace. Likewise, OWCs may be considered as fabricated rigid structures.

The final categorisation which Hagerman and Heller [324] apply to the WEC devices is to segregate them according to their fixing or mooring system. In this attempt, they create four different groupings; inertial structure, seafloor anchor, fixed structure, and inertial or fixed structure. From the image of their classification system displaying the various WEC devices, (see Figure 1.4), it appears that the inertial structures are slack line or catenary moored devices (either directly moored to the seabed or moored to a suspended plate, which is, in-turn moored to the seabed) and the seafloor anchored WECs are a various form of taut-leg moored devices. The fixed structures are some form of oscillating body displacing in either a heaving or surging motion. The final grouping in this classification system is the inertial or fixed structure, which comprises the various OWC WECs and overtopping devices.

This was one of the first attempts to present an all-encompassing categorisation of the various WEC devices concepts available at that time; and, indeed it was a commendable effort. However, this classification effort also highlights the difficulties with developing an established taxonomy system for all the various WEC device concepts. Indeed, since the classification system of Hagerman and Heller was developed, there have been many hundreds more various WEC concepts developed, many of which can fit into multiple groupings according to Hagerman and Heller's system [324].

Classification by WEC concepts, Falnes and Løvseth, 1991. Another comprehensive effort to classify the various WEC systems was presented by Falnes and

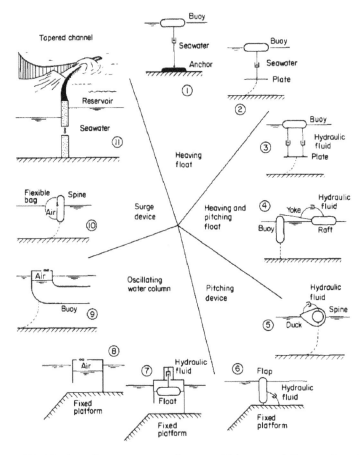

Figure 1.5: WEC classification according to Falnes and Løvseth, 1991, [230].

Løvseth in 1991 [230]. According to this method, the various WEC devices or concepts can be divided into five separate groups. These categories consisted of: heaving float devices, heaving and pitching float devices, pitching devices, oscillating water columns, and surge devices as shown in Figure 1.5. In creating this classification system, Falnes and Løvseth drew upon the earlier work of Hagerman and Heller [324] and attempted to present a clearer and better defined categorisation. They eliminated the heave and surge WEC group, and introduced a separate class for OWC type devices. The tapered channel device was still designated as a surge device similar to the manner in which Hagerman and Heller presented this device in their classification. As in the earlier system of Hagerman and Heller, it may have been assigned to this category as the waves surge over the front parapet wall into the reservoir. However, in contemporary classifications systems this type of tapered channel or overtopping device is usually assigned to a separate category as a overtopping device. The OWC type WEC was assigned to a category by itself, as is the common approach in many of the present day classification systems. Noticeably, there was still a lack of category for pressure differential systems. The most well known pressure differential WEC, the Archimedes Wave Swing, still had not been developed. This device, patented in 1993, was developed two full years after this classification system of Falnes and Løvseth was established [230]. This highlights the fluidity of such classification systems for new and evolving technology sectors; and illustrates the requirement for new categories to be added, existing categories amended, and device reclassification according to technology development.

Oscillating body devices were subdivided into three different categories; Heaving float, heaving and pitching float, and pitching devices. In the pitching type devices, Falnes and Løvseth highlighted two specific technologies. The first device was Salter's Edinburgh Duck, and the second devices was a bottom-hinged flap type device. The pitching Duck undoubtedly belongs to this pitching classification group, nevertheless, in this classificatio system it straddles the two groupings of pitching devices and heaving and pitching float devices. The bottom-hinged pitching-flap type device included in Falnes and Løvseth's taxonomy did not feature in the earlier classification system of Hagerman and Heller, instead they presented a top-hinged suspended flap type device which they asserted was a surging type of device. There is a present-day lack of consensus on the terminology for the flap device. Some researchers refer to this as a bottom hinged flap which operates in pitch mode; another commonly used nomenclature for this type of device is an Oscillating Wave Surge Converter (OWSC), which implies that the energy is generated by wave surging motion. Clearly, the device is mobilised by the surging action of the wave to displace in a pitching motion. This highlights another difficulty with the classification system – should the devices be classified according to their own motion or according the motion of the displacing force? There appears to be some disagreement on this point in the literature.

In the classification system of Falnes and Løvseth [230], the heaving float category presents two devices, each consisting of floating buoys. The first device is bottom tethered and second device appears to to be connected to a weighted plate which is, itself, slack-line-moored. The final category in this classification system is the heaving and pitching float, which includes two device designs, the first design is a floating buoy which is doubly tethered to a submerged plate, which is in turn, slack-line-moored to the seabed. The second design consists of a convoluted, coupled, floating buoy-raft system with the two components connected by means of a yoke. The buoy is moored to the seabed.

Falnes and Løvseth's classification system provided some clarity between the various WEC designs and their group designations, especially with the introduction of the additional OWC category; however, there still remained some ambiguity regarding a few of the devices, and which category they belong to. Clearly an alternative classification system was required.

Classification by working principles, Falcão, 2010. During the 1990s a decline in wave energy device development occurred, which may be attributed to the worldwide decline in oil prices during that time. The growth in the number of new devices and WEC design concepts being created decreased, and progress in the WEC sector slowed. Consequently, the existing classification systems persisted until there was an uptake in the development of new devices in the WEC sector at the end of the first decade of the 21st century. In 2010 a new classification system based on the operating principles of WEC devices was advanced by Falcão [24] (see Figure 1.6).

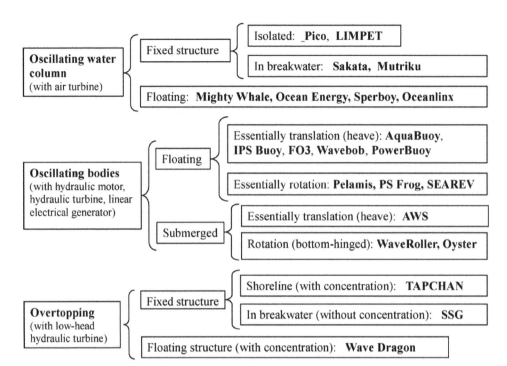

Figure 1.6: WEC classification according to Falcão, 2010 [24].

According to this new taxonomy, there were three main types of WEC devices; oscillating water columns, oscillating bodies, and overtopping type WECs. This was a deviation from the earlier and more convoluted classification systems of Hagerman and Heller [324, 323] and Falnes and Løvseth [230]. Each of the three groupings were further subdivided according to whether they were fixed, floating, or submerged WECs. By creating these subdivisions, Falcão, in effect, combined the device application (onshore, nearshore, offshore) and reaction source (i.e., their mooring or fixing mechanism) classifications.

According to this system, oscillating water columns that are fixed, are onshore or near shore devices. OWCs that are floating are either offshore or nearshore, and should be seafloor anchored. Oscillating bodies are separated into two categories; floating and submerged. Those devices that are considered as floating are deployed in either offshore or nearshore sites, and should be slack-line or catenary-moored. Furthermore, they can be considered as inertial structures. The floating, oscillating bodies were further subdivided into devices which predominantly operate in heave motion or the bodies that operate with rotation motion. The submerged oscillating bodies were subdivided into heaving type devices and rotating devices and both of these devices sub-categories are either taut-line moored or bottom fixed structures which should be sited in nearshore locations.

The final grouping of devices according to Falcão's system [24] are overtopping type devices. These are subdivided according to fixed structures or floating structures. The fixed structures can be positioned in onshore or near shore locations, such

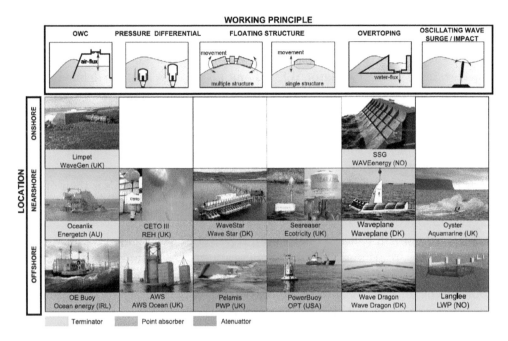

Figure 1.7: WEC classification according to Lopez et al., 2013 [471].

as integrated with a breakwater wall. The floating type overtopping device can be deployed to offshore or nearshore locations and can be either, catenary, slack-line, or taut-line moored. Falcão also presented specific examples of devices in each category of his classification system. This was the first time that a pressure differential device, i.e., the Archimedes Wave Swing, had explicitly appeared in the classification system, however, Falcão chose to integrate this device into the heaving, submerged oscillating body category.

Classification by application, orientation and operating principles, Lopez et al., 2013. Lopez et al., [471] presented an updated classification system for WECs in 2013 (see Figure 1.7). Their classification metrics were: application (referred to as location in their paper), orientation (Lopez et al., used the terms, device size, and directional wave characteristics), and operating principle. The location metric was divided into onshore, nearshore, and offshore. Onshore devices included those WECs which were bottom fixed, in shallow water of less than 10 meters depth. Breakwater integrated devices were also classed as onshore devices. Nearshore devices were those WECs that were situated within a few hundred meters from the shore, and were in waters of less than approximately 25 meters depth. Offshore devices were those WECs that were deployed in deep waters of more than 40 meters depth and were floating or submerged devices, moored to the seabed.

The second characteristic which Lopez et al. used to classify WECs was devices' size, and direction in relation to the incident wave. This metric was subdivided into three groupings: attenuator, point absorber, and terminator. Other authors have, in the past, referred to this method as classification by orientation. According to Lopez et al., attenuator type devices *are long structures compared to the incident wavelength and are placed parallel with respect to the wave direction.* The authors state that, they are usually constructed as *a series of cylindrical bodies that are linked together by a multi-degree-of-freedom, hinged joint,* [471]. Lopez et al. cite the Pelamis as a classical example of an attenuator WEC. Point absorber WECs are defined by Lopez et al., as devices that are significantly smaller than the incident wavelength and are capable of harvesting wave energy in multi-directional sea-states. They operate in heaving or pitching mode. According to Lopez et al., terminator devices *are similar to attenuators; they are long devices compared to the wave length but are orientated perpendicularly to the incident wave direction.*

The final metric which Lopez et al., [471] use to categorise WECs is their operating principle. Here, they follow a similar approach to Falcão [24]; they specify separate categories for OWCs, overtopping devices and floating structures (generally analogous to oscillating bodies in Falcão's taxonomy). However Lopez et al., also introduced two further groupings, pressure differential devices appear as a group for the first time and oscillating wave surge converters appear as an independent category. There seems to be some discrepancy between the classification system image the authors present and the text description of the classification system they present. In their classification matrix, shown here as Figure 1.7, OWCs and pressure differential devices are assigned to separate categories, however in the text, the authors claim that OWCs are a form of pressure differential device. This assertion is at odds with the commonly accepted, contemporary classification approaches which designate OWCs as an independent class of device based on operating principle. The authors also label the CETO 3 WEC as a pressure differential device, however this is a submerged point absorber, which relies on the orbital motion of the wave to displace it. The WaveStar device is labelled as a point absorber; however, the future-proposed design of the device, with its many floating buoys in close proximity to each other, could designated it to be classed as a terminator or attenuator depending on its orientation to the incident waves.

Classification by operating principle, orientation, PTO, and application, IRENA, 2014. The International renewable Energy Agency (IRENA) published a report detailing the various ways WECs can be classified in 2014 (see Figure 1.8(a)) [548]. One classification method according to IRENA is based on the operating principle, and using this approach, WECs can be grouped according to OWCs, oscillating bodies, and overtopping devices. Each of these groups can be further subdivided according to their reaction source, i.e., the manner in which the device is supported or moored. This system of classification is very similar to that of Falcão [24]. The second manner in which devices can be classified is according to their orientation. Using this method, WECs can be attenuator devices, terminator devices, or point absorbers. In the IRENA report, the authors also present a novel classification approach in which the device is categorised according to its PTO system. Correspondingly, there are four different groups in this classification approach, WECs which employ a pneumatic PTO system, those that use a hydro-driven PTO system, ones that use a hydraulic PTO, and those that employ a direct-drive PTO mechanism. These PTO systems are usually device specific; for example, the PTO system in an OWC is usually pneumatically driven, the turbine in an overtopping device relies on hydro power, hydraulic

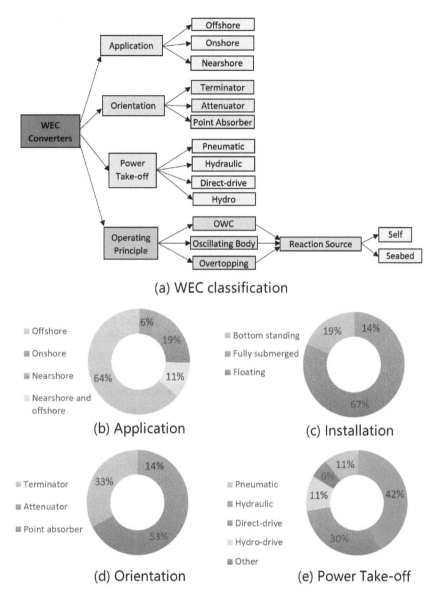

(a) WEC classification

(b) Application

(c) Installation

(d) Orientation

(e) Power Take-off

Figure 1.8: WEC classification according to IRENA, 2014 [548].

systems are used in attenuator devices such as the Pelamis WEC, and direct drive systems are usually employed in oscillating body type WECs. The final method by which to classify WECs according to the IRENA report, relies on classification by application. Using this method, WEC devices are categorised according to their deployment site, i.e., whether they are positioned onshore, nearshore, or offshore. The report also presented a percentage breakdown of device types under development by a selection of companies based on the various classification systems as of 2014, (see Figures 1.8(b–e)).

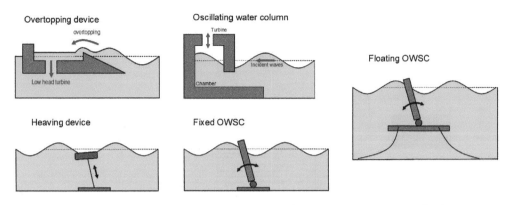

Figure 1.9: WEC classification according to Babarit, 2015 [42]

Classification by working principles, Babarit, 2015. A new classification system was presented by Babarit in 2015 [42] (see Figure 1.9). This taxonomy generally follows that of the earlier system proposed by Falcão [24], in that it presents three major classification groupings. The first group is overtopping devices, the second group is oscillating water columns, and the third group is oscillating bodies. However, Babarit proffers a major deviation from Falcão's system in the manner by which these groupings are further subdivided; he makes no distinction between fixed or floating devices. Babarit also suggests a new approach for the method in which oscillating body WECs are classified. In contrast to Falcão's approach which segregated oscillating bodies according to floating or submerged, Babarit, instead, separates them according the their degree of freedom motion. In so doing, he creates two subcategories of oscillating bodies, heaving devices and oscillating wave surge converters (OWSCs). The rational behind creating these subdivision categories is based upon the hydrodynamic performance and the theoretical maximum capture width discrepancies between these technologies. Babarit states that a surging or pitching oscillating body has a theoretical maximum capture width twice that of a heaving oscillating body based on the far field radiated wave generated by the device's motion. Accordingly, he divides the devices in line with their predominant degree-of-freedom motion direction. Babarit further subdivides the OWSCs into those that are bottom fixed, and those that are fixed to a floating reference, again based upon their performance. Fig. 1.9 shows the five main classification systems according to Babarit's taxonomy. It should be noted that Babarit also acknowledges that there are various other forms of WEC that do not fit neatly into his classification system and also allows for variants of each of the device categories shown in Figure 1.9; then, in total he ends up with 10 different categories. Arguably, he considers articulated body type devices such as Pelamis as variants of heaving oscillating bodies. This categorisation is inconsistent with the generally accepted designation of these devices as pitching or rotating type oscillating bodies or attenuators. Babarit also acknowledges that some of the newer type, second-generation WEC devices such as the Anaconda developed at the University of Southampton or the WEC S3® developed by SBM Offshore do not fit into existing classification systems.

With such emphasis on the WEC efficiency, it can be conjectured that Babarit's classification method is a device-efficiency based taxonomy. This highlights the

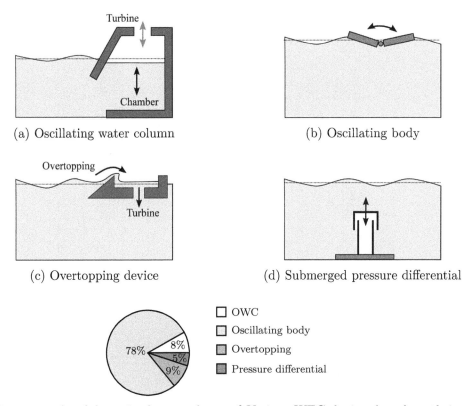

(a) Oscillating water column

(b) Oscillating body

(c) Overtopping device

(d) Submerged pressure differential

□ OWC
□ Oscillating body
▨ Overtopping
■ Pressure differential

(e) Percentage breakdown in the prevalence of Various WEC devices based on their operating principle

Figure 1.10: WEC technology classification according to working principles [698, 548].

ever-evolving nature of the classification system for WEC and as new technology concepts are developed, further categories of WEC based on different metrics may need to be considered.

The following sections describe the main categories of existing WEC technologies, according to their operating principle following Falcão's approach [165]. According to this system, the devices can be broadly grouped into oscillating body systems, oscillating water columns, and wave overtopping devices [38, 165, 698]. The authors add an additional category for pressure differential devices, as there is a growing interest in this technology (see Figure 1.10 for the schematic representation of these devices). With regard to the number of prototype devices in existence, as of 2017, it was estimated that 78% were oscillating body type devices, 8% were OWCs, 9% were overtopping type devices, and 5% were pressure differential WECs [698, 548] (see Figure 1.10). A number of example devices, both historical and contemporary, are examined in the following sections including discussions on the devices' PTO systems, their application, orientation, and reaction source. The classification parameters for some of these devices are also presented in Table 1.1. As highlighted in the preceding paragraphs, it should be noted that some devices can fall into a number of classifications, for example, some oscillating body devices may operate either as attenuators or as terminators, whilst some may be suitable for deployment in nearshore areas, whilst

Table 1.1: Classification of typical wave energy converters.

Application	Orientation	Device	Working principle	Installation	Oscillating mode	PTO	Power rating (kW), Scale	Test sea site
Onshore	Terminator	Mutriku	OWC	Bottom-fixed	-	Pneumatic	296	Bay of Biscay, Sain
		Limpet	OWC	Bottom-fixed	-	Pneumatic	500	Isle of Islay, Scotland
Nearshore	Point absorber	CETO5	Oscillating body heaving	Submerged & seafloor reference	Heave	Hydraulic	240	Graden Island, Australia
		Corpower WEC	Oscillating body heaving	Floating & seafloor reference	Heave	Direct mechanical drive	300	Agucadoura, Portugal
		mWave™	Pressure differential	Bottom-fixed	-	Pneumatic	1500	Wales, UK
	Terminator	UniWave200	OWC	Bottom-fixed	-	Pneumatic	200	King Island, Australia
		Oyster 800	Oscillating body OWSC	Bottom-fixed	Pitch	Hydraulic	800	EMEC, Scotland
		Zhoushan	Oscillating body pitching	Floating & seafloor reference	Heave	Hydraulic	500	Wanshan Island, China
Offshore	Point Absorber	Toftestallen Power Plant	Oscillating body heaving	bottom-fixed	Heave	Pneumatic	500	Toftestallen, Norway
		OE 12 buoy	OWC	Floating	-	Pneumatic	500	Hawaii, USA
	Terminator	Wave Dragon	Overtopping	Floating	-	Hydro turbine	20	Nissum Bredning, Denmark
	Attenuator	Pelamis	Oscillating body articulated	Floating	Pitch & yaw	Hydraulic	750	EMEC, Scotland

others are better suited to offshore environments. The authors therefore acknowledge that their categorisation of certain devices may be discordant with the opinions of other experts in the field of wave energy converter design.

1.1.3 Oscillating water column

One of the most promising technologies for the extraction of wave energy is the oscillating water column (OWC) device (see Figures 1.10a, 7.1a [344]) . This is a simple technology in which the working principle depends on a hollow chamber that is partially submerged below the ocean's free surface. The chamber has an opening below the minimum wave level, typically on the prevailing incident wave side. There are few moving parts, which determines that the technology is less prone to mechanical failure. The chamber traps a column of air above the water free surface, which is expelled, and inhaled through an orifice in the chamber in a cyclic manner. This is induced by the oscillatory action of the rising and falling free surface inside the chamber caused by the impinging incident waves. A directionally independent turbine, such as a Wells turbine, [277, 648] which is connected to the orifice, is thus powered. Energy can then be extracted on both the exhalation and inhalation phases of the cycle. Whilst OWCs are a conceptually simple device and much research has focused on their development, there are two main drawbacks that have beset their commercialisation. The first obstacle is the relatively low wave energy capture efficiency. Many different prototype designs have been proposed to alleviate this deficiency, including ongoing research focusing on optimising the structure's geometry to reduce the hydrodynamic energy losses, such that more energy can be transmitted from the impinging waves to pneumatic power that drives the turbine. The second main drawback associated with OWCs is related to their durability and survivability. As the function of OWCs is to harvest wave energy, it is natural that they should be deployed in regions with high wave energy density; however, this requires that they be situated in harsh environments. There have been numerous documented cases of OWC failures either by progressive deterioration of the structure and machinery [583] or, in some cases, there have been catastrophic failures during storm events [531, 807, 532, 403, 196].

Advances in OWC design is encouraging and their viability as a commercial renewable energy technology is improving. The design of OWCs can generally be categorised into floating offshore devices or fixed devices. The following subsections present additional information on each of these different classifications. Further information on OWCs and some design case studies are presented in Chapter 7.

1.1.3.1 Floating OWCs

The Kaimei was one of the first floating OWC WECs to investigate the production of electricity on a large scale. The device consisted of a large barge constructed in Japan under the guidance of Yoshio Masuda, and had eight OWC chambers mounted, each with a 125 kW rating [545, 425, 505, 504]. Since then, there have been a number of floating OWC WECs developed and deployed. Many of these devices have been proof or concept or prototype installations. These include The Mighty Whale, which entered service in 1998 [818, 359, 601], the Spar buoy which is a hollow cylindrical device that is vertically submerged in the ocean such that it can harvest wave energy [221, 289], and the Oceanlinx, which began testing off the coast of New South Wales, Australia, in 2010 [344]. The Ocean Energy OE12 Buoy, is a floating OWC developed by the Ocean Energy company based in Ireland (see Figure 1.11). The device has undergone a progressive development and testing program, from 1:50 scale tests to 1:4 scale trials over three years at Galway Bay test site and in Cork Harbour, Ireland. The 500 kW, full-scale device commenced testing in Hawaii in 2019. Larger versions of the device are planned, with a 2.5 MW utility scale OE50 buoy in development. In order to augment energy harvesting efficiency, a number of distinct designs have been developed for offshore OWC devices. The conventional design relies on a chamber with the opening orientated in the prevailing incident wave direction. Other designs rely on an opening at the leeward side of the chamber, which has been shown to amplify the wave resonance effects, thereby increasing the energy harvesting efficiency. This type of device is known as a Backward Bent Duct Buoy (BBDB) [506, 780, 560]. Studies have also focused on cylindrical chamber designs which dictate that the performance of the device is directionally independent of the incident waves [879, 714, 579, 585, 878]. A wave-powered navigational buoy developed by the National Institute of Ocean Technology (NIOT) in India and deployed at Kamarajar Port in Chennai is another example of a floating OWC [611, 612]. This device has been operational since 2017 and has proven to be robust in all weather conditions.

Figure 1.11: Ocean Energy OE12 Buoy, (image courtesy of Ocean Energy).

1.1.3.2 Fixed OWCs

Onshore or fixed OWCs have been around for many years [216, 219, 691, 774]. One of the first commercial examples of this technology was constructed in Toftestallen, Norway in 1985 [489]. Other notable early examples were the LIMPET power plant installed on the Isle of Islay of the west coast of Scotland in 2000 [75, 834, 254, 11], the Pico power plant [218, 222, 93, 90] in the Azores, which went into operation in 1999 and the OSPREY WEC, which was deployed off the coast of Scotland in 1995 [534]. The Mutriku power plant in Spain, which commenced operation in 2011, is one of the most famous examples of a land fixed OWC [779, 236, 235, 363]. This installation consists of 16 OWC chambers integrated into a rubble mound breakwater structure. The plant has a total capacity of 296 kW and has provided over 2.1 GWh of electricity to the local village of Mutriku since it became operational. The main advantages of the onshore OWC device are ease of construction and low installation costs. In some cases, the OWCs have been integrated into coastal defence systems to provide a dual functioning structure. The OWC at Mutriku, which is integrated into a breakwater wall, is an excellent example of such a construction. Furthermore, any maintenance works required over the design life duration of the device are made easier by the accessibility of the device. The main disadvantage associated with the onshore OWC is the reduced wave energy density in nearshore areas. However, research has concentrated on increasing the wave power at near shore locations through methods such as wave focusing and optimising the refracted and reflected wave energy at nearby structures [512, 657, 863]. Other studies focusing on optimising the chamber geometry have also been conducted, [179, 513, 485, 464]. Research into the effectiveness of multi-chambered OWCs [582, 869, 578, 582, 197] and the provision of a stepped seabed [580, 665, 663, 664] in the vicinity of the front wall of the OWC are further examples of such work.

Figure 1.12: Mutriku power plant. Image republished under the Creative Commons CC-BY-SA-4.0 licence from [236].

A bottom-standing OWC WEC was constructed near Jeju Island, South Korea, in 2016 [459]. The plant, developed by the Korea Research Institute of Ships and Ocean Engineering (KRISO), is located 1.5 km offshore and has a 500 kW capacity rating. The installation has two axial flow impulse type turbines attached that drive two 250 kW turbines. A 22.9 kV underwater cable connects the OWC to the power grid. This novel power plant has advanced control and maintenance technology installed, allowing it to operate only during the optimum sea states and to monitor continuously for any mechanical disruptions.

Wave Swell Energy installed an OWC WEC off the coast of Tasmania in January 2021 [89, 143]. The device, which was floated into position and ballasted such that it is an near-shore, bottom-mounted OWC, is known as UniWave200 . It has a 200 kW rating and operates with a unidirectional turbine in contrast with most OWC devices, which have bidirectional turbines installed. It is expected that this simpler, unidirectional turbine will be more reliable and prolong the design life of the OWC mechanical components.

Figure 1.13: UniWave200, a 200 kW rated OWC developed by Wave Swell Energy, (image courtesy of Wave Swell Energy).

1.1.4 Oscillating body systems

There are many different forms of oscillating body WECs [86] (also see Figures 1.4, 1.5, 1.6, 1.7, 1.9, and 1.10b). Usually these types of WEC are separated into floating bodies that convert the kinetic energy from their wave-induced motion into electrical energy and submerged bodies that act in the same way [700, 844]. Oscillating body devices can also be categorised according to their orientation to the incident waves, as shown in Figure 1.3. These converters are usually restricted to operating with a single degree of freedom: the devices commonly displace with heave, surge, or pitch motion. Point absorber WECs operating in heave motion are conceptually the simplest type of wave energy harvesting device. In their most basic form, they comprise of bottom-tethered floating buoys that oscillate vertically as the ocean waves interact with them. There are also hinged oscillating body systems with multiple degrees of freedom, which rely on a combination of pitch and surge motions to generate energy. These systems vary greatly in their design, with some devices being extremely large, such as the floating, 120 meter long Pelamis, [853] which is an example of an attenuator type WEC. Another class of oscillating body system is the submerged hinged WEC [125]. This form of submerged device is usually classed as a terminator type WEC, and usually comprises of a bottom-hinged flap that is connected to the seabed with the top of the structure free to move according to the wave direction. The bottom hinged flap can be either fully or mostly submerged, e.g., the WaveRoller WEC, or, designed to protrude above the sea level, e.g., the Oyster WEC. Then, as the waves interact with the flap, it moves back and forth, generating electricity either through a hydraulic PTO system or by driving a motor and thus powering a generator.

1.1.4.1 *Floating oscillating bodies*

Floating oscillating body absorbers are a common type of WEC device. They usually consist of a spherical or cylindrical floating buoy that is connected to a structure that guides the device to move in a prescribed motion. Usually they displace with a single degree-of-freedom such as heaving, pitching, surging, yaw, etc. Point absorbers, a particular type of floating oscillating body as discussed in Section 1.1.2 usually (but not always) displace with a heaving motion. They can be deployed offshore or they can be fixed to some coastline infrastructure such as harbour or breakwater walls. Those free-standing point absorber devices that are deployed offshore must be moored in position and their location necessitates underwater transmission cables to connect to the power grid. The installation and maintenance of these ancillary systems can be costly and complex undertakings. Chapter 4 discusses these point absorber type WECs in greater detail and presents some design case studies. Other floating oscillating body devices displace with a pitching motion, such the Edinburgh duck, or a combination of pitching and yaw motion, such as the Pelamis WEC.

Heaving

This type of WEC usually falls into the orientation category of a point absorber or quasi-point absorber, as presented in Figure 1.3. The Wavebob [825, 638, 824] was a floating point absorber developed in Ireland and patented in 1999; however, due to a lack of funding, the company responsible for developing the technology was discontinued in 2013. Another notable example of a floating point absorber was the AquaBuOY, developed by Finavera Renewables Ocean Energy (now defunct) [827, 808, 469, 438]. This device was comprised of a 3 meters diameter cylindrical buoy attached to a 21 meters vertical shaft. The electricity was generated by a hydraulically-driven turbine, which powered a generator. The PB3 PowerBuoy® (see Figure 1.14) is a WEC developed by Ocean Power Technologies, with an typical average power output of 8.4 kWh/day [87, 609, 377]. Some examples of full-scale prototype WECs include a device developed at Uppsala University [358] and tested off the west coast of Sweden in 2006, and the SeaBeav I [198, 199], a 10 kW point absorber developed by Oregon State University and tested in 2007. Frequently, the types of point absorber described previously, can be challenging to install due to the necessity for them to be tethered to the sea-bed, and maintenance costs are generally expensive when compared with shore-mounted WECs. Therefore, they may not be the preferred option, depending on certain bathymetry conditions.

Figure 1.14: PB3 PowerBuoy® WEC, developed by Ocean Power Technologies during commercial deployment in the North Sea, (image courtesy of Ocean Power Technologies).

SINN Power, founded in 2014, is another company that is focusing on the development of point absorbers for wave energy conversion [424, 387, 451]. Their device consists of a simple oblate-spheroid shaped buoy, which is connected to a vertical lifting rod. This shaft is attached to a structure such as a platform, elevated above the ocean surface or a breakwater wall. Then, as the waves interact with the floating buoy, the vertical lifting rod rises and falls. Using their patented PTO unit, the SINN PowerTrain 3.4, electricity can be generated. The company have been developing a modular type of floating platform, which is composed of a number of attached floating buoys. This platform design may be amenable to the incorporation of solar and wind power conversion technology. Another of their designs allows for the floating buoys to be attached to an existing edifice, such as a breakwater wall.

CorPower Ocean [167, 288], a Swedish company, are currently in the test phase of its pilot HiWave-5 project to develop an array of floating point absorbers, which will demonstrate the commercial feasibility of their concept. The first phase of the project concentrated on the full-scale testing of a single 9 meters by 18 meters C4 WEC device with a 300 kW power rating, taking their technology to TRL 7 [89]. The next phase of the project will be to complete full-scale testing on an array of 3 of the latest generation of C5 devices. The company anticipates a full commercial launch of their concept in 2024 [89].

An alternative design for floating oscillating bodies is to mount them on barges that are towed to a prescribed location and anchored in position. Due to the elongated shape of the floating buoy attached to the barge, these types of WEC are considered to be terminator type devices. The Wanshan [713] MW-level Wave Energy Demonstration Project is one such ongoing program with the goal of developing a megawatt level wave energy farm at Wanshan Island near Zhuhai, off the coast of China's Guangdong province [89]. As part of this project, two 500 kW WECs have already been designed by the Guangzhou Institute of Energy Conversion (GIEC) at the Chinese Academy of Sciences (CAS) and constructed by China Merchants Heavy Industry. The first of these 500 kW devices, the Zhoushan (Figure 1.15), was deployed in 2020 and the second device, the Changshan is currently undergoing open-sea tests, with construction having been completed in 2020. The design of the devices employs the *Sharp Eagle* technology [866] that has been under continuous development since 2011. This technology consists of a number of absorbing buoys connected to a hinged double floating body, which form a semi-submersible barge. Once deployed to the predetermined location, the main body of the device is ballasted with seawater to sink it to its semi-submersed working depth. It is then fixed in place with a number of anchors securing it to the seabed. A hydraulic system is used for the PTO [852].

Figure 1.15: The Zhoushan WEC, deployed in 2020 [89], (image courtesy of Guangzhou Institute of Energy Conversion)

Pitching

This type of device can generally be categorised, according to its orientation, as a terminator. One of earliest of these WEC devices was the Edinburgh Duck, developed by Stephen Salter in 1974 [307, 687, 845, 399]. It consisted of a pear-shaped body that housed a number of gyroscopes that swung back and forth as the device pitched due to wave action. These gyroscope motions activated a hydraulic pump that initiated a turbine to drive the generator. A prototype of the Edinburgh Duck was tested in 1976 off the east coast of Scotland. The device was rated to output 20 kW of power and had a calculated wave energy conversion efficiency of 90% [114]. Whilst very high energy conversion efficiencies were observed in the laboratory tests, these were never realised in the open seas testing. Interest in this concept was revived in 2009 by the Guangzhou Institute of Energy Conversion at the Chinese Academy of Sciences who constructed and deployed a device similar to Salter's Edinburgh Duck [855, 238]. This updated design had a 10 kW rating. A third generation device was deployed in 2013 with a 100 kW rating (see Figure 1.16). Another example of a pitching WEC was the PS Frog Mk 5, developed in 2005 at Lancaster University in the UK [9, 47]. This device operated as an inverted pendulum with a paddle-shaped section orientated perpendicularly to the wave direction. The paddle was connected by a vertical shaft to a submerged, ballasted chamber, which provided the righting moment. The device had a maximum mean predicted power output of just over 1.2 MW.

Figure 1.16: Pitching Duck type WEC constructed by the Guangzhou Institute of Energy Conversion, at the Chinese Academy of Sciences, (image courtesy of Guangzhou Institute of Energy Conversion).

One of the most successful pitching attenuator devices was the Pelamis WEC, developed by the UK-based Pelamis Wave Power company [156, 683, 605] (see Figure 1.17). The company carried out their first full-scale prototype tests of their P1 model in 2004 at the European Marine Energy Centre (EMEC) off the west coast of Orkney Island, north of Scotland. The device consisted of 4 tubular sections that were interconnected via articulated joints. The Pelamis WEC was positioned such that its spanwise direction was aligned with the prevailing wave direction. Hydraulic rams were attached at the hinged joints and, as the adjacent sections flexed relative to each other under the action of wave-induced motion, high pressure oil was pumped through the hydraulic motor to drive three generators. The P1 model was 120 meters long in total, 3.5 meters in diameter and rated at 750 kW power output. A second generation model, the Pelamis P2, was deployed to commence testing at EMEC in 2010. This updated version was composed of 5 sections and was 180 meters long and 4 meters in diameter. The Pelamis became the world's first grid-connected offshore WEC when it was connected to the UK grid in 2004. In 2008, three Pelamis P1 devices were installed near Póvoa do Varzin off the north coast of Portugal to undergo array deployment configuration testing. Further in-depth discussion, and design examples for pitching attenuator type devices are presented in Chapter 6.

Figure 1.17: Pelamis P2 WEC undergoing testing at EMEC [644].

Another such pitching attenuator device is the Blue X WEC developed by Mocean Energy. Construction on a prototype device was completed in 2021 and the WEC was deployed to begin sea trials at EMEC in April of 2021 [89]. The structure is composed of a 20 meters long, 38 tons, double-hulled construction connected by a revolute joint. The device is designed to be positioned with its longitudinal axis parallel to the wave direction. As the waves pass by the structure, it flexes about the joint, activating a direct drive PTO system to generate electricity.

A conceptually similar device has been developed by the Crestwing company in Denmark [25, 407, 408, 406]. Their technology, called the Tordenskiold, is composed of two large pontoons or floating hulls, which are connected by a revolute joint at the submerged seam between the two hulls. The overall structure is 30 meters long and 7.5 meters wide and has a mass of about 65 tons. When operational, the device is orientated with its long axis parallel to the wave propagation direction. A shaft, or push rod, connects the two hulls. As the revolute joint flexes due to wave interactions, the two hulls experience relative rotational motion. This activates the push rod, which forms a direct-drive PTO rack and pinion system and powers a generator to produce electricity. Initial scaled testing was conducted in a wave tank at Aalborg University during 2008 to 2009 and the results demonstrated that it had a 40–50%, operational efficiency. Construction of the full-scale model was completed in 2017 and initial testing commenced at Frederikshavn harbour in 2018. A second round of testing was carried out from February to November 2020 near the Hirsholm Islands, Denmark. Further testing and development of the PTO and generator system are planned before its commercial deployment.

1.1.4.2 Submerged oscillating bodies

Fully submerged oscillating body absorbers are somewhat less common than floating oscillating body absorbers. These submerged oscillating body devices rely on the orbital motion of the passing waves to generate electricity; this is in contrast to operating mode of pressure differential WECs (which are discussed later) which rely on the added mass of the wave crest passing over the device to activate it.

Heaving

The Carnegie Clean Energy company have developed a range of fully submerged wave energy point absorbers: one of their devices is known as the CETO 5 (see Figure 1.18). The main body of the device, the buoyant actuator, is moored to the seabed by a single cable, such that it is beneath the ocean surface but is still displaced by the wave action, predominantly in heave motion. The Albany Wave Energy Project, initiated in 2017, proposed the installation of a number of later generation CETO 6 units off the coast of Western Australia as a demonstration project. However, due to funding difficulties, the project was terminated. As of 2021, the Carnegie Clean Energy Company are focusing on developing intelligent PTO control systems

Figure 1.18: Fully submerged CETO 5 WEC concept, (image courtesy of Carnegie Clean Energy).

for the CETO 6 WEC device. Further discussion on the CETO devices is presented in Chapter 5.

Pitching

Another form of oscillating body WEC is the oscillating wave surge converter (OWSC) [690]. These are also sometimes known as bottom hinged devices [253] and are generally in the form of terminator devices. These OWSCs are usually fixed to the sea bottom and incorporate a paddle component, which is permitted to move in a rotational manner about the bottom fixing. There have been a number of these devices that have reached the prototype construction stage and the concept is not a new technology, with early patents for their design reaching back to 1954.

One of the original and best-known bottom-hinged WECs to reach full deployment status was the Oyster [661, 603, 412, 103] (see Figure 1.19). Developed by the Aquamarine Power company in Scotland in collaboration with a research team at Queen's University in Belfast between 2003 and 2009, the Oyster was one of the most promising WEC technologies. In 2009, a full-scale 315 kW demonstrator device was installed at EMEC near Orkney Island, north of Scotland. The hinged flap, which was almost fully submerged, was designed to be deployed in around 10 meters water depth. As the flap oscillated, it activated two hydraulic pistons that pumped pressurised water to drive an onshore hydroelectric turbine. Later, a second generation Oyster 800, with an 800 kW rating, was deployed at EMEC and was grid connected in 2012. The Oyster 800 continued generating electricity up until 2015, when Aquamarine Power ceased to trade. Prior to the dissolution of the company, plans were in development to construct a 2.4 MW wave energy farm, consisting of up to 50 Oyster devices, to supply electricity to 38,000 homes in Scotland. The project, however, was never realised, and Aquamarine Power went into administration.

Figure 1.19: Oyster, (image courtesy of Ramboll).

Exowave [89], is a Danish company that has also developed a type of bottom-fixed flapping WEC. However, their design concept differs in that their devices are smaller and modular in form. Designed to operate in fully-submerged conditions, a number of flapping plates can be connected together on the seabed in a truss-like configuration. The device is planned to undergo testing at the DanWEC test site [91, 767], off the coast of Denmark, in the second half of 2021.

In 2019, AW-Energy, a Finnish company, installed a 350 kW bottom-anchored, flapping-panel type device at Peniche, Portugal. The WEC, named WaveRoller, is designed to operate in nearshore areas at depths between 8 and 20 meters, where the wave shoaling effect is more intense [117, 468, 462, 257, 398]. This shoaling effect amplifies the horizontal component of the orbital motion of the subsurface water waves and the water particles move in a horizontal elliptic path as the nearshore water depth decreases. The horizontal forces from the waves cause the flapping panel to move back and forth. The device can operate in partially- or fully-submerged conditions. A hydraulic piston pump system is connected to a hydraulic motor, which powers an electric generator. The device, which was deployed in 2019, was successfully connected to the Portuguese national power-grid. AW-Energy are currently in the process of developing larger units, rated up to 1 MW. The company anticipates that their technology is suitable for deployment in large wave energy harvesting farms.

Some of the other notable submerged pitching oscillating body WECs include Aqua Power Technologies' multi-axis pitching type WEC and the BioWAVE [865, 253, 480] device developed by the Australian company, BioPower Systems. However, these two devices are not strictly limited to pitching motions and the BioWAVE device could also be classified as a bottom-hinged device.

Surging

Submerged oscillating body devices that operate in surge motion are generally aligned with their longitudinal axis orientated in the wave propagation direction. As such, they are a type of attenuator device. The Wavepiston [21, 654, 614] is a device developed by a Danish company of the same name, as shown in Figure 1.20. After four years of half-scale prototype testing at Hanstholm in the North Sea, a full-scale demonstration system, comprising a chain of 24 energy collector plates having an overall length of 200 meters was deployed at the Oceanic Platform of the Canary Islands (PLOCAN) testing facility around the beginning of 2021. The chain of collectors is anchored between two buoys in a direction parallel to the incident wave direction. Then, as the waves roll along the system, the collector plates move in a surging manner, pumping pressurised water to a turbine. The device is flexible, light-weight, modular and has a low environmental impact. Deployment is relatively simple and, as the device operates in near shore regions, the maintenance requirements are minimal and relatively straightforward.

1.1.5 Overtopping systems

An overtopping WEC is a conceptually simple structure consisting of a large basin that is isolated from the incident wave direction by a low parapet wall (see Figure 1.10c. Usually there is some form of wave collection system, typically

Figure 1.20: Wavepiston WEC device, (image courtesy of Wavepiston).

comprising walls that extend outwards from the reservoir in a widening, tapered manner to provide a channel that focuses the waves to the parapet wall. Then, as the waves approach the reservoir through the channel system, their amplitude increases through a wave-focusing action. These waves with increased amplitude spill over the parapet wall into the reservoir. The water stored internally in the reservoir is at a higher head relative to the external mean sea level. Due to this action, the wave energy has been transformed into potential energy. Then, a low head flow turbine is positioned at the reservoir outflow location and, as the stored water leaves the basin, the turbine is activated. In this manner, the stored potential energy is converted to fluid flow kinetic energy and finally into electrical energy. The basin should be sized in such a way that the rate of charging of the basin by the incident waves will not overwhelm its capacity, and it should also be proportionately large to ensure the smooth operation of the attached turbine system.

There have been a number of wave overtopping devices constructed around the world [533, 404]. These include the TAPCHAN power plant constructed in Norway in 1985 [258, 215] (see Figure 1.21). This installation utilised the natural topology of the coastline to create a large storage reservoir to hold the seawater. The power plant had a 350 kW power output.

The offshore Wave Dragon is a floating overtopping WEC, originally developed in Denmark in 2003, by a company of the same name. According to the manufacturer's technical specifications, there are a number of models available with the largest device measuring 390 meters by 220 meters and capable of delivering up to 12 MW of electricity [409, 757, 731, 734]. The device consists of a doubly curved front ramp upon which, waves surge up and spill into a reservoir located behind the ramp. The seawater drains from the reservoir back into the sea through a number of low-head hydro turbines. The device is slack-moored to the seabed. As of 2021, The company

Figure 1.21: TAPCHAN overtopping device constructed in Norway 1985 [649].

are preparing plans to deploy a 50 MW array of Wave Dragon WECs off the coast of Portugal.

A new design for an overtopping device was developed and constructed in 2008, wherein a number of reservoirs are positioned on top of each other so that, as the incident wave runs up on the front wall, it flows into the reservoirs through openings in the front wall. The Seawave Slot Cone Generator [805, 405] is an example of such a device that has undergone extensive testing, with a few prototypes either constructed or in various stages of development [495, 101, 137].

1.1.6 Pressure differential systems

Another type of WEC system is the pressure differential device. The device relies on pressure differentials to load and unload a plate or other element of the WEC cyclically as the wave passes over it. This type of energy converter is fully submerged under the ocean surface and is usually fixed to the sea-bed. The pressure differential device does not rely on wave-induced water motion to displace the WEC's mobile or deformable component but rather the increased pressure due to the added mass from the crest of the wave passing over the device. Generally, this type of WEC consists of two elements. There is a static lower part, which is fixed to the seabed, and an upper part that moves vertically, relative to the lower part. In this design, the upper part of the device is usually elevated by a spring mechanism. In some cases, the connecting element joining the upper and lower part of the pressure differential WEC is formed from a hollow section and the compressibility of a trapped air pocket within this hollow module can act as the spring. As the wave crest passes over the device, the pressure increases and the upper part is forced downwards, depressing the spring; then, as the wave trough passes over the device, the pressure is relieved and the top section of the device moves vertically upwards, unloading the spring. This cyclic loading and unloading and associated vertical movement of the upper section can be harnessed to generate energy. The Archimedes Wave Swing [167, 640, 792, 68] was a fully-submerged, pressure differential oscillating body that was deployed off the coast of Portugal in 2004. It had a 2 MW rated power output and utilised a direct-drive linear electrical generator. Bombora, a wave energy company based in Australia is currently developing a pressure differential WEC (see Figure 1.22). This device, the mWave™, differs from others in that a deformable membrane displaces as the wave crest and trough pass over it. It has less moving parts and is thus less susceptible to mechanical failure. The displacement of membrane forces air through a pneumatically driven PTO system.

Figure 1.22: Artist depiction of Bombora's mWave™ device, (image courtesy of Bombora).

1.1.7 Summary

In this section, the origins of WECs have been presented, the development of the various efforts to classify the many technologies has been described and many diverse types of WEC have been examined. However, even with the myriad different designs and concepts for wave energy extraction, the commercialisation of the sector remains relatively slow. In the next section, some of the advantages and disadvantages of wave energy over other forms of renewable energy are demonstrated. Additionally, some of the reasons for the slow development and uptake of wave energy will be examined. Specifically, the economic aspects of wave energy conversion and the TRL of the various devices are investigated.

1.2 ECONOMIC ASPECTS OF WAVE ENERGY CONVERTERS

It is important to realise that economic factors are some of the greatest drivers of technological advancements. Wave energy devices are not exempt from this rule. In this section, some of the economic factors and influences that have had an effect on the development and commercialisation of wave energy technology are examined.

1.2.1 Development status of renewable energy

WEC devices have been developed and employed for power generation across several decades, yet wave energy is still a relatively new form of renewable energy source compared with wind and solar energy [726]. In most instances, the previously developed WECs have been single devices designed to provide energy on a small scale. Historically, there have been some efforts to commercialise wave energy but, at that time, the contemporary technology was incompatible with its commercialisation. However, in more recent times, many of these WEC designs have been improved technologically; thus, there has been a significant, renewed effort to commercialise the sector of late.

In contrast with the efforts to commercialise wave power, both solar and wind energy have been exploited on an industrial scale for many years. The first commercial solar power plant was built in Maadi, Egypt, around 1912, to power a 50 kW engine, which was used to drive an irrigating pump supplying water to nearby cotton farms [34]. This installation relied on parabolic reflectors to concentrate the sun's rays onto boilers, which generated steam to drive the pump. The first photovoltaic power station with a rating of 1 megawatt-peak (MW_p) was constructed in 1982 in California. Since that time, there have been many thousands of utility scale solar parks constructed around the globe, with some rated up to 1 GW_p, such as the Mohammed bin Rashid Al Maktoum Solar Park in the United Arab Emirates [26]. From the early 2000s onwards, the growth of photovoltic power pants has seen a dramatic increase and the rise in photovoltaic capacity has been almost exponential. In 2019, almost 3% of the global electricity demand was met by photovoltaic solar power plants [367].

Wind power has a much older history. For millennia, sailors have been using the power of wind to propel their ships all around the globe. Windmills and windpumps have been in existence for more than two thousand years and have been used to convert wind energy into mechanical energy to drive various types of machinery on an industrialised scale. In terms of wind energy being converted to electrical energy, we can trace its origin back to 1887, when a Scottish Professor named James Blyth from Anderson's College, now the University of Strathclyde, used a rudimentary wind turbine to power the lighting in his home [642]. More recently, the global oil crisis in the 1970s had a large influence on the rapid development and utilisation of wind energy [797]. Many countries were determined to reduce their reliance on fossil fuels and extensive government funded research into alternative energy sources ensued. Wind power technology developed quickly and, by the early 1980s, grid-connected wind farms were producing electricity in California [716]. Throughout the 1990s and 2000s, with growing knowledge of the impact on fossil fuels on global warming, research continued into wind power and governments provided incentives to utility providers to increase their uptake of wind power. This has led to a rapid growth in the wind energy sector, with the United States, Europe, and China all being major installers and proponents of wind farms. In 2019, the globally installed capacity of wind farms was approximately 650 GW [433]; and these supplied approximately 5.3% of the world's electricity requirements [84].

With the rapid development and widespread deployment of both wind and solar energy, wave energy has lagged behind in its market penetration. The wide ranging and complex reasons for this will be presented and examined in the following sections. Nevertheless, since about the last decade of the 20th century, society has begun to realise that there are vast stores of energy in the world's oceans that can be used to generate electricity. Consequently, research into technologies that can exploit this energy in an efficient manner has gathered pace. In the last thirty or so years since interest in wave energy was ignited, there have been many thousands of WEC devices invented and patented; this trend continues and shows no sign of slowing down [368]. Some have argued that this explosion in the number of devices invented has been detrimental to the coherence of the wave energy sector as confusion abounds as to which devices are best. Nonetheless, as the sector matures, the technology will improve and those devices that are most appropriate to a given wave climate will be identified. The next section explores the history of wave energy and the developments that have brought the technology to its contemporary state.

1.2.2 Role of wave energy in the renewable energy mix

According to the Intergovernmental Panel on Climate Change (IPCC), Special Report on Renewable Energy Sources and Climate Change Mitigation, published in 2012 [189], the total theoretical global wave energy potential is estimated to be 32,000 TWh/yr. The global total electricity consumption in 2018 was 22,315 TWh, according to the International Energy Agency [366]. These consumption rates dropped significantly in 2020 and 2021. Wave energy alone has the theoretical potential to meet the world's electricity requirements, although the amount that can be harvested in practice is much lower. In 2020 it was reported that 3147 TWh of the world's electrical power came from all the combined renewable sources [85]. In 2019 it was estimated that the total combined capacity of all operational WEC devices around the globe amounted to only 2.31 MW [368]. This value falls well short of the 37 MW of wave energy capacity proposed by 2020 in the Joint Research Centre (JRC) Ocean Energy Systems (OES) Report published in 2016 [482]. Clearly, the wave energy sector has not reached maturity and requires investment and research stimulus to reach its potential.

1.2.2.1 Temporal and spatial availability

A substantial amount of wind energy must be put into the ocean surface before waves start to form. The same applies to the wind energy and its primary driving force, solar radiation, where a significant heating of the earth's surface must occur before wind is generated. As stated in [672], this process can be considered as a time integration of the stimulus and a similar response is observed when the main driving force is removed. For instance, when the wind ceases to blow, waves do not disappear immediately but begin to decrease gradually. These differences in the dynamic responses influence the availability of renewable sources. As a result, wave resources suitable for electricity generation are available up to 90% of the time, while solar and wind power plants can operate efficiently for only 20–30% of the time [616, 184].

Another advantage of wave energy is its natural seasonal variability, which coincides with the changing electricity demand throughout the year in many countries with temperate climates. This trend is presented in Figure 1.23 [655]. In addition, the hourly variability in wave energy is three times less than that of wind energy [66], which is significant for the potential integration and utilisation of this source of energy in the power system.

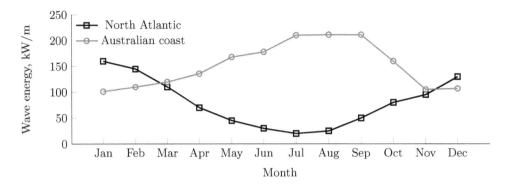

Figure 1.23: Monthly variability of the mean wave power in North Atlantic [655] and on the Australian coast [347].

The distribution of the annual mean wave energy density around the globe varies significantly, as shown in Figure 1.24 [140]. Areas of the globe in moderate to high latitudes, between 30° and 60° north and south, respectively, have the highest energy densities; whilst equatorial seas have low energy densities. These higher wave energy densities can be attributed to the consistent trade winds, which stream in a westerly direction in these regions. Furthermore, offshore regions generally have a higher wave energy density than near-shore locations. This irregular distribution of mean annual wave energy density has led some countries to invest more resources in the

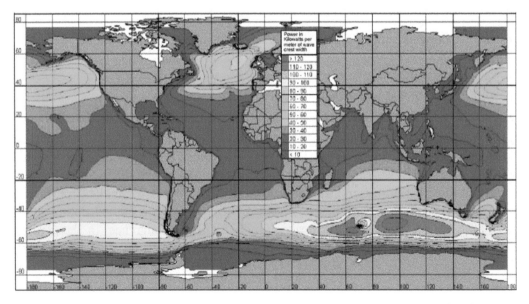

Figure 1.24: Annual mean wave energy density around the globe [140].

technology's development that will benefit them directly. For example, some of the countries on the western seaboard of Europe, where wave energy density is significant, have played a leading role in pushing forward the sector [484].

The predictability of power generation from renewable sources, such as wind, wave or solar, is an important factor for the energy sector that helps to manage the grid load and imbalances in the electrical system. Data has shown that it is possible to achieve an accurate wave forecast for 36 hours, while wind speed and directionality can only be predicted accurately 12 hours in advance [66]. Other research demonstrates that the output from WECs is up to 35% more predictable than power generated from wind turbines [127]. In summary, wave power has been shown to be more available, less variable and more predictable than wind or solar energy.

1.2.2.2 Environmental impact

All sources of renewable energy are considered to be "clean" due to the much lower levels of emissions and pollution than for conventional fossil fuels. However, even clean sources of energy have environmental impacts during some stages of their life-cycle including raw material extraction, construction, transport, installation, operation, maintenance, and end-of-life disposal [296]. The life-cycle assessment is a standardised technique that allows for the tracking of the pollutant flows associated with a particular energy technology. The resultant environmental impacts of selected renewable sources are presented in Table 1.2.

Table 1.2: Environmental impacts of renewable energy technologies. Data has been adapted from [773, 739, 695].

Environmental effect	Wind	Solar PV	Geo-thermal	Biomass	Hydro-power	Wave
Greenhouse gas emission [gCO_2/kWh]	11	48	38	230	24	17
Land use [m^2/MWh]	69	12	0.4	488	122	< 0.5
Air pollution, NOx [mg/kWh]	25	110	0	550	30	0
Waste generation		+	+			
Acoustic noise	+					+
Endangered species	+				+	+

It should be noted that wave power conversion technology is at a pre-commercial stage and the amount of information available on its environmental impacts is minimal. Based on the existing data, wave energy has one of the lowest levels of life-cycle greenhouse gas emissions and requires only a limited onshore area for power generators (chosen technologies only) and grid connection equipment. There are some concerns that WECs produce noise and vibrations under water that may have the potential for negative effects on marine fauna [836]. However, the published data on acoustic environmental impacts of WECs is sparse and this requires further investigation, especially when large-scale arrays of WECs are installed and operational.

1.2.2.3 Technology readiness level

At this point, the concept of technology readiness level (TRL) is introduced. This is a scale from 1 to 9, conceived by NASA in 1974, which can be used to classify the developmental maturity of a new technology [749] (see Figure1.25).

TECHNOLOGY READINESS LEVELS

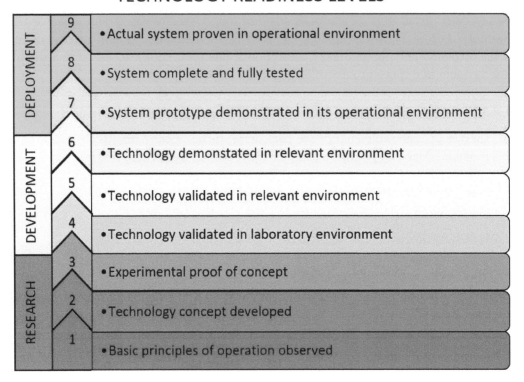

Figure 1.25: Technology readiness levels

Levels 1–3 represent the research phase in the creation of the new technology, levels 4–6 describe the development phase and levels 7–9 are used to describe the deployment and testing phases of the device. At the bottom end of the scale, a proposed new technology, whose basic principles are being observed and investigated, is characterised by a TRL rating of 1. TRL 9 represents a technology that has undergone widespread deployment and is commonly in commercial use. According to the JRC Ocean Energy Status Report, published in 2016, the TRL for wave energy devices ranges between 1 and 8 (see Figure 1.26) [482]; however, it can be argued that some technologies, such as the grid-connected array of three Pelamis WECs and the long-established OWC type Mutriku power plant, demonstrate proven operational aptitude for wave energy harvesting. Thus, these devices could be argued to be at TRL 9.

There are a number of different WEC concepts and within each concept category there are a vast number of designs. These different designs range from experimental proof of concept level right through to deployed prototypes and, in some cases, grid-connected devices. Therefore, the TRLs for WECs exhibit a large spread across the range. Figure 1.26 exhibits the TRL classification range for different WEC concepts; the red point indicates the maximum TRL that a constructed device has reached [482].

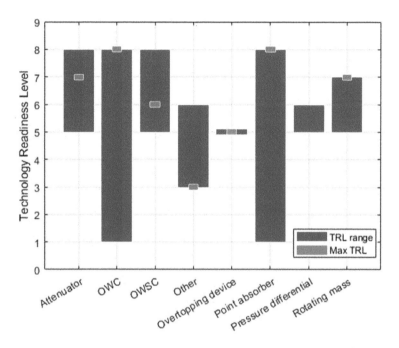

Figure 1.26: Wave energy device technology readiness levels (TRL) [482].

Examples of the various forms of renewable energy demonstrated in Section 1.2.1 clearly show that both wind and solar power technologies have reached TRL 9, as presented in Table 1.3, while wave energy devices are still predominantly in the development phase at TRLs 5, 6, and 7, with many additional conceptual designs at lower TRLs [482].

There is a myriad of reasons why this is the situation. As wind and solar power are already proven technologies in widespread, commercial, and grid-connected deployment, there may be a reluctance to allocate research investment into a new, competing technology. Additionally, whilst wind energy and solar energy harvester technologies have a limited number of basic operating principles, there have been a much greater number of diverse designs proposed for wave energy harvester technologies. These competing designs, all vying for market leadership in their own right, may have had a compound, adverse impact on the development of wave power as a source of renewable energy. The development of a wave power plant is a multi-disciplinary project, which unites such fields of knowledge as hydrodynamics, mechanical and structural engineering, electrical engineering and control, materials sciences, marketing and economics. Some of the other major barriers that are preventing wave technology from advancing faster include:

(i) wave-specific barriers [132]:

Table 1.3: Maturity of selected renewable energy technologies (as of 2020). Sources: [364] updated; own analysis.

Technology	Sector	Demonstration	Commercialisation		
			Inception	Take-off	Consolidation
Biomass	Electricity and heat	Thermal gasification		Anaerobic digestion	Modern boilers and stoves
	Transport	Advanced biofuels		Conventional biofuels	
Geothermal		Enhanced geothermal		Conventional geothermal	
Hydro					Hydro
Marine		Wave — Tidal and stream			
Solar	Heat	Solar cooling		Solar water heaters	
	Electricity	PV – 3rd generation	CSP tower	CSP trough	PV cristalline and thin film
Wind			Offshore wind		Onshore wind

(a) irregularity of sea states (amplitude and phase), which make it difficult to control the device and achieve the maximum efficiency (see Chapter 3);

(b) high peak-to-average loading ratios on the structure, which affect the cost of the converter and increase the cost of generated electricity;

(c) wave power plants are subject to low frequency oscillations, while effective off-the-shelf PTO systems are designed for high-frequency motions;

(ii) barriers common to offshore industries [370]:

(a) uncertainties about environmental regulations;

(b) insufficient infrastructure: offshore grid connections, and operation and maintenance facilities;

(c) the lack of processes for planning and licensing of marine activities.

In order to overcome these challenges to the wave energy sector, a number of initiatives have been introduced, both nationally (e.g., via the European Union) and globally, to assist wave developers in advancing their technologies and moving towards commercialisation. These include the MaRINET2 project that offers access to experimental facilities, Ocean Energy Systems' projects that contribute to the state-of-the art in wave energy modelling, assessment of environmental issues, and development of internationally accepted metrics for measuring ocean energy technology progress [484]. Moreover, there is growing financial support for marine renewables at government level in Ireland, the UK (notably in Scotland), the EU, and Australia.

Notwithstanding the challenges to the commercialisation of wave energy outlined above, harvesting energy from ocean waves has many advantages over other forms of renewable energy [132]. In contrast with both wind and solar power, it is a relatively constant source of energy with comparatively little temporal variation. Solar power plants cannot operate at night-time, and during overcast days they do not operate efficiently. Wind power faces the same problems with regards to temporal variability in wind speeds. Solar power-plants require large tracts of land to install photovoltaic arrays, often up to 4 hectares per MW [811]. Onshore wind turbine arrays also require vast areas of land. Modern onshore wind turbines have an average capacity of about 2 MW and a rotor diameter of about 100 meters. The turbines are usually spaced at a minimum of 6 rotor diameters from each other in the streamwise direction and 4 to 5 rotor diameters in the cross-wind direction [795]. This necessitates an extensive land area for the installation of a wind turbine array. Photovoltaic arrays require the mining of rare earth minerals, which has a significant, detrimental environmental impact. Furthermore, construction of solar power plants and onshore wind farms are often opposed by local residents, as they can have a negative visual (and sometimes auditory in the case of wind farms) impact on the landscape and its ecology, a phenomenon referred to as the "not in my backyard (NIMBY)" effect [832]. Additionally, whilst offshore wind turbines generate significantly more power than their onshore counterparts, they are significantly more expensive to install and maintain. Most of these listed disadvantages associated with other forms of renewable energy are relatively minor for WEC devices.

Clearly, the deployment of WEC technologies has fallen behind other forms of renewable energies. However, the outlook is not all bleak; in the last couple of years there has been a resurgent interest in wave energy. There have been a number of newly-constructed wave energy devices and research continues apace. During the 20th and the early 21st centuries, European academic institutes and enterprises were at the forefront of the wave energy technology sector [484]. However, in 2020, China's State Council Information Office released a white paper titled *Energy in China's New Era* [768]. This document set out future policy for China's energy sector. It stated that China will *accelerate its transformation towards green and low carbon development in the economy and society*. The document goes on to state that China will *also reinforce R&D and pilot demonstrations on harnessing ocean power such as tidal and wave energy*. As the world's largest electricity consumer and producer, this is a significant development and China has indeed accelerated its WEC development program. Up to 2020, China has completed sea trials on more than 40 wave energy devices [89], with their flagship device, the Zhoushan WEC, being the first of two 500 kW point absorbers to begin sea trials in 2020.

These renewed efforts to stimulate the wave energy sector are essential for bringing the industry to maturity and advancing the technology along the TRL scale to reach commercial viability.

1.2.2.4   Niche market for wave energy

As the wave energy sector has not reached industrial maturity, the cost of the generated electricity from ocean waves is considerably higher than that of competing technologies (see Figure 1.27). These higher costs are also negatively affecting investment attractiveness for the wave power sector and may be influencing the relatively slow progress of wave energy technology through the TRL spectrum [477]. However, as the technology improves, these costs are set to reduce.

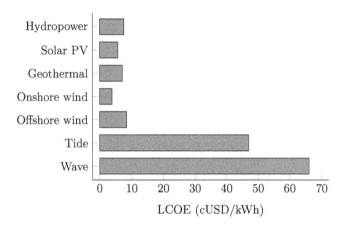

Figure 1.27: LCOE for chosen renewable energy sources (as of 2020) [369, 478].

While many renewable energy projects are large-scale, ocean power has the potential to provide off-grid solutions for remote island communities where diesel generators are still in use. Moreover, there are industries and applications that rely on expensive energy sources and require relatively small power capacities (< 100 kW) that could become niche markets for wave energy utilisation. These include fish farms, offshore platforms, ocean remote sensing, and coastal protection applications.

1.2.3   Performance measures

The choice of metric to be used for WEC performance assessment is largely determined by the TRL of a particular prototype. Table 1.4 demonstrates which performance measures are usually used at different stages of the wave energy development.

1.2.3.1   Power production assessment

Absorbed, or captured power, [W], is the average mechanical power absorbed by a WEC from a regular wave or a sea state. Depending on the PTO machinery used,

Table 1.4: Stages of the wave energy development process (adapted from [357]) and associated performance measures used at each stage.

Stage	Description	TRL	Performance measures
Stage 0	Concept creation	1	CW, RCW
Stage 1	Concept development	2 3	CW, RCW, AEP
Stage 2	Design optimisation	4	AEP, η_{eff}, DEL, LCOE, ACE, q-factor
Stage 3	Scaled demonstration	5 6	AEP, peak loads
Stage 4	Commercial scale single device demonstration	7 8	LCOE, NPV, PBP
Stage 5	Commercial scale array demonstration	9	LCOE, NPV, PBP

the calculation of the average power varies:

$$\text{Absorbed power [W]} = \frac{1}{T}\int_0^T (\text{force} \times \text{velocity})\,dt$$

$$= \frac{1}{T}\int_0^T (\text{torque} \times \text{angular velocity})\,dt$$

$$= \frac{1}{T}\int_0^T (\text{pressure difference} \times \text{flow rate})\,dt. \tag{1.1}$$

More commonly, the power absorption properties of a WEC are demonstrated in terms of the power matrix, as shown in Figure 1.28, indicating the average power capture in each sea state.

Significant wave height [m]	3	4	5	6	7	8	9	10	11	12	13	14	15	16	17
7.5					1424	1508	1457	1324	1174	1004	864	746	651	556	497
7					1264	1355	1291	1164	1017	907	756	664	580	507	440
6.5					1111	1189	1122	1014	903	779	660	587	509	443	387
6					963	1027	974	897	803	697	583	515	444	388	341
5.5					822	879	851	754	659	588	497	445	387	339	293
5				606	703	745	703	645	565	496	430	380	328	287	251
4.5				504	585	608	594	542	458	416	365	324	280	243	215
4				413	477	495	482	427	377	337	303	272	227	197	172
3.5				330	378	388	384	340	303	268	245	219	187	148	144
3			175	250	292	308	291	266	236	215	184	166	140	129	110
2.5			123	178	210	225	206	191	175	156	137	122	106	93	79
2	9	41	82	118	139	149	138	131	120	105	94	82	73	61	51
1.5	5	22	47	69	81	86	83	77	72	63	56	47	44	32	30
1	2	9	21	31	38	38	40	36	34	29	26	23	20	16	14
0.5	1	3	6	8	10	11	10	10	9	8	7	6	5	3	3
	3	4	5	6	7	8	9	10	11	12	13	14	15	16	17

Peak wave period [s]

Figure 1.28: Example power matrix (in [kW]) of the three-tether WEC from Chapter 5.

Capture width (CW) [98], sometimes called capture length, is defined as the ratio of the average power absorbed by a WEC to the average incident wave power:

$$\text{CW [m]} = \frac{\text{Absorbed/captured power [W]}}{\text{Available wave power [W/m]}}. \tag{1.2}$$

In other words, the capture width of the WEC can be explained as the width of the wave front over which the device absorbs 100% of the incident wave power. Depending on the purpose of the analysis, the capture width can be evaluated independently for each regular wave frequency, for each sea state, or as an average value for a particular deployment site.

Relative capture width (RCW), or capture width ratio (CWR), represents the hydrodynamic efficiency of a device, defined as the ratio of the capture width to the characteristic width of the device:

$$\text{RCW} = \frac{\text{Absorbed/captured power [W]}}{\text{Available wave power [W/m]} \times \text{Characteristic width [m]}}. \tag{1.3}$$

The RCW is a dimensionless parameter and may be interpreted as a fraction of the wave power absorbed by a WEC. If the relative capture width is greater than 1, then the device absorbs more power than is contained in a wave of its width.

The annual energy production (AEP) is probably the most important power performance metric that is widely used as an input to evaluate the techno-economic attractiveness of the wave power projects. For a given wave climate, characterised by the scatter diagram representing the probability of occurrence of each sea state, the yearly mean annual energy production is estimated as:

$$\text{AEP [Wh]} = \frac{8760 \text{ [h/year]}}{100 [\%]} \times \sum_{H_s, T_p} \left(\text{Power matrix [W]} \times \text{Scatter diagram [\%]} \right). \tag{1.4}$$

The wave farm interaction factor, sometimes called the park effect [41] or q-factor, quantifies the effect of wave interactions on power absorption in a wave array. It is calculated as the ratio of the power absorbed by the wave farm to the sum of the power extracted by each isolated WEC:

$$q = \frac{\text{Power absorbed by the wave array [W]}}{\sum_{i=1}^{N} \text{Power absorbed by an isolated } i\text{-th WEC [W]}}. \tag{1.5}$$

The values of the q-factor that are greater than 1 mean that the WECs experience constructive hydrodynamic interaction within the wave farm.

Modified q_{mod}-factor; Babarit [40] argued that the usual q-factor masks the real amount of absorbed power, and a modified factor q_{mod} was suggested for use instead. In contrast with the usual q-factor, which characterises the wave farm performance as a whole, q_{mod} is evaluated for each i-th device independently, as:

$$q_{mod}^{i} = \frac{P_i - P_{isolated}}{\max_{T_w} P_{isolated}(T_w)}. \tag{1.6}$$

where P_i is the power absorbed by i-th WEC in a farm, $P_{isolated}$ is the power absorbed by an isolated i-th WEC, and $\max_{T_w} P_{isolated}(T_w)$ is the maximum absorbed power by an isolated WEC across wave periods T_w. For this performance measurement, the positive values of q_{mod} indicate a constructive effect for the wave interaction on the power production of this WEC, while negative values correspond with a destructive effect.

The **Capacity density** of a wave energy farm demonstrates the amount of energy delivered to the grid per ocean area, per year:

$$\text{Capacity density } [\text{kWh/km}^2] = \frac{\text{AEP of a wave farm } [\text{kWh}]}{\text{Wave farm area } [\text{km}^2]}. \tag{1.7}$$

This measure is widely used to assess and compare installed offshore wind farms.

1.2.3.2 Techno-economic analysis

The **levelised cost of energy (LCOE)** is the most important measure for energy investment, and is usually used to compare the feasibility of various renewable energy projects. The LCOE is defined as [144]:

$$\begin{aligned} \text{LCOE } [\$/\text{kWh}] &= \frac{\text{Total costs over project lifetime}}{\text{Total energy produced over project lifetime}} \\ &= \frac{\text{CAPEX} + \sum_{t=1}^{n} \dfrac{\text{OPEX}_t}{(1+r)^t}}{\sum_{t=1}^{n} \dfrac{\text{AEP}_t}{(1+r)^t}} \end{aligned} \tag{1.8}$$

where CAPEX is the capital expenditures of the project, OPEX_t corresponds to the operations and maintenance expenditures in the year t, r is the discount, and n is the expected lifetime of the project.

As it is quite difficult to evaluate LCOE for new technologies, several alternative measures have been proposed.

Average climate capture width per characteristic capital expenditure (ACE) was used by the United States Department of Energy to compare the performance of WECs for the Wave Energy Prize [791, 790, 163]:

$$\text{ACE } [\text{m}/\$] = \frac{\text{Capture width } [\text{m}]}{\text{Characteristic capital expenditure } [\$]}, \tag{1.9}$$

where

$$\begin{aligned} \text{Characteristic capital expenditure } [\$] =& \\ \text{Total surface area } [\text{m}^2] \times \text{Structural thickness } [\text{m}] \times& \\ \text{Material density } [\text{kg/m}^3] \times \text{Cost of material } [\$/\text{kg}]& \end{aligned} \tag{1.10}$$

Babarit et al., [51] proposed the use a characteristic mass of the WEC as a proxy for the project costs. Thus, the annual energy production normalised by the characteristic mass of the system was used as a techno-economic metric to compare different WEC prototypes.

It has been proven [163] that both ACE and AEP per unit mass demonstrate a clear correlation with LCOE and can be used to optimise the wave energy system:

$$\text{LCOE [\$/kWh]} \approx \text{RDC} \times \left(\frac{\text{AEP [kWh]}}{\text{Characteristic mass [kg]}} \right)^{-0.5}, \tag{1.11}$$

$$\text{LCOE [\$/kWh]} \approx \text{RDC} \times (\text{ACE})^{-0.5}. \tag{1.12}$$

The **net present value (NPV)** is an economic measure to evaluate the profitability of a system, which is calculated as [144]:

$$\text{NPV} = \sum_{t=1}^{n} \frac{(\text{Net cash flow})_t}{(1+r)^t}. \tag{1.13}$$

The NVP indicates how much value is added to the investment. It is only worth investing in the project if the NVP is positive.

The **payback period (PBP)** is the time needed for the cumulative cash flow to be positive, i.e., the period it takes for the project investment to earn itself back.

$$\text{PBP} = \frac{\text{Initial investment}}{\text{Net annual cash flow}}. \tag{1.14}$$

According to the Ocean Energy Technology Development Report [478], published in 2020, the LCOE produced from wave energy in 2015 in Europe was in the region of 0.47 EUR/kWh to 1.4 EUR/kWh, with a reference value of 0.72 EUR/kWh. This reference value dropped to 0.56 EUR/kWh with the installation of an extra 8 MW capacity of wave energy by the year 2018. However, since then, the LCOE of wave energy has remained unchanged. As additional devices are developed and are grid-connected, the LCOE is predicted to drop further due to economies of scale cost-reductions, and improvements in WEC technology efficiency. It is anticipated that this will set in motion a chain of events whereby more investors are attracted, thus accelerating the technological developments and further increasing the energy harvesting efficiency. Figure 1.29 shows a projection for the LCOE produced from wave energy arrays as the sector develops. A target value for the LCOE for wave energy of 0.2 EUR/kWh has been set for the year 2025, according to OES.

A number of different solutions are available to decrease the LCOE of wave energy. The CAPEX can be reduced by developing technologies that are easier and cheaper to build and deploy, and the OPEX can be reduced by ensuring more durable structures with extended design life duration and lower maintenance costs. Additionally, improvements can be made at the WEC design stage to develop more efficient devices that exploit more energy from the incident waves. Sections 1.3.1 and 1.3.2 provide some examples of methods to reduce the CAPEX and OPEX expenditures for WEC devices. In order to extract more energy from the wave environment, it is also important to optimise both the device geometry and the power take-off (PTO) system.

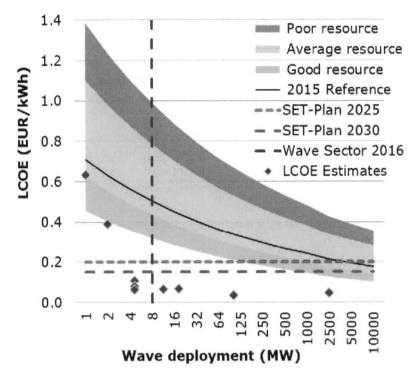

Figure 1.29: Wave energy LCOE trends in Europe [478].

At present, there is a significant amount on research on the optimisation of PTO systems using control theory to enhance the performance of WECs [10, 461, 274, 624, 706, 308]. Some of the key considerations include balancing the peak-to-average power ratio to ensure smoother operation of the device and maximising the average power extracted by the PTO system. Chapter 3 discusses these topics in greater detail.

1.3 LATEST DEVELOPMENTS IN WAVE ENERGY HARVESTING RESEARCH

The European Commission (EC) assessment of National Energy and Climate Plans (NECPs), has stated that wave energy technologies are still in the development phase and at TRL 6–7 [207]. Wave energy is not expected to contribute a sizable proportion of the total renewable energy sector until at least 2030. There is still much ongoing research into developing more durable structures, more efficient devices and reducing the LCOE sourced from wave energy. The following subsections briefly discuss these topics in more detail.

1.3.1 Integration with pre-existing and planned infrastructure

Until the 2020s, the main markets for wave energy harvesting installations that had already been constructed or deployed had been small coastal communities [490]. The Mutriku Power Plant, which is described below in more detail, is one of the best known examples of a breakwater-integrated WEC. Other, similar proposed plants are in various stages of development. These wave energy installations, some of which are in array configurations, are designed using a diverse range of devices, including oscillating buoy type converters or offshore and onshore OWCs. It is anticipated that these farms will not only supply the local grid requirements, but they will have a large enough output to serve electrical requirements on a larger, regional scale. Another approach currently under consideration is the use of wave energy devices to support offshore ocean industries, such as providing the electrical requirements for oil and gas drilling platforms [595], aquaculture and offshore ocean research stations. There is also much interest in the integration of wave energy harvesting devices with other renewable energy systems, such as offshore wind energy farms [643, 39] and solar energy platforms [596, 656]. If proven to be mechanically complementary, such dual function installations should be significantly more cost-effective to operate over extended periods of time.

There have been a number of instances of WEC technology being integrated into either pre-existing coastline infrastructures or newly-built coastal protection systems [360, 574, 626, 320, 119, 868, 183, 102, 867]. In some cases, the WEC devices have also been built on naturally-formed shoreline features [544]. The main benefits from integrating WEC technology into shoreline infrastructure include the ease of installation, ease of maintenance and consequently reduced CAPEX and OPEX costs. Other benefits include lower environmental impact and, in certain cases, the infrastructure may act to concentrate the wave energy, thereby increasing the localised energy density. The main disadvantage of such structures is that the wave energy density in near shore areas is almost always lower than the wave energy density in offshore regions.

One of the earliest examples of a WEC device being integrated into an artificial breakwater wall was the Sakata wave power plant, built around 1989 [594, 487, 751, 593]. This 60 kW demonstration project consisted of a specially constructed, vertical concrete breakwater wall that included a 20 meters long hollow caisson, which formed the OWC chamber. The plant remained operational for a number of years, proving to be a valuable test facility and providing a large amount of data for future installations. The efficiency of the plant was relatively low, with a wave-to-wire efficiency of just 3.26%. Nevertheless, the project successfully demonstrated the harvesting ability of a breakwater-integrated OWC system.

The power plant at Mutriku, on Spain's northern coast, is one of the most widely known such OWC-breakwater structures [779, 236, 235, 363]. Construction on the plant commenced in 2006 and the plant was commissioned in 2011. It consists of a rubble mound breakwater wall housing 16 individual OWC air chambers measuring 4.5 meters wide by 3.1 meters deep by 10 meters high. The plant is rated at 296 kW and has been grid-connected, providing power to the local town of Mutriku since it was commissioned. The wave-to-wire efficiency is again quite low at 2.56%; however, the plant is deemed a commercial success due to the longevity of its operation. The Mutriku power plant has been the subject of numerous studies and there is a significant amount of data published about it. Currently, the plant provides a single OWC chamber for testing of new wave energy harvesting technology and machinery, such as novel turbine designs.

Other forms of wave energy harvesting devices have also been integrated into the shoreline environment. The TAPCHAN overtopping WEC, built in Norway in the 1980s, was a device that exploited the shoreline topology by utilising a naturally-formed onshore reservoir to store seawater at a higher energy potential than the surrounding sea level [258, 215]. This stored water was used to drive a low head turbine to produce electricity. Another example of a breakwater-WEC integrated structure is the ReWEC3 device installed at the port of Civitavecchia, Italy, in 2012 [31, 30, 498, 29]. The WEC component is a U-shaped OWC. It is estimated that the incorporation of the OWC only added 5% CAPEX costs to the initial breakwater installation price. It includes 136 chambers spread over a total length of 576 meters, making it one of the largest WEC devices ever constructed. The device has an esti-mated efficiency of 26.33%, making it significantly more efficient than the Mutriku device, and one of the most efficient near-shore OWCs yet constructed. Other notable WEC-breakwater systems include the OBREC overtopping device [372, 138, 607], installed in the San Vincenzo breakwater in Napoli, Italy, in 2015 and the point ab-sorber, wall mounted type device developed by SINN Power and undergoing full-scale testing at the breakwater wall in Iraklino, Greece, since 2015 [424, 387, 451].

1.3.2 Prolonging the design life of the WEC technology

As shown in Figure 1.24, the annual mean average wave density is region-specific and varies significantly around the globe [140]. One of the challenges to the wave energy sector is to extract as much energy from waves, whilst at the same time ensuring the survivability of the harvesting device. Certainly, it is more desirable to position the WEC in a region with high wave energy density; however, these areas experience some of the harshest weather conditions on earth. Thus, the conundrum is how to maximise the energy harvest, whilst at the same time ensuring the device remains in service for the duration of its design life. Presently, this is one of the main concerns in the wave energy sector. Previous experience has demonstrated that this is a major challenge to overcome [677, 403, 556, 486]. There have been a number of notable failures of WEC devices in the past, due either to catastrophic failure during storm events or from progressive collapse or fatigue failure due to structural deterioration in the harsh environment. For example, the OWC power plant at Mutriku [779, 236, 235, 363] sustained damage during storms in 2007 and 2008, during its construction. The plant also experienced severe structural damage to many of the OWC chambers in 2009 during storms. Four of the sixteen chambers were destroyed and the front wall was damaged in many of the other chambers.

Another notable example of a WEC device that was beset with problems is the Pico power plant, built in the Azore Islands in 1999 [218, 222, 93, 90]. Soon after completion, the plant experienced flooding, which rendered the turbine machinery in-operational, thus limited testing was permitted during this time. The plant was refurbished in 2004–2006 and became operational with limited power production in 2005. However mechanical problems persisted until 2008 when repairs allowed the plant to run at its rated power capacity of 400 kW. In 2012, the plant commenced fully autonomous operation but one year later electrical problems occurred and the plant was closed in 2016. However, it continued to supply power in an intermittent, limited capacity to the island grid until early 2018. There was renewed interest in renovating and restarting the project in 2017; however, the plant suffered extreme damage and partial collapse in 2018, excluding any possibility of rendering the plant operational again.

In 2006, an onshore oscillating buoy developed by the Guangzhou Institute of Energy Conversion catastrophically failed after only 2 days of being operational. A critical shaft broke and the buoy section fell into the sea and was lost in a typhoon approximately two weeks later [855].The Toftestallen shoreline-mounted OWC power plant built in Norway in 1985 and rated at 500 kW was destroyed three years after its construction when a storm damaged its foundation fixings, causing the device to sink [489].

A number of other devices have also been submerged during the deployment or testing phase. These include the first generation OSPREY device [534], a 2 MW OWC type device, which was lost in 1995 during the testing phase and the Trident device, which was lost during deployment in 2009. The Oceanlinx MK3 prototype [344], a floating OWC WEC deployed at Port Kembala in Australia, sank in 2010 during adverse weather conditions as it was being towed to calmer waters. The device had only been deployed for three months; however, testing had been completed and the device had been grid-connected for two of those months.

These examples of WEC device failures demonstrate the requirement to conduct further research on the strengthening and resilience optimisation of WEC structures to ensure that they remain serviceable for their design life duration. The Framework for Ocean Energy Technology published by OES in 2021 [357] defines nine evaluation areas that are key to the success of a wave energy project. Included in these evaluation criteria are *reliability*, *survivability*, and *maintainability*. These are all key factors in ensuring the WEC device remains operational for the duration of its design life. According to the OES document, reliability is defined as the *probability that an item can perform a necessary function under given conditions for a given time interval*; survivability is a measure of the ability of a subsystem or device to experience an event (*Survival Event*) outside the expected design conditions and not sustain damage or loss of functionality beyond an acceptable level, allowing a return to an acceptable level of operation after the event has passed; and maintainability is defined as the *ability to be retained in, or to be restored to a state to perform as required, under given conditions of use and maintenance*. Clearly all three of these evaluation areas determine whether the WEC remains functional during its design life. In terms of reliability [769], it is inevitable that some systems will fail during the design life of the device. However, it is important to minimise these failures through good design and, where these failures do occur, it is critical to ensure that they can be rectified as quickly as possible with minimal consequences to the power generation output. Some reliability issues may not impact the daily operation of the WEC device and may be repaired during pre-scheduled maintenance works. With regards to survivability [652, 651], the OES document states *Activities relating to Survivability predominantly occur during the earliest stages while significant design decisions are being made. They focus on understanding the events which may cause damage [410, 806] or loss of functionality for the device and develop means to mitigate these risks. Extensive understanding of the operational requirements at the intended commercial deployment site and open water test sites, as well as the environmental conditions and device behaviour are fundamental.* The document goes on to list the factors upon which survivability depends:

- Likelihood of experiencing an event which results in components, subsystems, or devices operating beyond their expected design conditions.

- Likelihood of being able to predict or detect the survival event and take suitable protective action.

- Likelihood of resisting the event, having taken suitable protective action.

- Likelihood of resisting the event, not having taken suitable protective action.

Clearly, the survivability of a WEC device is heavily dependent on the initial design of that device and knowledge of its operational environment. Device maintenance is another critical consideration that influences the operational duration of the WEC device. Indeed, some components will experience a greater amount of wear and tear and consequently have a shorter design life than the overall installation. These components will require servicing or replacement at regular intervals. Biofouling and the harsh marine environment contributing to accelerated corrosive effects dictate that device maintenance will be a periodic necessity. These maintenance works should be considered and a plan for their implementation must be included at the WEC device design stage. Other important considerations for maintenance include the environmental conditions in which the maintenance can be conducted, for example, can the work be carried out in bad weather?; and is it necessary to conduct the maintenance on site or can the component be moved onshore to carry out necessary repairs?; etc. Finally, machine down-time and the cost of maintenance must be considered. A well maintained device will have an increased prospect of remaining in service for its design life duration. Much research is currently ongoing to extend the design life of WEC devices and ensure their survivability [652, 651]. Common themes of interest include mooring line analyses [381, 336, 110, 380, 848], the provision of wave impact pressure relief channels on the device super-structure and integration with shoreline devices that can provide a more robust support for the WEC technology [410, 806]. Research into improving the efficiency of WEC devices to allow them to be deployed in calmer waters whilst still being commercially feasible is also an important topic which is receiving much interest [118, 400]. Finally, research on wave-structure interactions to improve our knowledge of the nonlinear and complex load transfer mechanisms [831, 652, 321] is continuing apace and will ensure the increased durability of WEC devices deployed in the future.

1.4 FUTURE DEVELOPMENTS IN WAVE ENERGY CONVERTER SYSTEMS

The Low Carbon Energy Observatory report, published by the European Commission in 2020 [477], has classified the priorities for innovation of key components of WEC devices. Technological developments concerning PTO optimisation and improvement, moorings and foundations and array dynamics, interactions, and optimisation are all classified as high priority; whilst environmental impacts and monitoring are classified as medium priority. A workshop on the identification of future emerging technologies in the ocean energy sector, organised by the JRC, the European Commission's science and knowledge service, in 2018 [480], also expanded on these innovation priority areas. According to this document, wave energy technology can be divided into first and second generation devices. First generation devices, including Pelamis , Wavebob, Oceanlinx, Wave Dragon etc., are characterised by a mathematical analysis design that attempts to extract the entirety of the energy contained in the waves. In contrast, second generation devices are those WECs that extract energy from waves in a different way from the mathematically-inspired concepts followed by the first generation devices. Instead, these second generations devices rely on innovative, deformable materials, triboelectric activated polymer materials or multiple degree of freedom oscillations. These second generation designs are currently at a very low TRL level and are mostly in the research phase at TRL 1-3.

The wave energy sector as a whole is currently regarded as being at TRL 6-7. Whilst there have been a number of successful demonstrator projects, and in some cases WEC technology has been operational over long durations, the total electrical output remains low. Thus, wave energy is still regarded as being at the pre-commercial stage. It is estimated that between 2015 and 2020 alone, there were more than one thousand wave energy device patents granted across Europe, the USA, and Asia. With this rapid growth in the sector, more efficient devices are being developed; however, further innovations need to be realised to elevate the wave energy sector on the TRL scale. The following sections discuss three of the most important future developments that are necessary to ensure the commercialisation of the wave energy sector.

1.4.1 Efficiency augmentation

At present, there is a large amount of work being undertaken to augment the efficiency of WECs. This work is being conducted at both academic institutions and commercial enterprises. The workshop on identification of future emerging technologies in the ocean energy sector has identified a number of challenges related to the operational efficiency of WECs. These include:

- Poor efficiencies in the conversion from captured mechanical power from wave motion to electrical power (influence by the PTO system, control, and power electronics);

- Limitations of PTO loading during extreme events (influenced by power electronics);

- Fluctuations of power output;

- The WECs predominantly harness energy from the vertical or surge motion in waves while multiple degree of freedom motion modes are not yet well developed.

With regard to the first generation WECs, a number of new, innovative designs have been developed to optimise the power capture and increase the efficiency performance. WaveNet [305, 488, 715] is a multiple point absorber that allows wave energy to be captured from 5 out of the 6 degrees of freedom, The Symphony WEC [104, 430] design is based on the Archimedes WaveSwing device [167, 640, 792, 68] but incorporates a newly-patented water turbine design, which is predicted to be more efficient. HACE [374] is a company that is developing a multi-chambered OWC with a trefoil plan footprint, which is operational in all wave directions. Seabreath [501, 532, 681] is another OWC device that can produce a continuous, unidirectional airflow due to its two-chambered design. This unidirectional air flow is proposed to be more efficient than the traditional OWC designs. Hann-Ocean Energy have developed a novel twin-chambered OWC system named Drakoo [849] that directs water through a hydro turbine connected to the chamber. The water flow is constant and unidirectional, resulting in smooth operation of the hydro turbine.

WEC devices that can be considered as purely second generation include the WEC S3® developed by SBM Offshore. This device consists of a floating, flexible membrane with embedded electro-active polymers, which are capable of direct energy conversion from wave to electricity. [481, 375, 48, 53, 136]. The electricity is generated by the wave-induced motion of the flexible membrane. The LILYPAD WEC [136, 481] is a floating double membrane device with the upper membrane floating on the ocean surface and the lower, weighted membrane fully submerged. The upper and lower membranes are connected by a system of extensible hose pumps and ties. As waves interact with the upper membrane, hydraulic pressure is built up in the system of hose pumps, activating a turbo generator. Other notable second generation WEC devices that are at an early stage of development as of 2021 are the Costas Wave device [322], developed at Aalborg University, and the Anaconda, developed at the University of Southampton in the UK [346, 112, 332]. At the subsystems level, a number of innovations are also being made, most notably to the PTO systems. Table 1.1 highlights some of the companies, their devices, and the WEC category in which these innovative PTO systems are installed.

Research work on control systems [352, 835] for WEC devices is also ongoing, with specific emphasis on passive control [814, 673, 271], reactive control [354, 241, 733, 7], phase control [96, 688], and latching control [762, 847, 770, 452]. Sandia National Laboratory in the US, Wave Energy Scotland, and the Centre of Ocean Research (COER) at the National University of Ireland, Maynooth, are all involved in augmenting WEC control systems in particular. In Chapter 3 a more comprehensive discussion of WEC control systems is presented.

1.4.2 Integration with other renewable energy harvesting systems and technologies

The synergistic integration of WEC systems with alternative marine structures can present significant advantages. The CAPEX can be reduced by combining WECs with existing shoreline infrastructure. Additionally, if new coastal infrastructure projects are planned, such as breakwater walls, port and harbour edifices, desalination plants, causeways etc., the carefully designed integration of a WEC system can generate longer-term construction cost-sharing benefits. Other benefits include a reduction in the design life, maintenance costs due to ease of access to carry out ongoing works, reduced environmental impact due to the consolidation of multiple, varied purpose structures into a single installation, and improvement to each of the component structures' design life duration due to complementary strengthening of each constituent substructure.

The integration of WECs with different forms of renewable energy devices has also received some attention since the early 2020s. There are a number of projects considering the feasibility of constructing floating solar panel arrays [596, 656] with WEC structures integrated into the platform. Similar ideas are being considered for the integration of WECs into floating wind turbine platforms [643, 39]. An overview of some of these projects, which are either at the research or testing stage, is presented in the following paragraphs.

There are a number of examples of WEC-breakwater integrated devices either already constructed or in the planning phase. The WEC-harbour wall built at Sakata [594, 487, 751, 593] in Japan was one of the first such WEC-infrastructure integrated installation constructed. Currently, there is ongoing research at the University of Stellenbosch in South Africa to construct a breakwater-integrated multi-chambered OWC named ShoreSWEC [386, 385, 170]. Research into the device development builds upon studies conducted at the same university in the 1970s and 1980s; however, the program does not seem to have advanced very much since then. The Siadar Wave Power Station [821, 13], originally proposed to be built at Siadar Bay, close to the Isle of Lewis in Scotland, was another WEC-breakwater integrated installation that was not realised. Had it been built, the plant was rated to produce 4 MW of electricity. The proposed project was cancelled in 2012. The Overtopping BReakwater for Energy Conversion (OBREC) device is a prototype, overtopping type device that has been integrated into a rubble mound breakwater wall at Naples Harbour in Italy, and has been operational since 2016 [806, 557]. The ReWEC3 OWC-integrated breakwater has also been constructed in Italy at the harbour of Civitavecchia, with construction completed in 2015 and a total power output of 2.6 MW [31]. The Seawave Slot Cone Generator (SSG) is another example of an overtopping-type device that is designed to be integrated into a breakwater wall but which is still in the development phase as of 2021 [805, 405]. Feasibility studies are also ongoing for a number of breakwater-WEC integrated structures in the Finistère region on the north-west coast of France [465]. KRISO, the Korean Research Institute of Ships and Ocean Engineering is also investigating dual functioning WEC installations, including a 30 kW OWC WEC integrated into a breakwater wall to provide energy to off-grid islands [608, 401, 402].

Design concepts to integrate WECs with other installations providing alternate forms of renewable energy are less advanced. The proposed W2Power device [434, 109, 333, 514], patented and launched by Pelagic Power and currently owned by EnerOcean, was originally conceived as an integrated energy harvesting platform comprising of a pair of wind turbines and a number of point absorbers. The platform

Figure 1.30: (a) the EnerOcean W2Power device undergoing testing at the Flowave tank at Edinburgh University, (image courtesy of EnerOcean), [514, 333], (b) photo of the prototype 1:6 scale EnerOcean W2Power device, (image courtesy of EnerOcean), (c) conceptual image of the DualSub device, (image courtesy of Marine Power Systems), (d) conceptual image of the modular SINN Power wave-wind-solar platform, [732].

is triangular in plan view with the wind turbines positioned outside of the area of the triangle. The point absorbers are distributed along the platform's edges. In its latest released design (2019), the W2Power platform is rated for 12 MW of wind power using two 6 MW-class turbines, and additionally could produce up to 3 MW of wave power in strong wave climate conditions. Figure 1.30 (a) shows the 1:40 scale model platform incorporating WECs undergoing testing at the Flowave tank at Edinburgh University, and Figure 1.30 (b) is a photo of the prototype 1:6 scale device that was successfully sea tested over a period of 4 months in 2019, at the PLOCAN test site off the island of Gran Canaria. As of 2021, the W2Power floating wind turbine (without WECs) is considered to be at TRL 6. An array of W2Power wind turbines is in the planning phase which will take the technology beyond TRL 7. As an integrated technology to capture both wind and wave energy, the device has reached TRL 5 at the conclusion of the Flowave tank tests conducted in 2015.

Marine Power Systems (MPS), a company based in Wales, is also investigating the possibility of combining wind energy with wave energy. Their technology, named

DualSub (see Figure 1.30 (c)) [337], combines a platform-supported wind turbine with a system of submersed buoys that generates electricity by displacing in an orbital manner due to wave motion. The technology has already been subjected to some scaled testing and is projected to be capable of producing up to 20 MW of power at full-scale.

The Ocean Hybrid Platform (OHP) [133, 175, 639] developed by SINN Power is a modular, integrated wave-wind-solar platform which is currently undergoing testing at the port of Iraklio in Greece (see Figure 1.30 (d)).

The Renewable Energy Integration Demonstrator – Singapore (REIDS) is a Singapore-based R3D (Research, Development, Demonstration, Deployment) project that is investigating the integration of wave energy with other forms of renewable energy to power island communities in south-east Asia. The Wavegem platform has been undergoing testing at the marine technology SEM-REV test site off the coast of France since 2019 [516, 133]. This platform, developed by the GEPS Techno company, consists of a hybrid wave-solar power production system with a future projected capacity of up to 1 MW of power. Another integration concept that has been investigated and developed to prototype level is the combination of a WEC with an aquaculture cage. In 2020, the Penghu demonstration project [576, 747], developed by the Guangzhou Institute of Energy Conversion (GIEC), completed 18 months of testing. The device is an integrated semi-submersible platform that combines aquaculture and wave energy harvesting. The device has been patented in China, the EU, and Japan.

Currently there is a large body of research being produced on the viability of wave-powered desalination plants. In most cases, the total wave energy produced is consumed by the desalination plant; however, in a number of cases, the energy produced is in excess of the plant requirements and this energy has the potential to be fed back into the grid system. The fully submerged point absorber, CETO 5 WEC, was proposed to power a desalination plant on Garden Island off the west coast of Australia [647]. The Danish company, Wavepiston has begun the full-scale testing phase of their technology, also named Wavepiston, at the PLOCAN test site [21, 654, 614]. The device is capable of harnessing wave energy whilst simultaneously pumping water to a desalination plant. The system is projected to have a peak power output of 200 kW and will be capable of powering up to 140 households.

Finally, there is limited research ongoing into the integration of wave power devices with the offshore oil and gas industry [439, 157, 377]. Lundin Energy, in collaboration with the Swedish company, Ocean Harvesting Technologies, is currently undertaking a one year study to investigate the possibility of combining wave energy technology with oil and gas platforms to facilitate the electrification of daily running operations. Mocean Energy are also working with ocean technology companies EC-OG, Chrysaor, Modus, OGTCV and Baker Hughes to develop a wave-powered renewable energy system capable of operating oil and gas subsea equipment. Bombora is another such company that are investigating the feasibility of using their devices to supply clean energy for the operational requirements of offshore oil and gas platforms. This research into efforts to decarbonise the non-renewable sector will have clear benefits in terms of reducing operational pollutant emissions and meeting future climate change targets.

1.4.3 Wave energy farms

As WEC device development matures and progresses along the TRL scale, the commercialisation of the technology is becoming an important consideration. Economies of scale cost reductions dictate that WEC arrays will become more cost effective as component device construction, deployment, maintenance, and operational costs reduce when more WECs are integrated into an array. At present there are a number of small arrays in operation. Some of these, such as the array of OWCs operating at the Mutriku power plant, are integrated into a single structure; whilst other arrays, such as the three Pelamis devices, were deployed to operate independently, albeit in close proximity. In this section, some of these WEC arrays that are already operational or at the planning stage, are discussed.

Seabased, a company that develops floating buoy type WECs, has supplied a number of their devices to WEC farms around Europe and in Africa [115, 682, 880]. The Ada Foah wave farm in Ghana was installed in 2016 and initially consisted of 6 L12 floating buoy type devices generating 400 kW of power. Following on from the success of this project, the Ghanaian government signed a contract with Seabased in 2019 to provide enough WECs to install a wave energy farm with a capacity of 100 MW. The Sotenäs wave energy farm [535, 721] off the coast of Sweden consisted of 36 WEC buoys supplied by Seabased. The plant had a capacity of almost 3 MW. The plant ceased operations in 2019. The Aguçadoura test facility, off the north coast of Portugal was the location selected to install three Pelamis WECs in 2008 [156, 195, 345, 500]. The array was commissioned in 2008 but was closed down just two months later when technical problems with the devices were identified. The array had a total installed capacity of 2.25 MW (see Figure 1.31).

Figure 1.31: Pelamis array of 3 WECs [165].

The Swedish wave energy company, CorPower Ocean, has secured a 10 year marine licence to install and begin testing an array of four full-scale C4 WEC devices at the Aguçadoura site [293]. The project, HiWave-5, aims to demonstrate that the technology has reached TRL 8. The Australian company, Bombora, intends to begin testing on a small scale commercial array of 1.5 MW mWave™ devices at the Albany test site in southwest Australia [107, 128]. Depending on the localised site wave climate, the testing could comprise up to six devices. The Perth Wave Energy Project at Garden Island, Western Australia, consists of three CETO 6 type submerged buoys developed by Carnegie Clean Energy [493, 646, 844]. The power plant, which also works as a dual-purpose desalination facility, has a peak rated capacity of 5 MW. The Wanshan 1 MW Wave Energy Demonstration project is currently being installed in China and consists of two 500 kW point absorbertype WEC devices, named the Zhoushan and the Changshan [90]. SINN power are also currently testing their breakwater fixed multi point absorber WEC array at Iraklio Port, Greece [424, 387, 451]. Eco Wave Power [598, 175, 810] have developed a similar technology to SINN Power. Their devices are also wall-mounted floating buoys which are designed to be attached to breakwater walls or harbour walls, etc. The company has received a permit to install ten point absorbers to be attached to a breakwater wall at the port of Jaffa in Israel as part of the EWP-EDF One project. The combined capacity of the installation will be 100 kW. The company has also installed a WEC array of point absorbers at an old World War II ammunition jetty on the east coast of Gibraltar. The plant currently provides 100 kW to the electric grid but plans to expand the plant to 5 MW are in place.

One of the greatest challenges in the development of wave energy array farms is to understand the complex wave-structure interaction effects, especially in the diffraction and multi-point source radiated wave field. In the wave energy array sector, this interactive wave field response across multiple WECs is known as the park effect [41, 303, 281]. This subject and other WEC array topics are explored in further detail in Chapter 9.

II

Fundamental Theories Applicable to Wave Energy Harvesting

CHAPTER 2

Fluid dynamics and wave-structure interactions

Malin Göteman[1], Robert Mayon[2], Yingyi Liu[3],
Siming Zheng[4], Rongquan Wang[2]

[1]Uppsala University, malin.goteman@angstrom.uu.se,
[2]Dalian University of Technology, rmayon@dlut.edu.cn,
[3]Kyushu University, liuyingyi@riam.kyushu-u.ac.jp,
[4]University of Plymouth, siming.zheng@plymouth.ac.uk

2.1 FLUIDS AND WAVE-STRUCTURE INTERACTIONS

To understand the energy absorption from ocean waves, it is necessary to understand the source itself – the fluid and the ocean waves – and how it interacts with structures, such as a wave energy converter. Here, the properties and description of ocean waves will first be discussed based on fundamental as well as stochastic principles, after which the hydrodynamic forces due to the waves and their interaction with bodies in the fluid will be discussed.

2.1.1 Ocean waves

Several types of ocean waves exist. Tsunami waves are very long, fast waves caused by an earthquake or landslide, and capillary waves are small ripples on the water surface, generated by the wind and dominated by surface tension effects. In wave energy applications, the waves of interest are *wind-generated gravity surface waves*, i.e., waves resulting from wind blowing at the ocean surface, and dominated by gravity and inertial forces. Wind-generated ocean waves are thus a renewable energy source, distilled in two steps from the solar energy incident on the earth, which causes wind and in the next step waves. As such, ocean waves contain more energy per unit volume than wind and solar energy, and the wave energy resource roughly resembles the characteristics of wind energy and is largest at high latitudes, as shown in Figure 1.24.

DOI: 10.1201/9781003198956-2

Figure 2.1: In general, ocean waves consist of waves of many frequencies, travelling simultaneously in different directions (left). Due to the dispersive property of ocean waves, waves propagating over long distances separate according to frequency, producing swells (right).

As water waves, ocean waves can in their most fundamental form be described by the Navier-Stokes equations for incompressible fluids, as will be discussed in Section 2.1.1.1. Numerical solutions of these equations form the basis for the computational fluid dynamics methods that are discussed in detail in Section 2.3. For many applications, however, simplified theories based on neglecting viscosity, turbulence, and non-linear effects of the fluid can be used, resulting in linear potential flow theory. These assumptions and their limits of validity will be discussed in Section 2.1.1.2. As we will see, the solutions describe surface waves that satisfy the *dispersion relation*, meaning that the waves are dispersive and travel with speeds proportional to their wave lengths. This gives rise to the formation of swells, which can be seen in Figure 2.1 and will be discussed in more detail in Section 2.1.1.3. From the linearity of the problem, a sum of solutions is again a solution to the linear potential flow theory, and ocean waves can, to a good approximation, be described as superpositions of sinusoidal waves with different frequencies and phases. Such irregular waves can be described by stochastical parameters and wave spectra, to be introduced in Section 2.1.1.4.

2.1.1.1 *Navier-Stokes equations*

The fundamental equations describing fluids are the Navier-Stokes equations together with the continuity equation. The continuity equation stems from the physical principle that the mass of the fluid element must be conserved, and can be expressed as:

$$\frac{\partial \rho}{\partial t} + \nabla \cdot (\rho \mathbf{u}) = 0 \tag{2.1}$$

where the first term represents the change in fluid density ρ and the second term the mass flow, and $\mathbf{u}$ denotes the fluid velocity.

Water can (to a good approximation) be considered incompressible [242], and to distinguish water from other fluids such as gases, an incompressibility constraint can be imposed,

$$\nabla \cdot \mathbf{u} = 0. \tag{2.2}$$

Finally, the Navier-Stokes equation is derived from the fundamental principle of momentum conservation. In other words, it is derived by applying Newton's second law

$$\mathbf{F} = m\mathbf{a} \tag{2.3}$$

to a small fluid volume, and expressing on the left-hand-side all the forces acting on the fluid element, and expressing the right-hand-side as the density of the fluid times its acceleration, integrated over the fluid volume. The forces acting on the fluid are external forces (usually only gravity) and internal forces acting on the surface of the fluid element by neighbouring fluid elements. The internal force is called stress force and can be divided into pressure p (acting perpendicular to the surface) and shear stress (acting parallel). For incompressible fluids, the shear stress can be rewritten in terms of viscosity ν, and the Navier-Stokes equation takes the form

$$\frac{\partial \mathbf{u}}{\partial t} + \mathbf{u} \cdot \nabla \mathbf{u} = -\frac{1}{\rho}\nabla p + \nu \nabla^2 \mathbf{u} + \frac{1}{\rho}\mathbf{F}_{\text{ext}} \tag{2.4}$$

where $\mathbf{F}_{\text{ext}}$ is the external net force acting on the fluid, usually only the gravity force $\mathbf{F}_{\text{ext}} = \nabla(-\rho g z)$. For inviscid flow, the Navier-Stokes equations reduce to the Euler equations.

An incompressible fluid is thus governed by the continuity Eq. (2.1), an incompressibility constraint (2.2), and the Navier-Stokes Eq. (2.4). These form a system of non-linear partial differential equations, and finding analytical solutions is difficult, or impossible. In fact, to prove the existence of smooth solutions in three dimensions is one of the most important open problems in mathematics and physics, and a reward of 1 million USD has been offered by the Clay Mathematics Institute for a solution or counterexample.

To find solutions, there are two main approaches. The first, which will be discussed in detail in Section 2.3, is to discretise the problem and solve it using approximate numerical methods, such as Reynolds-averaged Navier-Stokes (RANS) computational fluid dynamics (CFD) methods. The second approach is to make assumptions to simplify the equations, such that analytical or numerical solutions can be readily found. This approach will now be discussed in detail in the remainder of this section and in Section 2.2.2.

2.1.1.2 Linear potential flow theory

Consider a fluid which is irrotational ($\nabla \times \mathbf{u} = 0$). This assumption is equivalent (at least for simply connected domains) to the fluid velocity being conservative, i.e., it is the gradient of some fluid velocity potential, $\mathbf{u} = \nabla \Phi$. Together with the incompressibility constraint in Eq. (2.2), this implies that the fluid potential satisfies the Laplace equation,

$$\nabla^2 \Phi = 0. \tag{2.5}$$

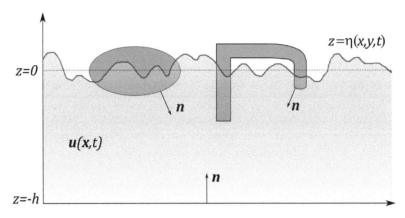

Figure 2.2: Notation of the fluid domain.

Furthermore, assume that the fluid is ideal (i.e., it has neglectable viscosity $\nu = 0$) and that the external force in the Navier-Stokes Eq. (2.4) is gravity $\mathbf{F}_{\text{ext}} = (0,0,-\rho g) = \nabla(-\rho g z)$, then the Navier-Stokes equation takes the form

$$\frac{\partial}{\partial t}\nabla\Phi + \frac{1}{2}\nabla\left((\nabla\Phi)^2\right) = -\frac{1}{\rho}\nabla p + \frac{1}{\rho}\nabla(-\rho g z). \tag{2.6}$$

By collecting the terms on one side and integrating over the spatial dimensions, the Bernoulli equation is obtained,

$$\frac{\partial}{\partial t}\Phi + \frac{1}{2}(\nabla\Phi)^2 + \frac{1}{\rho}p + gz = \text{const.} \tag{2.7}$$

A priori, the constant on the right hand side may be time-dependent $C = C(t)$, but this time dependency may be embedded in a field redefinition $\Phi' = \Phi - \int_{t_0}^{t} C(\tau)d\tau$. When multiplied with the density, the second term can be seen to represent the kinetic energy $\rho u^2/2$; whereas the term $\rho g z$ represents the potential energy. As we will come back to in Section 2.1.2, when computing hydrodynamic forces, the Bernoulli equation can be used to compute the pressure in the fluid domain.

At the free surface, the pressure equals the atmospheric pressure, $p = p_{\text{atm}}$. To find the constant C on the right hand side of the Bernoulli equation, consider the static case where there are no waves, i.e., there is no fluid velocity $\mathbf{u} = 0$, ϕ is constant and $z = 0$ at the free surface, see Figure 2.2. This implies that $C = \frac{1}{\rho}p_{\text{atm}}$, which further implies that generally, at the free surface $z = \eta(x,y,t)$, the dynamic boundary constraint holds,

$$\frac{\partial}{\partial t}\Phi + \frac{1}{2}(\nabla\Phi)^2 + g\eta = 0 \quad \text{at } z = \eta(x,y,t). \tag{2.8}$$

At any fixed, rigid boundary in the fluid domain, such as the sea bed, the fluid is constrained such that it cannot penetrate the boundary. In other words, the fluid velocity in the direction of the rigid boundary must vanish,

$$\frac{\partial\Phi}{\partial\mathbf{n}} = 0, \tag{2.9}$$

where $\mathbf{n}$ is the normal of the boundary pointing in towards the fluid domain. Note that in many applications in CFD, the non-penetration constraint (2.9) is replaced by a more conservative no-slip condition, where the relative fluid velocity is zero in all directions at rigid walls. In addition, the fluid particles should stay in the water, i.e., the kinematic constraint is imposed such that the vertical velocity of a fluid particle at the free surface, $u_z = \partial\Phi/\partial z$ should equal the vertical velocity of the surface itself, $\dot\eta(t)$,

$$\frac{\partial\Phi}{\partial z} = \frac{\partial\eta}{\partial x}\frac{\partial\Phi}{\partial x} + \frac{\partial\eta}{\partial y}\frac{\partial\Phi}{\partial y} + \frac{\partial\eta}{\partial t} \quad \text{at } z = \eta(x,y,t), \tag{2.10}$$

where it was also used that the fluid velocity components can be written in terms of the velocity potential, $\dot x(t) = u_x = \partial\Phi/\partial x$.

In total, the fluid is governed by the Laplace Eq. (2.5) together with the boundary constraints (2.8) and (2.10) at the free surface and (2.9) at any rigid boundary. If the rigid boundary is moving, a more general form of the boundary constraint (2.9) is valid where the velocity of the fluid equals the velocity of the moving boundary.

Linearisation Potential flow theory relies on the assumptions of irrotational, inviscid, and incompressible fluid, but is still described by non-linear partial differential equations. To simplify the non-linear boundary constraints (2.8) and (2.10) at the free surface, a further assumption on non-steep waves is required, i.e., that the wave height H is small in relation to the wave length λ, $H \ll \lambda$. A small parameter

$$\epsilon = ka \ll 1 \tag{2.11}$$

is often defined, where $k = 2\pi/\lambda$ is the wave number and $a = H/2$ the wave amplitude. The fluid potential and the surface elevation can then be expanded as perturbations around the still free surface $z \approx 0$ and the first-order approximation taken. In practice, this amounts to neglecting all non-linear terms in Eqs. (2.8) and (2.10) since they are small. The resulting linear boundary constraints at the free surface are

$$\frac{\partial\Phi}{\partial t} + g\eta = 0 \quad \text{at } z = 0 \tag{2.12}$$

$$\frac{\partial\Phi}{\partial z} - \frac{\partial\eta}{\partial t} = 0 \quad \text{at } z = 0 \tag{2.13}$$

which can be combined into

$$\frac{\partial^2\Phi}{\partial t^2} + g\frac{\partial\Phi}{\partial z} = 0 \quad \text{at } z = 0. \tag{2.14}$$

Together with the fixed rigid body boundary constraint (2.9) and the Laplace Eq. (2.5) in the full domain, they define linear potential flow theory, also known as Airy wave theory.

As defined by Eq. (2.11), the linearisation relies on a perturbation in a small parameter, which only remains small as long as the waves are non-steep. As the waves become steeper, higher-order non-linear terms should be included in the approximation. This is shown in Figure 2.3, which specifies the validity of linear and

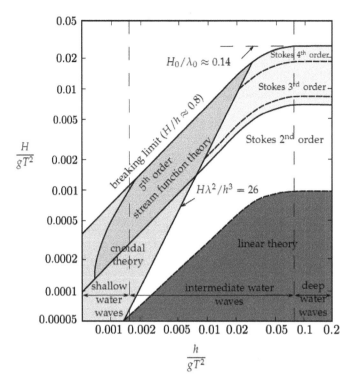

Figure 2.3: Wave model suitability, adapted from [276].

higher-order theories at different water depths and at different wave heights. The validity of different theories describing ocean waves as well as their corresponding solvers will be further discussed in Section 2.5.

2.1.1.3 Dispersive waves

A solution to the Laplace Eq. (2.5) and the linear boundary constraints at the free surface (2.12)–(2.13) and at fixed rigid boundaries (2.9) can be found by separation of variables. Here, the derivation will be considered for a wave propagating only in the x-direction, but the generalisation to wave propagation in both the x and y directions follows analogously.

Consider waves propagating along the x-direction and make the ansatz for a harmonic wave with wave number k and angular frequency ω,

$$\Phi(x,z,t) = Z(z)\sin(kx - \omega t). \tag{2.15}$$

Inserting the ansatz into the Laplace equation gives

$$0 = \nabla^2 \Phi = \left[-k^2 Z(z) + Z''(z)\right]\sin(kx - \omega t), \tag{2.16}$$

implying that the ansatz satisfies the Laplace equation provided that the vertical function $Z(z)$ takes the form $Z(z) = Ae^{kz} + Be^{-kz}$ for some constants A and B. The boundary constraint at the seabed $z = -h$ further implies that

$$0 = \frac{\partial \Phi}{\partial z}\Big|_{z=-h} = \Big[Ake^{-kh} - Bke^{kh}\Big]\sin(kx - \omega t) \tag{2.17}$$

which implies that the two constants are related as $A/B = e^{2kh}$. Finally, the potential must satisfy the boundary constraints at the free surface, which in combined forms take the expression

$$0 = \frac{\partial^2 \Phi}{\partial t^2} + g\frac{\partial \Phi}{\partial z}\Big|_{z=0} = \Big[-\omega^2(A + B) + gk(A - B)\Big]\sin(kx - \omega t) \tag{2.18}$$

from which follows that

$$\omega^2 = gk\frac{A - B}{A + B} = gk\frac{e^{2kh} - 1}{e^{2kh} + 1} = gk\tanh(kh). \tag{2.19}$$

With the relationship between the constants A and B inserted, the vertical function takes the form $Z(z) = Be^{kz}(e^{k(z+h)} + e^{-k(z+h)}) = C\cosh(k(z+h))$, where C is some new constant. Often, the constant C is chosen such that $Z(0) = gH/(2\omega)$, which gives the final linear solution: a fluid potential of the form

$$\Phi(x, z, t) = \frac{gH}{2\omega}\frac{\cosh(k(z+h))}{\cosh(kh)}\sin(kx - \omega t) \tag{2.20}$$

satisfies the Laplace Eq. (2.5) and the linear boundary constraints (2.12)–(2.13) at the free surface and the sea bed (2.9), provided that the *dispersion equation* holds,

$$\omega^2 = gk\tanh(kh). \tag{2.21}$$

The dispersion equation relates the frequency of a wave with its wave length, and implies that waves of different wave lengths in general travel with different speeds, i.e., that the phase speed $v_p = \omega/k \propto \lambda$. Waves for which this is true are denoted dispersive. This can be compared with electromagnetic waves in vacuum that are non-dispersive. The dispersion equation for electromagnetic waves in vacuum is $\omega = ck$, where c is the speed of light, implying that the phase speed is always equal to the constant speed of light, $v_p = \omega/k = c$.

The surface elevation corresponding to the fluid potential (2.20) is given by the free surface boundary constraint (2.12),

$$\eta(x, t) = -\frac{1}{g}\frac{\partial \Phi}{\partial t}\Big|_{z=0} = A\cos(kx - \omega t) \tag{2.22}$$

which is a harmonic plane wave with amplitude $A = H/2$, propagating along the x-direction.

The energy available in a harmonic wave is the sum of its kinetic and potential energy. In deep water, it equals

$$E = \frac{1}{8}\rho g H^2 = \rho g\langle \eta^2 \rangle, \tag{2.23}$$

where ρ is the density, H the wave height, and $\langle \eta^2 \rangle$ is the time average of the surface elevation (2.22) squared, also called the variance.

In addition to the energy available in ocean waves per unit ocean area, the transportation of energy, or *energy flux*, is a relevant parameter for wave energy applications. The average energy transported by the per-unit frontage of the incident wave can be calculated as [226, 839]

$$J = \frac{\rho g^2}{4\omega}|A|^2, \tag{2.24}$$

in deep water, where A is the complex wave elevation, and

$$J = \frac{\rho g^2}{4\omega}D(kh)|A|^2, \tag{2.25}$$

in finite-depth water, where $D(kh)$ is the depth function

$$D(kh) = \tanh(kh)\left[1 + \frac{2kh}{\sinh(2kh)}\right]. \tag{2.26}$$

which is a function of the depth h and asymptotic to 1 in deep water. Note that J will be used in Chapters 3, 5, and 8 in the definitions of capture width, interaction factors, etc.

2.1.1.4   *Wave spectra and wave parameters*

From the linearity of the Laplace equation and the boundary constraint, solutions can be superposed into new solutions. In general, ocean waves consist of sums of many harmonic waves of the form (2.20), travelling in many directions simultaneously. Since ocean waves are dispersive, when they travel over long distances they will separate according to their frequency. When reaching a distant point, the resulting superposition will consist of waves with similar frequencies, the so-called swells, see Figure 2.1.

Consider an irregular ocean surface composed of many harmonic waves (2.22) of amplitudes A_n and angular frequency $\omega_n = nd\omega$. At a certain point (x, y), the time-dependent surface elevation is given by $\eta(t) = \sum_n A_n \cos(\omega_n t + \varphi_n)$, where φ_n are phases. The variance of the surface elevation for the irregular waves is

$$\langle \eta^2 \rangle = \frac{1}{T}\int_0^T \left[\sum_{n=0}^{\infty} A_n \cos(\omega_n t + \varphi_n)\right]\left[\sum_{n=0}^{\infty} A_m \cos(\omega_m t + \varphi_m)\right]dt = \sum_{n=0}^{\infty}\frac{1}{2}A_n^2, \tag{2.27}$$

where the orthonormality of trigonometric functions was used, implying that the only contribution from the product of the cosine functions is when $m = n$. By comparing (2.23) and (2.27), the conclusion can be drawn that the energy of the nth wave component is given in terms of the amplitude of the wave component as $E_n = \frac{1}{2}\rho g A_n^2$.

At this point a new function $S_\eta(f_n)$, called the spectral density function, can be defined as $S_\eta(f_n)df = \frac{1}{2}A_n^2$ with units m^{2}s, so that

$$\langle \eta^2 \rangle = \sum_{n=0}^{\infty}\frac{1}{2}A_n^2 = \frac{1}{\rho g}\sum_{n=0}^{\infty}E_n = \sum_{n=0}^{\infty}S_\eta(f_n)df, \tag{2.28}$$

where $f_n = \omega_n/2\pi$ are the frequencies. The spectral density function is proportional to the energy density for each wave component f_n. When the frequency step becomes infinitesimally small $df \to 0$, the sum can be written as an integral, and the variance, and thus the energy of the irregular waves, can be expressed as

$$E = \rho g \langle \eta^2 \rangle = \rho g \int_0^\infty S_\eta(\omega)d\omega. \tag{2.29}$$

A harmonic wave is characterised by its wave height, wave length, and period (and the latter two parameters are related by the dispersion equation). Clearly, irregular waves cannot be described by a single wave height and period. Instead, these parameters must be defined statistically, which is undertaken in terms of so-called spectral moments. The nth spectral moment is defined in terms of the spectral density function and the wave frequency as

$$m_n = \int_0^\infty f^n S_\eta(f)df. \tag{2.30}$$

From comparison with Eq. (2.29), it is clear that the zeroth spectral moment is proportional to the energy of the irregular waves,

$$m_0 = \int_0^\infty S_\eta(f)df = \langle \eta^2 \rangle = \frac{1}{\rho g}E. \tag{2.31}$$

The significant wave height is defined in terms of the zeroth spectral moment as

$$H_s = 4\sqrt{m_0}, \tag{2.32}$$

sometimes also denoted H_{m0}. Its definition has historical reasons. An older definition of significant wave height is the average wave height of the highest 1/3rd of the waves, denoted $H_{1/3}$. This was the wave height "estimated by a well-trained observer". The new definition in terms of spectral moments agrees with the historical one within a few percent, $H_s \approx H_{1/3}$.

The energy period of irregular waves is defined as

$$T_e = \frac{m_{-1}}{m_0}. \tag{2.33}$$

It is the period with which a single harmonic wave would have the same energy as the irregular waves. Another useful statistical property is the average up- or down-crossing periods. An up-crossing period is the time between when the surface elevation passes from $z < 0$ to $z > 0$, to the next such event. The up-crossing period is defined as the average of those periods, and can be written in terms of spectral moments as $T_z = \sqrt{m_0/m_2}$. The average crest period $T_c = \sqrt{m_2/m_4}$ is defined as the length of the wave record divided by the number of crests within that record, and the peak period $T_p = 1/f_p$ is defined as the period for which the spectral density has a peak, $S'(f_p) = 0$.

From the definition of the significant wave height and Eq. (2.31), the energy in irregular ocean waves beneath a unit area is given by

$$E = \frac{1}{16}\rho g H_s^2, \tag{2.34}$$

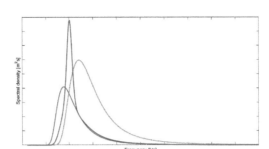

Figure 2.4: Different wave spectra of the form in Eq. (2.35) for different values of the constants A and B.

which can be compared with the analogous expression for harmonic waves in Eq. (2.23).

As derived above, the irregular ocean waves can be described stochastically in terms of their wave spectra and related wave parameters, such as significant wave height and peak or energy period, and the energy available in ocean waves is directly proportional to the integral of the spectra over all wave frequencies. In general, the different wave components may travel in different directions simultaneously, and the wave spectrum is dependent on direction, $S(\theta, f)$. For a more thorough description of ocean waves and directional wave spectra, the reader is referred to a textbook specific for the subject, such as [587].

From considerations of a finite valued energy and energy flux, and based on empirical studies, it has been found that a typical spectrum can be written in the form

$$S(\omega) = \frac{A}{\omega^5} e^{-B/\omega^4} \tag{2.35}$$

where A and B are constants that can be determined in different ways, and $\omega = 2\pi f$ is the angular frequency of the wave component. The typical form of such a function can be seen in Figure 2.4. Empirical formulas have been derived from large data sets of realistic wave measurements to find the explicit form of the spectral density functions, and different wave spectra exist that describe waves in different ocean basins to varying degrees of accuracy.

The Bretschneider spectrum was derived in conditions where the wind has only a limited distance to generate the waves [88],

$$S_{\text{Bret}}(\omega) = \frac{5\omega_p^4 H_s^2}{16} \frac{1}{\omega^5} \exp\left(-\frac{5}{4}\left(\frac{\omega_p}{\omega}\right)^4\right) \tag{2.36}$$

where ω_p is the peak angular frequency. Based on the assumption of a fully developed sea, where equilibrium has been reached between the waves and the wind, the Pierson-Moskowitz spectrum was presented on the form [635]

$$S_{\text{PM}}(\omega) = \frac{\alpha g^2}{\omega^5} \exp\left(-\beta\left(\frac{\omega_0}{\omega}\right)^4\right) \tag{2.37}$$

where $\alpha = 0.0081$, $\beta = 0.74$, $\omega_0 = g/U_{19.5}$, and $U_{19.5}$ is the wind speed measured at a height of 19.5 m above the sea surface. Another often-used parameter is the wind speed at 10 m above the surface, which in most situations is related to the wind speed at 19.5 m as $U_{19.5} \approx 1.026 U_{10}$, where a drag coefficient of $1.3 \cdot 10^{-3}$ was assumed. The peak frequency of the Pierson-Moskowitz spectrum is the one for which $S'(\omega_p) = 0$, which is $\omega_p = 0.844 g/U_{19.5}$ rad/s^{-1}.

In the JOint North Sea WAve Project, sharper peaks were observed than were predicted by the Pierson-Moskowitz spectrum, and it was found that the waves were never fully in equilibrium, they continued to develop through non-linear wave interactions, implying also that wave speed can become larger than the wind speed. A modified spectrum with a peak enhancement factor was developed, the so-called JONSWAP spectrum [339],

$$S_{\text{JONSWAP}}(\omega) = \frac{\alpha g^2}{\omega^5} \exp\left(-\beta \left(\frac{\omega_p}{\omega}\right)^4\right) \gamma^r \qquad (2.38)$$

where the parameters were defined as $\alpha = 0.076(U_{10}^2/gF)^{0.22}$, where U_{10} is the wind speed at 10 m height, $\beta = 1.25$, $\gamma = 3.3$, and $r = e^{-(\omega-\omega_p)^2/(2\sigma^2\omega_p^2)}$ where the parameter $\sigma = 0.07$ for $\omega \leq \omega_p$ (angular frequencies smaller than or equal to the peak angular frequency) and $\sigma = 0.09$ for $\omega > \omega_p$.

The *fetch F* is defined as the distance over which the wind blows with constant velocity. As discussed above, the waves of relevance for wave energy are generated by winds, which is reflected in the form of the empirical spectra. The stronger the wind, the longer the duration of the wind, and the longer the distance over the ocean surface the wind has blown (the so-called fetch), the bigger the resulting waves will be.

2.1.2 Wave-structure interaction

Once the wave field is obtained, the wave-structure interaction and the dynamics of a structure due to the hydrodynamic forces can be studied.

2.1.2.1 Hydrodynamic forces

Once the equations describing the ocean waves have been solved, the pressure in the fluid can be determined. In the case of the full Navier-Stokes Eq. (2.4), the fluid velocity and the pressure are obtained numerically using a CFD method. In the case of (linear) potential flow theory, solutions to the Laplace Eq. (2.5) and the boundary constraints (2.9), (2.12)–(2.13) can be solved either by analytical or numerical methods, and the pressure can be obtained from the Bernoulli Eq. (2.7).

The force and force moment on a structure in a fluid is given by the fluid pressure integrated along the wetted surface S of the structure as

$$\mathbf{F} = \iint_S p d\mathbf{S} \qquad (2.39)$$

where for the force vector $d\mathbf{S} = \mathbf{n}dS$, $\mathbf{n}$ being the normal vector pointing out from the structure into the fluid, and for the force moment vector $d\mathbf{S} = \mathbf{r} \times \mathbf{n}dS$, where $\mathbf{r}$ is the moment arm pointing from the axis of rotation to the surface element dS.

From this point, the fluid will be restricted to the case of potential flow theory. The total fluid pressure is the sum of the hydrodynamic pressure due to the waves and the hydrostatic pressure. From Eq. (2.7), the hydrodynamic pressure can be obtained as

$$p_{\mathrm{dyn}} = -\rho \left(\frac{\partial \Phi}{\partial t} + \frac{1}{2}(\nabla \Phi)^2 \right). \tag{2.40}$$

As discussed in the previous section, in linear potential flow theory the non-linear term can be neglected and the dynamic pressure can be described in terms of the first term. Furthermore, due to the linearity, the fluid velocity potential can be divided into potentials corresponding to incident, scattered and radiated waves,

$$\Phi = \Phi_I + \Phi_S + \Phi_R. \tag{2.41}$$

The scattered waves appear when waves are scattered off a fixed structure, and the radiated waves are due to the structure's own motion in the water.[1]

The force and force moments of Eq. (2.39) can then be separated into forces resulting from incident and scattered waves off a fixed structure, denoted excitation force $\mathbf{F}_e$, and forces resulting from the motion of the structure in the absence of incident waves, denoted radiation force $\mathbf{F}_r$.

$$\mathbf{F} = \underbrace{-\rho \iint_S \left(\frac{\partial \Phi_I}{\partial t} + \frac{\partial \Phi_S}{\partial t} \right) d\mathbf{S}}_{\mathbf{F}_e} + \underbrace{-\rho \iint_S \frac{\partial \Phi_R}{\partial t} d\mathbf{S}}_{\mathbf{F}_r} \tag{2.42}$$

and analogously for the force moment. In the frequency domain description of the problem, the expressions simplify further, which will be discussed in Section 3.1.2.

The radiation force is usually written as a sum of a term proportional to the acceleration of the structure, added mass, and a term proportional to its velocity, radiation damping. (In the frequency domain, these correspond to the imaginary and real parts of the radiation force, respectively.) The terms *added mass* and *radiation damping* (or radiation resistance) indicate the physical interpretation of the terms: the added mass can be understood as the added inertial force due to the mass of the volume of water following the body's oscillation, whereas the radiation damping can be understood as the damping of the oscillatory motion due to the emitted energy in terms of radiated waves.

2.1.2.2 Hydrostatic forces

The hydrostatic forces are due to the fluid loading acting on a body when placed in still water. It originates from the static-pressure term $-\rho g z$ in Eq. (2.6), because

---

[1]Note that the term *diffracted* waves is sometimes used to describe the sum of incident and scattered waves, $\Phi_D = \Phi_I + \Phi_S$, whereas sometimes the opposite meaning of diffracted and scattered waves is used, i.e., $\Phi_S = \Phi_I + \Phi_D$.

the body's wet surface experiences a varying hydrostatic pressure as a result of its oscillation. The hydrostatic forces can be determined using the same method for the hydrodynamic forces, and can be expressed as

$$\mathbf{F}_s = -\rho g \iint_S z \mathrm{d}\mathbf{S}, \tag{2.43}$$

and analogously for the force moment.

2.1.2.3 Hydrodynamic responses

The motion of a rigid body can be characterised by six components corresponding to six degrees of freedom (DoF) or modes of (oscillatory) motion. For a vessel like an elongated body (directed parallel to the x axis, as indicated in Figure 2.5), the six modes numbered 1 to 6 are named surge, sway, heave, roll, pitch, and yaw, respectively.

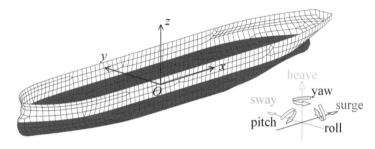

Figure 2.5: A vessel in still water with its wet surface marked in blue colour. A rigid body has six modes of motion: surge, sway, heave, roll, pitch, and yaw.

A six-dimensional generalised velocity vector $\dot{\mathbf{x}}$ is introduced with components

$$(u_1, u_2, u_3) = (U_x, U_y, U_z) = \bar{U}, \tag{2.44}$$

$$(u_4, u_5, u_6) = (\Omega_x, \Omega_y, \Omega_z) = \bar{\Omega}, \tag{2.45}$$

where $\bar{U}$ is the velocity of a reference point and $\bar{\Omega}$ is the angular velocity vector corresponding to rotation about the reference point. Note that the components numbered 1 to 3 have SI units of m/s; whereas the remaining components numbered 4 to 6 have SI units of rad/s.

From Newton's second law, the dynamic equation for an oscillating body may be written as

$$\mathbf{M}\ddot{\mathbf{x}} = \mathbf{F}_e + \mathbf{F}_r + \mathbf{F}_s + \mathbf{F}_p + \mathbf{F}_c + \mathbf{F}_m + \mathbf{F}_v, \tag{2.46}$$

where $\mathbf{M}$ denotes the inertia of the oscillating body. $\mathbf{F}_e$ and $\mathbf{F}_r$ represent the six-dimensional generalised vectors of the wave excitation forces/moments and wave radiation forces/moments, respectively (see Eq. (2.42)). $\mathbf{F}_s$ denotes the hydrostatic buoyancy force vector (see Eq.(2.43)).

Apart from the fluid force, including both the total wave force, $\mathbf{F}_e + \mathbf{F}_r$, and the hydrostatic buoyancy force, $\mathbf{F}_s$, some additional forces are considered as the four last terms in Eq. (2.46). $\mathbf{F}_p$ denotes a load force due to some purpose. It could be a power take-off (PTO) system necessary for the conversion of wave energy and/or a control force intended, for example, to reduce the oscillation of a floating body or to enhance wave power absorption of the wave energy device. A load force induced by interconnected constraint may also be considered as a component of this kind of force; where $\mathbf{F}_m$ represents the mooring force and $\mathbf{F}_v$ represents an unavoidable viscous effect.

2.2 LINEAR POTENTIAL FLOW THEORY SOLVERS

For many wave energy applications, ocean waves can, to a good approximation, be described by linear potential flow theory, as was discussed in Section 2.1.1.2, where the equations governing linear potential flow theory were derived. One solution representing a harmonic wave was derived in Section 2.1.1.3 to illustrate the dispersion equation. Here, more general solutions to the theory will be derived, using analytical methods in Section 2.2.1 and the numerical boundary element method in Section 2.2.2.

2.2.1 Analytical solutions

To solve the equations of motion (2.46), the dynamical forces must be determined. From Eq. (2.42), the forces can be obtained as integrals of the fluid velocity potentials. The fluid potentials should satisfy the Laplace Eq. (2.5) and the boundary constraints at fluid boundaries, including the free surface. Again, the problem will be restricted to the linear potential flow theory, where the boundary constraints at the free surface take the form (2.14).

A common strategy is to solve the potentials in the frequency domain using separation of variables. Using Fourier transform, the fluid potential can be considered in the frequency domain,

$$\phi(\mathbf{x}) = \int_{-\infty}^{\infty} \Phi(\mathbf{x}, t) e^{i\omega t} dt, \tag{2.47}$$

where the frequency dependence in $\phi(\mathbf{x})$ is implicit. In the frequency domain, the time derivative translates to multiplication with the frequency,

$$\frac{\partial}{\partial t} \Phi(\mathbf{x}, t) = \int_{-\infty}^{\infty} [-i\omega\phi(\mathbf{x})] e^{-i\omega t} dt, \quad \frac{\partial}{\partial t} \leftrightarrow -i\omega. \tag{2.48}$$

To find a potential satisfying both the Laplace equation and the linear boundary constraints, an ansatz can be made in the form

$$\phi(x, y, z) = W(x, y) Z(z). \tag{2.49}$$

Inserted back into the Laplace equation and separating the horizontal and vertical coordinates reveals that

$$\frac{1}{W} \left(\frac{\partial^2 W}{\partial x^2} + \frac{\partial^2 W}{\partial y^2} \right) = -\frac{1}{Z} \frac{\partial^2 Z}{\partial z^2} = \alpha \tag{2.50}$$

where α is a constant. Thus, the vertical eigenfunctions satisfy the differential equation $Z''(z) + \alpha Z(z) = 0$ and take the form of trigonometric functions,

$$Z(z) = C\cos(\alpha z + \beta). \tag{2.51}$$

When inserting this expression into the boundary constraint at the sea bed (2.9),

$$0 = \frac{\partial \phi}{\partial z}\Big|_{z=-h} = -W(x,y)C\sin(-\alpha h + \beta), \tag{2.52}$$

which is satisfied if $\beta = \alpha h$. Furthermore, the linear boundary constraint at the free surface (2.14) in the frequency domain takes the form

$$0 = -\omega^2 \phi + g\frac{\partial \phi}{\partial z}\Big|_{z=0} = -[\omega^2 + g\alpha\tan(\alpha h)]W(x,y)Z(z) \tag{2.53}$$

which is satisfied if $\omega^2 = -g\alpha\tan(\alpha h)$. To find which values of the constant α satisfy this relationship, the case when α is a real, positive number is considered. In that case, for each value of the angular frequency ω, there are infinitely many solutions $\alpha = k_m > 0$ to the equation

$$\omega^2 = -gk_m\tan(k_m h) \tag{2.54}$$

where each solution lies within the range $k_m \in \left(\frac{\pi}{h}\left(m - \frac{1}{2}\right), \frac{m\pi}{h}\right)$. Negative and real values of α correspond to the same solutions. In the case when α is a complex valued number, we write $\alpha = -ik$, giving the expression

$$\omega^2 = gk\tanh(kh) \tag{2.55}$$

which was also derived in Eq. (2.21). Eqs. (2.54) and (2.55) are called the dispersion equation. For each value of the angular frequency ω, there is a unique solution $k > 0$ satisfying the dispersion Eq. (2.55). The two cases of real and complex solutions can be combined into one, $Z(z) = C_m\cos(k_m(z+h))$, with $k_0 = -ik$. Typically, the constants C_m are chosen such that the vertical eigenfunctions are orthogonal,

$$Z_m(z) = \frac{\cos(k_m(z+h))}{\cos(k_m h)}, \qquad m \geq 0, \quad k_0 = -ik. \tag{2.56}$$

Note that in the case of $m = 0$ (i.e., the complex root of the dispersion solution), the vertical function takes the form of a hyperbolic cosine, $Z_0(z) \propto \cosh(k(z+h))$, also shown in Eq. (2.20).

The general solution to the horizontal part $W(x,y)$ of the fluid potential will take different forms depending on the geometry of the problem. In Section 2.1.1.3, it could be seen that if the function were independent of y, a solution could be found in a simple form of a trigonometric function, representing a plane wave propagating along the x-direction.

Many fluid systems relevant for wave energy can be described using cylindrical geometry; for instance, waves scattered off pillars with circular cross sections, or waves radiated from a stone dropped in a pond or from an oscillating cylindrical buoy. In such cases, it is appropriate to seek the horizontal function in cylindrical coordinates, as a product between a radial and an angular function, $W(x,y) = R(r)\Theta(\theta)$. When inserted into the Laplace Eq. (2.50) in cylindrical coordinates and separating the radial and angular parts, the following equations are obtained,

$$\frac{r^2}{R(r)}\left[R''(r) + \frac{1}{r}R'(r) - k_n^2 R(r)\right] = -\frac{1}{\Theta(\theta)}\Theta''(\theta) = n^2 \tag{2.57}$$

where n is a constant and k_n are the solutions to the dispersion Eqs. (2.54)–(2.55) obtained earlier. For the angular function, this is satisfied by a function of the form

$$\Theta(\theta) = Ce^{in\theta}, \tag{2.58}$$

where C is a constant and n an integer to satisfy the periodicity condition $\Theta(\theta + 2\pi) = \Theta(\theta)$. The radial function represents a modified Bessel equation, which is solved by modified Bessel functions,

$$R(r) = AK_n(k_m r) + BI_n(k_m r). \tag{2.59}$$

For the fundamental solution $k_0 = -ik$ of the dispersion equation, the modified Bessel functions correspond to a Hankel function of the first kind, $K_n(k_0 r) \propto H_n^{(1)}(kr)$, and a Bessel function of the first kind, $I_n(k_0 r) \propto J_n(kr)$, respectively.

To summarise, the fluid velocity potential that satisfies the Laplace equation and the linear boundary constraints in a cylindrical coordinate system takes the form

$$\phi(r,\theta,z) = \sum_{m=-\infty}^{\infty}\sum_{n=-\infty}^{\infty}[A_{mn}K_n(k_m r) + B_{mn}I_n(k_m r)]\,e^{in\theta}Z(z), \tag{2.60}$$

where the vertical eigenfunction takes the expression in (2.56). The case $k_0 = -ik$ corresponds with the dispersion Eq. (2.55) to propagating modes represented by Hankel functions and Bessel functions of the first kind, whereas the higher modes $k_m > 0$ correspond to the dispersion Eq. (2.54) with $m > 0$ and evanescent modes represented by modified Bessel functions.

In presence of one or several bodies, body boundary constraints and requirement of continuity of the velocity potentials over fluid domains can be used to determine the unknown coefficients A_{mn} and B_{mn} in Eq. (2.60). In some cases, such as a bottom-mounted cylinder, this can be done exactly, whereas in the case of a floating truncated cylinder the solution can be found semi-analytically by truncating the infinite sums in Eq. (2.60) and solving the resulting system of linear equations.

For more in-depth information and for solutions in other geometries and coordinate systems, the reader is referred to the excellent textbook [450]. The analytical solution will be extended to arrays of floating structures in Chapter 8.

2.2.2 Boundary element method

In addition to the analytical techniques, e.g., the method of separation of variables, a more general way is to apply a standard boundary element method (BEM). Without being limited to regular geometries, BEM has the advantage that all the unknowns are restricted on specified boundaries so that the preprocessing work such as meshing can be extremely straightforward. In BEM, the boundary integral equations are derived via Green's theorem within a confined or unconfined space. The boundary integral equations can be numerically solved by discretising the boundaries into a number of geometrical and physical elements. These elements can be approximated using Lagrange interpolating polynomials, B-splines, etc., depending on the requisite accuracy.

2.2.2.1 Boundary integral equation

The complete linear frequency-domain problem is comprised by the Laplace Eq. (2.5), the free-surface boundary conditions (2.12) and (2.13), the rigid body boundary condition (2.9), plus an appropriate wave radiation condition such as the Sommerfeld condition (2.61) in the far-field,

$$\lim_{R \to \infty} \left[\sqrt{kR} \left(\frac{\partial \phi}{\partial R} - \mathrm{i}k\phi \right) \right] = 0, \tag{2.61}$$

where R refers to the distance away from the body. Based on Green's second theorem, the radiated and diffracted wave velocity potentials on the immersed body surface S_B can be solved by a set of boundary integral equations. The integral equations can be with respect to either a distribution of sources

$$2\pi\sigma(\boldsymbol{x}) + \iint_{S_\mathrm{B}} \sigma(\boldsymbol{\xi}) \frac{\partial G(\boldsymbol{\xi}; \boldsymbol{x})}{\partial n_{\boldsymbol{x}}} \mathrm{d}S_{\boldsymbol{\xi}} = V_\mathrm{n}, \tag{2.62}$$

or mixed sources and dipoles

$$2\pi\phi(\boldsymbol{x}) + \iint_{S_\mathrm{B}} \phi(\boldsymbol{\xi}) \frac{\partial G(\boldsymbol{\xi}; \boldsymbol{x})}{\partial n_{\boldsymbol{\xi}}} \mathrm{d}S_{\boldsymbol{\xi}} = \iint_{S_\mathrm{B}} V_\mathrm{n}(\boldsymbol{\xi}) G(\boldsymbol{\xi}; \boldsymbol{x}) \mathrm{d}S_{\boldsymbol{\xi}}, \tag{2.63}$$

where $\boldsymbol{\xi}$ refers to the source point on the body surface and $\boldsymbol{x}$ the field point in the fluid domain or on the body surface, and V_n denotes the respective normal velocity on the body surface. It has been proved that the mixed source-dipole distribution method is more accurate than the source distribution method [154]. Eq. (2.12) can be discretised using splines or polynomials. Taking the most popular constant panel method for example (i.e., the zeroth-order Lagrange interpolating polynomials), the discrete form of Eq. (2.12) is

$$2\pi\phi(\boldsymbol{x}_i) + \sum_{j=1}^{N} D_{ij}\phi(\boldsymbol{x}_j) = \sum_{j=1}^{N} S_{ij} V_\mathrm{n}(\boldsymbol{x}_j), (i = 1, 2, ..., N), \tag{2.64}$$

where N is the number of panels on the immersed body surface. The integration of sources and dipoles over each panel can be represented by

$$S_{ij} = \iint_{S_{\mathrm{B},j}} G(\boldsymbol{\xi}; \boldsymbol{x}_i) \mathrm{d}S_{\boldsymbol{\xi}}, \tag{2.65}$$

$$D_{ij} = \iint_{S_{\mathrm{B},j}} \frac{\partial G\left(\boldsymbol{\xi};\boldsymbol{x}_i\right)}{\partial n_{\xi}} \mathrm{d}S_{\boldsymbol{\xi}}, \tag{2.66}$$

where $S_{\mathrm{B},j}$ denotes the jth panel surface.

2.2.2.2 Removal of irregular frequencies

Directly solving of Eq. (2.63) leads to substantial errors in the neighbourhood of the so-called "irregular frequencies". This phenomenon is caused by the waterplane section of the members of floating bodies that intersects the free water surface. The irregular frequencies actually coincide with the eigenfrequencies of the corresponding sloshing modes of the interior tank (assuming flow filling inside the tank).

There are several approaches to prevent these numerical errors. Ref. [881] presents the modified Green function method and the extended integral equation method (the latter has later been applied in WAMIT). Ref. [446] gives a comprehensive comparison between the extended integral equation method and the overdetermined integral equation method, and concludes that the overdetermined integral equation method is more computationally efficient, as it only requires a few discrete points on the waterplane area, in contrast with hundreds and thousands of waterplane panels in the extended integration method.

The overdetermined integral equation method assumes that the potentials on the interior water plane are zero. By applying Green's theorem in the interior domain of the floating body, an additional boundary integral equation is introduced in a combined application with Eq. (2.63):

$$\iint_{S_{\mathrm{B}}} \phi(\boldsymbol{x}) \frac{\partial G(\boldsymbol{\xi};\boldsymbol{x})}{\partial n_{\xi}} \mathrm{d}S_{\boldsymbol{\xi}} = \iint_{S_{\mathrm{B}}} V_{\mathrm{n},}(\boldsymbol{\xi}) G(\boldsymbol{\xi};\boldsymbol{x}) \mathrm{d}S_{\boldsymbol{\xi}}, (\boldsymbol{x} \in S_{\mathrm{WP}}, \boldsymbol{\xi} \in S_{\mathrm{B}}), \tag{2.67}$$

where S_{WP} denotes the interior waterplane area. By choosing several discrete points (say, M points) on S_{WP}, a set of over-determined linear algebraic equations can be constructed, which finally leads to the following linear algebraic system:

$$\sum_{n=1}^{N} \left\{ \sum_{m=1}^{M+N} A_{mn} A_{mp} \right\} \phi_k\left(\boldsymbol{x}_n\right) = \sum_{m=1}^{M+N} A_{mp} B_k\left(\boldsymbol{x}_m\right), (p=1,2,...,N), \tag{2.68}$$

Eq. (2.68) can regularly be solved without any problem. In addition to the advantage of less computational cost, the overdetermined integration method avoids evaluation of the logarithmic singularity of free-surface Green's function occurring in the limiting case when the panel is on the free surface. Interested readers can refer to Refs. [429, 454, 446] for the implementation details of this method.

2.2.2.3 Calculating free-surface Green's functions

The computational burden of BEM mainly lies in two aspects, namely the computation of free-surface Green's functions and solution of the resultant linear algebraic system. This holds true for both frequency-domain and time-domain BEMs. The infinite interval of the integral, the oscillating nature, and the singularity behaviour of the integrand are the three key difficulty points in calculating free-surface Green's functions.

There have been numerous works on developing efficient and accurate algorithms for free-surface Green's functions, for the deepwater condition (e.g., [572, 843]), and for the finite-depth water condition ([572, 120, 458, 473]). In general, the calculation strategies can be categorised into several types (in particular for finite-depth Green's function): (1) extracting slow-varying components from the Green function and using a Chebyshev or multi-dimensional polynomial method to approximate them (e.g., [572, 120, 473]); (2) applying asymptotic or power series expansions, such as eigenfunction expansions, rapid convergent series, or a combination with other numerical acceleration algorithms in different subregions (e.g., [634, 448, 456]); (3) decomposing the principal-value integral into two parts by subtracting a special term from the integrand and applying a direct Gauss-Laguerre quadrature to the numerical integration (e.g., [200, 455]). In order to reduce the repeated effort in implementation of these algorithms, [761] and [447] released their open-source codes for the deepwater Green function, and [458] released an open-source code for the finite-depth Green function.

2.2.2.4   Resolving linear algebraic system

As we may know, a direct solver such as Gauss elimination is generally robust but requires $O(N^3)$ computations (N denotes the matrix size), while some iterative methods can reduce the effort to $O(N^2)$ operations. The linear algebraic system, constituted by Eq. (2.63) for a submerged body or Eq. (2.63) and Eq. (2.68) for a floating body, is a full-rank dense complex system. For a large-scale computation of three-dimensional offshore structures, a direct inversion or inefficient iteration of such a large, dense system of linear equations with $O(N^4)$ unknowns for a set of wave frequencies is seemingly prohibitively time consuming even with modern computers. Note that no matter which method to remove irregular frequencies is employed, the condition number of the resultant linear algebraic system always increases. This means that an iterative method, such as the GMRES method, is no longer appropriate to solve equations like Eq. (2.68), because of the ill conditions. In addition, considering the usual cases when the radiation-diffraction problem needs to be solved with multiple wave headings for each single wave frequency, the computation effort of using either a direct method or an iterative method is still not acceptable because it needs to solve the linear equations for every wave heading in succession.

Considering all the reasons above, several improved options are recommended to solve the problem. The first one is to apply an LU decomposition (e.g., the "ZGETRF" subroutine of LAPACK) for the left-hand side matrix, since it needs to decompose the matrix only once for each wave frequency. This decomposition can be applied to calculate the wave forces for a distribution of wave headings via a forward and backward substitution (e.g., the "ZGETRS" subroutine of LAPACK) for triangular matrices (L and U), which can be solved directly without using the Gaussian elimination process. The second option is to use a preconditioned iterative method. However, one needs to be especially careful in choosing an appropriate preconditioner.

2.2.2.5 Parallelisation on multi-core machines

The problems to be solved are often of a very large size such that resolving the resultant linear systems requires huge computational resources. Nowadays, with the facility of a fast multi-core computer, it is natural to maximise the advantages of the current hardware technology in our computations. A large portion of the programs that people write and run daily are serial programs, especially in the marine hydrodynamic field to the best of the author's knowledge. Applying a parallelisation technique enables a better performance of the numerical code on multiple processors. Taking into consideration that in a typical case, the number of panels involved in the hydrodynamic computation is usually below ten thousand to thirty thousand and that off-the-shelf computation machines contain multiple processors, the open multi-processing (OpenMP) technique is considered to be an appropriate option for the BEM solver in marine hydrodynamics. Furthermore, in the case of a large-scale computation on a cluster of machines, the message passing interface (MPI) standard is recommended for use in programming. However, OpenMP needs much less effort than MPI as the latter can request substantial modifications on the code architecture.

2.2.2.6 Useful references and tools

Inspired by Hess and Smith (1964) [351], many numerical solvers based on BEM arose from the mid 1980s. In general, these solvers can be categorised as commercial, in-house, and open-source. In Table 2.1, a comparison is given regarding the mainstream BEM solvers that are available to users in the ocean engineering and renewable energy community. Note that in-house codes (DIFFRACT [188], WAFDUT [763], etc.) are not listed here as they cannot be publicly accessed. Moreover, some derived versions of these codes are neither included, such as WADAM [180] (a descendant of WAMIT [431]), OpenWARP [492] and Capytaine [16] (descendants of NEMOH [46]) since the main functionalities are basically the same as their parents'.

In addition, there are also some other resources that the users can find to assist their hydrodynamic analysis via BEM, such as mesh processing (BEMRosseta, BEMIO, Gmsh), post-processing (BEMRosseta), etc.

2.3 COMPUTATIONAL FLUID DYNAMICS

2.3.1 Governing equations

The Navier-Stokes and the continuity equations, expressed in Eq. (2.4) and (2.1) for incompressible flow, are the fundamental equations used in computational fluid dynamics (CFD) to generate a model of the viscous fluid flow.

In addition, an energy equation can be derived using the principle that energy must be conserved within a closed system, implying that further flow dependency characteristics such as compressibility or thermal effects can also be modelled using this system of equations.

Together with the energy equation, Eqs. (2.1) and (2.4) form a system of coupled partial differential equations that can be used to describe the motion of a viscous flow. Nevertheless, no general closed-form solution to this system of partial differential

Table 2.1: Comparison of the public-available mainstream BEM solvers in the frequency-domain.

Properties	WAMIT	AQWA	Hydrostar	Nemoh	HAMS
Approach	Potential and source formulation	Source formulation	Source formulation	Source formulation	Potential formulation
Discretisation	Constant panel or B-splines	Constant panel	Constant panel	Constant panel	Constant panel
Nonlinearity	Full QTF	Mean drift	Full QTF	Mean drift	Linear
Forward speed	Encounter frequency	Encounter frequency	Encounter frequency	No	No
Radiation-diffraction	Yes	Yes	Yes	Yes	Yes
Removal of irregular frequencies	Yes	Yes	Yes	No	Yes
Body symmetry	Yes	Yes	Yes	Yes	Yes
RAO calculation	Yes	Yes	Yes	Yes	Yes
Free-surface elevation	Yes	Yes	Yes	Yes	Yes
Multi-body modelling	Yes	Yes	Yes	Yes	No
Parallelisation	OpenMP	No	MPI	No	OpenMP
Operating System	Windows/Linux	Windows	Windows	Windows/Linux	Windows/Linux
Access	Commercial	Commercial	Commercial	Open-source	Open-source

equations has been formally proved: in fact it is considered one of the outstanding challenges in mathematics. In Sections 2.1 and 2.2, analytical and numerical solutions to a simplified version of the Navier-Stokes equations were presented, based on approximations of inviscous and irrotational flow. Another approach, which will be discussed in this section, is to consider the full Navier-Stokes equations, solving them by numerical methods. In this approach, the equations are discretised on a grid that represents the fluid domain under consideration and a set of initial conditions, and boundary conditions are stipulated at the domain edges. In this way, a predictive model of the flow is generated.

2.3.2   Volume of fluid method for free-surface flows

As the majority of WECs are surface-piercing devices that convert the free surface kinetic energy to electrical energy, the method by which the free surface is modelled is of paramount importance. The most common technique that is used to compute the free surface motion is the volume of fluid (VOF) method. The method is a numerical technique used to model complex free-surface flows. It is a particularly suitable approach for those simulations in which the free surface boundary undergoes large deformations. In CFD analyses, the transformation of the flow and the subsequent evolution of the free surface is achieved by using a discretisation method to solve a transport equation for the fluid in each cell [292, 356, 511]. The VOF method was primarily developed to overcome the inherent low-resolution problem that occurs at a free surface boundary interface in a multi-phase flow analysis, which arises due to convective flux averaging of flow properties across cell boundaries. This may occur when pure Eulerian or arbitrary Lagrangian-Eulerian (ALE) techniques are employed.

In the volume of fluid method, a function $F(x, y, t)$ is introduced at each grid cell in the domain. The value of this function is defined as unity at any cell entirely occupied by the fluid and zero at any cell completely devoid of fluid. Thus, cells with an intermediate value $0 < F < 1$ are those cells crossed by the free surface boundary. Hence, a cell that has at least one empty neighbour cell ($F = 0$) is by definition a free surface cell. This method allows for the determination of the fluid proportion in each cell through the storage of only a single variable. Because the VOF method only requires the value of the F function to determine the fluid spatial representation, the computational storage requirements are minimised.

Although those cells that contain the free surface are determined from the F function, the orientation of the surface requires additional computation. By calculating the derivatives of the F function at each cell boundary, the free surface normal can be established [684]. The normal direction to the free surface is then the direction in which the F function varies most rapidly (i.e., ∇F). From the value of the F function and the direction of the normal to the boundary, a line cutting the cell can be drawn which represents the free surface boundary. The temporal evolution of the F function and thus the advection of the flow in two dimensional space is governed by the following transport equation:

$$\frac{\partial F}{\partial t} + u\frac{\partial F}{\partial x} + v\frac{\partial F}{\partial y} = 0, \tag{2.69}$$

where u and v are the velocities in the component x and y directions, respectively. Then, as the simulation proceeds in time, the value of function F moves with the fluid. The fluxes across each cell in the fluid domain are then obtained from Eq. (2.69). The value of F is recorded and the simulation is advanced in time by the amount ∂t, at which stage the value of $F(x, y, t)$ is re-computed. OpenFOAM is one of the most commonly used CFD codes for the analysis of wave energy converters. The code employs the VOF method to compute the free surface evolution. In the OpenFOAM InterFOAM solver, the advection of the free surface is controlled by the explicit multidimensional universal limiter with an explicit solution (MULES) algorithm, which is a variation of the flux corrected transport (FCT) scheme [176]. The MULES algorithm relies in a straightforward upwind scheme to the computed advection in the interfacial cells [172]. This scheme ensures the fluxes into or out of a cell are limited to maintain the boundedness of the VOF method, thus ensuring the stability of the numerical code.

The main benefit of the VOF method is that it allows for the analysis of multiple fluid flow interactions (multi-phase flow) within a single simulation. However, it is not a trivial process to incorporate these interactions. A frequently observed problem is the smearing of the fluid boundary. This occurs as a result of diffusion of the transport equation over the mesh cell in which the boundary is located. One method by which this problem can be addressed is by specifying a localised high-resolution mesh in the region where the fluid interface will occur. However, prior knowledge of the evolution of the interfacial region is required for this technique.

Multiphase CFD models can be implemented using either an homogeneous or an inhomogeneous approach. According to the Ansys CFX guide [22], for a given transport process, the homogeneous model assumes that the transported quantities (with the exception of the volume fraction or F function value) for that process are the same for all phases. Thus, a common flow field for velocities, temperature, pressures, etc. is applied to all fluids within the domain. This simplifies the underlying code, which defines the interaction of the fluids at the boundary as a single mass conservation, and a momentum conservation equation is applied to both fluids as opposed to individual conservation equations being applied to each fluid in an inhomogeneous model approach at the fluid interface. The conservation of mass and momentum equations is formulated by summing the averaged fluid properties according to their constituent proportion in the boundary cell. For a two-phase flow, density ρ in the boundary cells is given by:

$$\rho = \sum_{a=1}^{2} r_a \rho_a \tag{2.70}$$

where r_a is the volumetric fraction of each constituent fluid in the free surface boundary cell. The conservation of mass equation for an incompressible fluid in tensor notation is:

$$\frac{\partial(p)}{\partial t} + \frac{\partial(\rho u_i)}{\partial x_i} = 0 \tag{2.71}$$

and the conservation of momentum for an incompressible fluid is:

$$\frac{\partial(\rho u_i)}{\partial t} + \frac{\partial(\rho u_i u_j)}{\partial x_j} = -\frac{\partial p}{\partial x_i} + \frac{\partial \tau_{i,j}}{\partial x_j} + f_i \qquad (2.72)$$

where ρ is given in Eq. (2.70) In the domain cells where $F = 1$ or $F = 0$, these equations reduce to the mass and momentum conservation equations for a single-phase fluid. The homogeneous model approach is incorporated into the OpenFOAM multiphase solvers. This limits the relative motion between individual fluids at the boundary interface cells to zero. This is analogous to a "no-slip" boundary condition between the individual fluids at the cells in which the fluid interface is located.

2.3.3 Computational fluid dynamics software

There is a wide variety of CFD software programs used to model wave interactions with energy harvesting devices. Generally, these programs can be classified as proprietary or open-source codes. The most common proprietary packages include Ansys Fluent, Ansys CFX, STAR-CCM+, and FLOW-3D. OpenFOAM is the most common open-source code, with some additional "in-house" codes being developed for specific modelling scenarios. The choice of which software package to employ depends on a number of factors. The first consideration is whether the user has access to the commercial software. The license fees associated with these commercial packages can often incur significant expenses. Another disadvantage of the commercial packages is the difficulties in modifying the underlying code. Frequently, the user has no access to the source code and thus any modifications to the program are prohibited. However, these commercially licenced packages are often more user-friendly, for example with simple-to-use graphical user interfaces (GUIs). In contrast with the commercial codes, the open-source codes do not require expensive licence fees and the source code is readily editable. These points are major advantages of open-source codes. However, many open-source codes do not have a dedicated GUI and many of the commands must be run from the command prompt window or from the terminal window depending on which operating system the software is installed upon. The lack of a GUI is often a disconcerting experience for new users who have little knowledge of CFD programming and who view the experience with trepidation. Nonetheless, it should be remembered that even whilst conducting CFD research on commercial software with a guiding GUI, the researcher should have a deep theoretical knowledge of the mathematics and fundamental physics underlying the simulation scenarios they recreate. The most popular software for the investigation of wave energy converters is the OpenFOAM program. This C++ based code was utilised in approximately 40% of all the published papers on wave energy employing a CFD modelling approach in 2018, and since that time the use of this software code has become even more prevalent. In the following sections, the steps required to create a numerical wave tank and generate a CFD simulation for the analysis of wave energy converters are presented and briefly described. The open-source code OpenFOAM is used for this case study; however, the process is similar for other software programs.

2.3.4 Creating a computational fluid dynamics simulation

The first step in creating a CFD model for the hydrodynamic analysis of a WEC is to generate the numerical wave tank that defines the limits of the numerical model domain. Following on, the domain is discretised by a suitably designed mesh that is capable of capturing the flow physics accurately, in a high-fidelity manner. Suitable initial conditions and boundary conditions are established. The correct numerical schemes should be specified for the solution of governing equations and the mathematical methods and algorithms by which these solutions are computed must be defined.

2.3.4.1 Numerical wave tank definition

The numerical wave tank is a computational representation of a physical region, either in the natural environment or in a controlled experimental setting where free-surface gravity waves are manifest. The wave tank may take the form of a flume, whereby one of the tank dimensions (length) is significantly longer than the other two dimensions (width and depth) or a wave basin in which two of the dimensions (length and width) are significantly longer than the third dimension (depth). The tank has a wave generator at one end and may have a wave absorption region at the other end to damp the incident wave energy. Usually, the free surface is open to the atmosphere. The tank should be wide enough that the sidewalls do not influence the study. The characteristics and requirements described in the preceding sentences relate to a physical wave tank; however, the numerical wave tank must also satisfy this criterion. Once the extents of the numerical wave tank are established, the next step is to discretise the domain with a mesh.

2.3.4.2 Mesh generation

The software user should first decide if they will utilise the program's built-in mesh generation functionality or if it is more advantageous to use a dedicated meshing software and import the generated mesh into the CFD program. This may depend on the complexity of the mesh and whether the WEC (that is immersed in the wave tank domain) has a highly convoluted geometry that may not be accurately rendered by the built-in meshing utility in the CFD program. Another consideration that influences the meshing process is whether the mesh will be static or dynamic. In most cases, WECs rely on the motion of some floating or submerged device to generate electricity. Therefore, to model the motion of these devices, the mesh must displace or distort in an appropriate manner to capture the device's motion. This requirement of a deformable mesh is not applicable to bottom-mounted or fixed OWCs.

Meshing software The majority of CFD software, whether commercial or open-source, incorporates a domain meshing functionality. However, as outlined in the previous section, there are situations in which the numerical modeller may choose to use a dedicated mesh generating software. There are a number of mesh generating and mesh editing programs for this purpose, for example, Pointwise, CF-Mesh+, SALOME etc. OpenFOAM, one of the most commonly used CFD softwares, has a dedicated mesh generator module named *SnappyHexMesh* which is capable of creating a domain mesh fitted to highly complex geometries.

SnappyHexMesh in OpenFOAM In order to ensure high fidelity results from a numerical simulation, the solid body WEC device that is used for the energy extraction must be represented accurately in the model. This can sometimes present challenges, as many WEC devices conform to an irregular and complex geometry. The difficulty then arises in the effort to construct a domain mesh that can capture the intricacies of the WEC surface topology. In OpenFOAM the modeller may first create a background mesh that encompasses the entire domain. The WEC geometry is created in independent CAD software and the file is stored as an *.STL or *.OBJ file. These are lightweight file formats that describe the structure's surface geometry in the form of vertex coordinates and surface or vertex normals which are obtained when the surface geometry is tessellated into a number of facets during the *.STL or *.OBJ file conversion process. The solid-body position in the CFD domain is determined and the background mesh is chiselled away in order to fit the solid body surface. A cell castellation process is performed to fit the cells to the solid surface. This process only occurs in those cells that contain an edge (in two-dimensional simulations) or surface (in three-dimensional simulations) boundary on the solid body. The castellation process may be iteratively performed to better capture the surface topology. Then, those cells that lie entirely within the volume of the solid body are removed and the cell vertices are snapped to the *.STL or *.OBJ surface geometry. Additional layers of cells can be added along the solid surface boundary to improve the local mesh quality at the WEC surface.

SnappyHexMesh has become a widely used tool for the generation of CFD domain meshes around complex and irregular shaped objects.

Dynamic mesh methods With the exception of oscillating water columns and overtopping structures, most devices used for extracting wave energy displace due to the wave-structure interaction. When conducting numerical simulations the motion of these devices must be captured accurately. This can sometimes pose a challenge to the stability of simulations. There are three main methods for the implementation of dynamic meshes in CFD simulations:

1. dynamic mesh morphing,

2. sliding interface mesh,

3. overset mesh.

Dynamic mesh morphing is a technique employed to allow for the displacement of a solid body. This technique is usually selected for simulations in which the solid body displacements are small or those simulations in which a single degree of freedom displacements occur. Grid connectivity is maintained as the solid body displaces. This means that the cell edges connecting the grid nodes remain unchanged. If the displacement of the solid body is excessive the grid may distort excessively resulting in a low-quality mesh with high aspect ratio cells or highly skewed grid cells that will reduce the accuracy of the solution or in extreme cases cause the simulation to fail. This method is seldom used in the simulation of WECs due to the relatively high amplitude, multi-degree of freedom motions to which due to their relatively high amplitude, multi-degree of freedom motions. In the OpenFOAM software, a *sixDoFRigidBody-Motion* solver is included which can manipulate the mesh according to the dynamic mesh morphing method. Using this technique, an inner region must be around the solid displacement body and an outer region some distance away from the body must be defined. These regions are usually defined by a radial distance away from the body. The mesh in both the inner and outer regions does not distort as the solid body displaces, but the mesh in the inner region will deform. In OpenFOAM the mesh displacement in the intermediate region is controlled by the spherical linear interpolation (SLERP) algorithm, based on the distance from the intermediate region cell to the moving body. This algorithm allows the mesh quality to be strictly controlled. The body displacement is diffused into the domain according to the Laplace equation:

$$\nabla \cdot (k\nabla u) = 0, \tag{2.73}$$

where k is the diffusivity coefficient and u is the velocity of the moving body.

The main disadvantage of the mesh morphing method is that solid body displacements should be small and the modeller should have prior knowledge of the amplitudes of displacement before simulations are conducted. Additionally, the outer radius which defines the extent of the mesh distortion for a moving body cannot overlap with another mesh distortion region around a second body. For this reason, this method is not suitable for use in simulations with multiple adjacent displacing bodies. This dynamic mesh modelling technique is especially applicable for simulating the motion of small amplitude heaving buoys.

The sliding interface mesh is an alternative dynamic meshing method that can be used to allow for large displacements of solid bodies in numerical simulations. In this technique, grid connectivity is not maintained. The user defines a background mesh that is applied throughout the domain and a local mesh that is defined in the region of the solid body. The local mesh is allowed to slide or rotate, relative to the background mesh. The interface between the two mesh regions can either be a straight line to allow for a single degree of freedom motion such as heave or it may be a curved interface to allow for rotational motion of the solid body. The solution to the governing equations is computed separately in each of the mesh regions and the field variable data such as pressures or velocities are transferred across the interface using an arbitrary mesh interface (AMI) technique. As the inner mesh region slides relative to the outer mesh region the cells become misaligned at the interfacial boundary the AMI algorithm computes the input weights from each face of the intersecting cells based on the fraction of the overlapping areas at the boundary. For two boundary adjacent cells, the sum of the weights should approach unity in order to preserve simulation stability.

The sliding interface mesh method has some advantages over the mesh morphing technique. The method can process large displacements especially in rotational degrees of freedom. Also since the dynamic mesh region is limited to the area around the moving solid body, this method is more suitable for simulation cases with multiple moving bodies in close proximity. Since the domain meshes in both the inner and outer do not deform, the cells preserve their shape, therefore errors arising for deformed cells are minimised. The most conspicuous disadvantage of the sliding interface mesh method is that this technique is mainly limited to a single degree of freedom motion of the solid body.

The overset grid approach is a powerful technique in dynamic meshing methods. In this method, two independent, disconnected meshes are defined within the domain. The first mesh is a background mesh that is applied throughout the entire domain. The second mesh is a body-fitted mesh that is defined around the WEC solid body and overlays the background mesh. The two meshes are permitted to move relative to each other and neither independent mesh deforms or distorts, keeping its original structure. Using this technique, the shifting cells within the domain (in both the background and overset mesh) are classified according to their characteristics and locations within the domain at each time step. The cells are categorised as blocked cells, fringe cells, donor cells, and acceptor cells. Blocked cells are those cells that are within the volume of the solid body. These are inactive cells, sometimes called hole cells as they form a hole in both the body-fitted and the background mesh. In the background mesh, the fringe cells are those cells that are adjacent to the hole. In the body-fitted mesh, the fringe cells are those cells that are at the outer boundary of the grid. Donor and acceptor cells are those additional cells that are in the region common to both the background and body-fitted meshes. An interpolation process is used to map the boundary values to the fringe cells and interpolation of values is performed between the donor and acceptor cells on both the body-fitted and the background meshes. The main advantage of the overset grid method is that it allows for large motions in multiple degrees of freedom whilst the mesh quality does not degrade. The main disadvantage of the method is that it requires significantly more processing time due to the computationally demanding interpolation procedure.

2.3.4.3 Boundary and initial conditions

In order to reduce the number of unknown terms in the system of governing equations, boundary conditions are specified at the edges of the fluid domain. There are a great number of disparate boundary conditions to simulate different physical circumstances. It is beyond the scope of this text to investigate all of these boundary conditions individually, but it is of critical importance that the CFD user is familiar with the physical significance and the mathematical implementation of the boundary conditions that are specified at any domain boundary. Most commonly-used boundary conditions in the case of wave energy harvesting include no-slip boundaries at solid surface boundaries or wave absorption boundary conditions at domain edges to prevent wave reflection. Additionally, initial conditions are specified either at a boundary edge or internally at some location within the domain. For the purpose of modelling wave energy harvesting systems, these initial conditions usually consist of some form of wave generation condition at one or more boundary edges.

2.4 WAVE TANK EXPERIMENTS

Numerical solvers, as described in earlier sections, are very useful to model wave-structure interaction problems in various conditions, in particular in the stages of WEC development. To gain trust in the simulations and to capture physical effects accurately, numerical solvers need to be complemented by physical experiments. With the increase in TRL of the WECs, wave tank experiments on the scaled physical model are required to gain confidence in the WEC's performance in a controlled, repeatable, and high-fidelity laboratory environment, prior to sea trials that are uncontrollable, uncertain, and both time- and cost-consuming.

2.4.1 Objectives

The objectives of wave tank experiments vary with the TRL of the WEC. For WECs at an early stage of development with low TRL (e.g., TRL$\leq$4), wave tank experiments are used to evaluate their power absorption performance, PTO control strategies, and numerical models of wave-structure interaction in typical operational sea states. The evaluation outcomes support further design optimisation of the WECs at the early conceptual stage where design flexibility is high. For mature WEC technologies with higher TRL (e.g., TRL$\geq$5), wave tank experiments can be used to assess their hydrodynamic responses, structural and mooring forces, and PTO efficiency, loads, and reliability in operational and extreme sea states at particular sea sites. Such assessments aim to further improve soundness in the engineering solutions and de-risk the subsequent sea trials.

2.4.2 Wave generation

Making desired and repeatable waves is fundamental to wave tank experiments. There are two main, different types of wavemakers used at the wave tank, according to the water depth of the experiments.

Piston wavemakers are used to simulate shallow water scenarios, where the water depth is roughly smaller than half a wavelength. Here the orbital particle motion is compressed into an ellipse and there is significant horizontal motion on the floor of the tank. This type of paddle is used to generate waves for modelling coastal structures, harbours, and shore-mounted wave energy devices.

Flap paddles are used to produce deep water waves where the orbital particle motion decays exponentially with depth and there is negligible motion at the bottom. Typical applications are the modelling of floating structures in deep water and the investigation of the physics of ocean waves. Often the hinge of the paddle is mounted on a ledge some distance above the tank floor.

To generate the desired wave spectra $S_{\eta d}(\omega)$, the property function $T(\omega)$ of the wavemaker should be confirmed, which is determined by mechanical transfer function $T_1(\omega)$, hydrodynamic transfer function $T_2(\omega)$, and deformation function of the wave

$T_3(\omega)$. $T_1(\omega) = e/R(\omega)$ is the transfer function between the analogue voltage $R(\omega)$ and paddle position e, $T_2(\omega, h) = A_0/e$ is the transfer function between the paddle position e and wave amplitude A_0, and $T_3(\omega)$ is the function of the wave deformation along the wave tank, determined by the wave conditions, water depth, and the boundary conditions. Thus, the transfer function of the wavemaker system $T(\omega)$ (unit: m/V) is given as follows:

$$T(\omega) = T_1(\omega)T_2(\omega, h)T_3(\omega). \tag{2.74}$$

For the piston type wavemaker $T_2(\omega, h)$ is given by [856]:

$$T_2(\omega, h) = \frac{4\sinh^2 kh}{2kh + \sinh 2kh}. \tag{2.75}$$

For the flap type wavemaker $T_2(\omega, h)$ is given by [856]:

$$T_2(\omega, h) = \frac{4\sinh kh}{kh}\left(\frac{1 - \cosh kh + kh\sinh kh}{2kh + \sinh 2kh}\right). \tag{2.76}$$

Thus, the spectral of wave maker is given by:

$$S_V(\omega) = \frac{S_{\eta d}(\omega)}{|T(\omega)|^2}. \tag{2.77}$$

The voltage signal of the wave maker can be obtained from a spectral analysis of the generated wave. In another word, the time signal of the wave can be represented as a superposition of a series of sinusoidal waves, and thus the voltage signal of the wave maker is [856]:

$$V(n\Delta t) = \sum_{i=1}^{M} \sqrt{2S_V(\hat{\omega}_i)\Delta\omega_i} \cos(\hat{\omega}_i n\Delta t + \beta_i) \tag{2.78}$$

where $V(n\Delta t)$ has the unit volt (V), and $n\Delta t(n = 0, 1, 2...)$ is the temporal discrete points.

2.4.3 Wave tank dimensions

The width of the tank depends on the proposed model tests. 2D model tests are usually carried out in a narrow straightforward wave tank, i.e., a wave flume, with the model fully blocking the width of the flume. The wave flume wall is usually transparent to enable good visibility of the wave-structure interaction. This type of model is relatively easy to analyse as the waves and flow act in a plane. For 3D model tests, a wider wave tank is needed so that the waves can pass around the model's sides. Generally, the most realistic mixed sea waves have to be modelled in a wide tank with multiple, individually-controlled paddles. A full range of waves and wave spectra can be generated by software controlling the paddles.

The wave tank is divided into three distinct zones (see Figure 2.6) and each zone should be sufficiently long. Firstly, there is a paddle and enough space for the evanescent waves to decay. Waves from a well-controlled paddle need to travel approximately twice the hinge depth of the paddle to become fully developed. Secondly, the model zone depends on the size and motion of the model. Enough length is needed in this zone to guarantee the sample time and avoid the influence of the reflected wave from the paddle. For wide tanks, the combination of width and length determines the angle of the waves that approach the model. Thirdly, there is the wave absorbing beach, which has to be at least half the length of the design wavelength to achieve 90% absorption.

The depth of the wave tank depends on the experimental water depth. Sufficient depth in the wave tank is needed to generate the desired wave conditions and the reflected phenomenon should also be considered to avoid the top overflow. The wave characters depend highly on the water depth. The relationship between the water depth and wave nonlinearity can be seen in Figure 2.3. For experiments using optical equipment (especially for this device to be located at the side of the wave tank), the depth of the wave tank is a very important factor to be considered since most wave tanks have steel beams. A wave tank with proper depth and width can avoid any beam sheltering of the model.

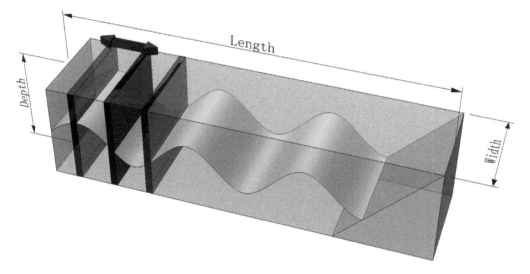

Figure 2.6: Schematic of wave tank

2.4.4 Scaling

Nearly all physical modelling of WECs is conducted at scale. A properly-scaled physical model is expected to behave in a manner similar to the prototype it is intended to emulate. To relate the responses of a scaled model with the prototype, the scaling laws or principles of similitude are considered.

2.4.4.1 Scaling law

Geometric similarity, kinematic similarity, and dynamic similarity are the three basic principles of similarity that should be taken into account for the study of fluid-structure interaction. After applying these three principles of similitude to a WEC problem, six relevant non-dimensional numbers, i.e., the *Froude, Reynolds, Cauchy, Weber, Euler,* and *Strouhal* numbers [306], can be obtained, of which the first four represent the ratio of the inertial forces to the gravity, viscous, elastic, and surface tension forces, respectively; whilst the Euler number corresponds to the ratio of the pressure forces to the inertial forces, and the Strouhal number is associated with the ratio of the temporal inertial forces to the convective inertia forces. To achieve complete similitude, all six of these criteria should be met, which is, however, not possible unless the scale factor is 1.0, i.e., the model is not scaled.

When testing WECs at scale, a decision over which criterion to uphold must be made. As inertial and gravitational forces are normally predominant for the scaling of wave interactions with WECs, most WEC tests are scaled following Froude's scaling law, which can be given by:

$$F_r = \frac{U}{\sqrt{gl}}, \tag{2.79}$$

where U is the fluid velocity, g the gravitational acceleration, and l the characteristic length.

Table 2.2 presents more explicitly the direct application of Froude's scaling law for the scaled model-related characteristics and results. The term *power density* refers to the power per unit length.

2.4.4.2 Scale issues

For some special circumstances, it may be necessary to use distorted models in the framework of Froude's scaling law. For example, when modelling a WEC floating in deep water, the horizontal lengths may be tens of m long, while the water depth may be hundreds of m. To model this system in a laboratory requires a horizontal scale large enough to avoid significant surface tension effects, and a vertical scale factor small enough to fit the model into the available space.

For some WECs, different parts of the model could be associated with different scaling laws. For example, the power take-off mechanism of oscillating water column (OWC) devices relies on the aerodynamics as the air in the OWC chamber is forced back and forth through a turbine. The hydrodynamics and aerodynamics of the OWC model are relevant to the Froude and the Reynolds laws, respectively, and their combination makes the scaling considerations for OWCs more complex than standard Froude's scaling law. More information on the topic of OWC scaling can be found in [823].

2.5   MODELLING OF INTERACTION BETWEEN OCEAN WAVE AND WAVE ENERGY CONVERTER

Complementary to costly model-scale experiments, numerical modelling is often used to understand the performance of the wave energy converter. The majority of studies on wave energy conversion have used linear potential flow theory as the basis for modelling hydrodynamics arising from wave-body interaction [622], as discussed earlier in this chapter. Although this simplifies/linearises the wave-structure interaction

Table 2.2: Scale factors for the Froude's scaling law.

Quantity	Dimension	Scale factor
Wave height/length	[L]	C_l
Wave period	[T]	$C_l^{0.5}$
Wave frequency	$[T^{-1}]$	$C_l^{-0.5}$
Power density	$[MLT^{-3}]$	$C_l^{2.5}$
Power	$[ML^2T^{-3}]$	$C_l^{3.5}$
Energy	$[ML^2T^{-2}]$	C_l^4
Linear displacement	[L]	C_l
Angular displacement	[-]	1
Linear velocity	$[LT^{-1}]$	$C_l^{0.5}$
Angular velocity	$[T^{-1}]$	$C_l^{-0.5}$
Linear acceleration	$[LT^{-2}]$	1
Angular acceleration	$[T^{-2}]$	C_l^{-1}
Mass	[M]	C_l^3
Volume	$[L^3]$	C_l^3
Pressure	$[ML^{-1}T^{-2}]$	C_l
Force	$[MLT^{-2}]$	C_l^3
Torque	$[ML^2T^{-2}]$	C_l^4
Linear stiffness	$[MT^{-2}]$	C_l^2
Angular stiffness	$[ML^2T^{-2}]$	C_l^4
Linear damping	$[MT^{-1}]$	$C_l^{2.5}$
Angular damping	$[ML^2T^{-1}]$	$C_l^{4.5}$

problem, as well as speeding up the simulation process, the utilisation of a linear solver (which assumes that the wave steepness and body motion are both small) usually contradicts the large motion arising from the wave energy converter and thus does not ensure the fidelity of the simulation results in medium to large wave conditions. As shown in Figure 2.3, deep-water waves with a steepness less than 0.01 can be approximated by a sinusoidal wave by applying linear wave theory [248]. However, wave energy converters are usually located in either intermediate water or shallow water, where waves are usually described by Stokes's wave theory or cnoidal wave theory. The operational conditions of WECs exceed the limits of linear wave theory and cover Stokes waves of the 2nd and 3rd orders [620]. At the extreme opposite, the

Navier-Stokes equation based CFD methods capture the full nonlinear hydrodynamics in the wave-body interaction problem leading to high fidelity simulation results regardless of the system's operational conditions. However, CFD is often only used to assess the survivability of the device, due to its high computational requirement.

A compromise between the linear method and the Navier-Stokes solver also exists [620], often termed weakly nonlinear solvers. These solvers are built based on the potential flow theory but with weak nonlinearities augmented to consider additional nonlinear effects. Examples for these nonlinearities are the consideration of the instantaneous body position in the wave field, extrapolation of wave kinematics above the mean sea level, and consideration of quadratic transfer functions for second-order wave-excitation forces. The computational time requirements for these codes are usually higher than for the purely linear codes, but they are often less expensive than CFD, whilst still able to capture notable nonlinearities. A typical example of a weakly nonlinear solver is the weak-scatterer potential flow method proposed for the seakeeping analysis of ships with forward speed [613], which is formulated based on the assumption that the perturbation wave field generated by the body oscillation is small compared with the incident wave field, such as the free surface conditions can be linearised at the incident wave elevation level. The weak-scatterer method takes into account the unsteady and nonlinear hydrodynamic loads associated with dynamic wave-body interactions. A further simplified version of the weak-scatterer method is the body-exact potential method, which assumes the free surface conditions can be linearised around the mean free surface elevation. This solver was proposed to account for the body motion induced nonlinearities but is only valid when small steepness waves are present. It is a compromise between the weak-scatterer method and the linear method. Table 2.3 summarises the main differences between the linear method, body-exact solver, weak-scatterer solver, and CFD [178]. Despite the fact that a large number of weakly nonlinear solvers have been proposed for WEC investigations, their capabilities in capturing nonlinearities arising from wave-WEC interactions in various wave conditions remain questionable. The WEC research and development community lacks and urgently needs a universal high-fidelity but low-cost model to improve confidence levels in numerical predictions of power production and load estimates, which are important quantities for the development of reliable and cost-efficient WECs.

To gain confidence in using numerical models and assessing the accuracy of these codes, the International Energy Agency Ocean Energy Systems (IEA OES) Task 10 project [830] compared a large number of linear, weakly nonlinear, and highly nonlinear codes on a simple WEC system of a heaving buoy. The main outcome of this project demonstrated that, in steep waves, the weakly nonlinear codes are able to capture higher-order peaks in the WEC response because of the consideration of nonlinear hydrostatics and Froude-Krylov forcing, and therefore they predict a reduced mean power output of the WEC that is closer to the reality. Highly accurate experiments were carried out on a heave decay test for a sphere [419], whose results were compared with the linear potential flow, nonlinear potential flow, and RANS CFD methods. For low drop heights, all methods performed very well, with deviations less than 1 mm for all models, but for larger drop heights and applications

Table 2.3: Main differences between the four wave-body interaction modelling methods. See [178] for more details.

	Linear potential flow	Body-exact potential flow solver	Weak-scatterer potential solver	Navier-Stokes-based CFD solver
Assumption	Irrotational and inviscid fluid, small body motion, $\epsilon \ll 1$	Irrotational and inviscid fluid, $\epsilon \ll 1$	Irrotational and inviscid fluid, $\Phi_P = \mathcal{O}(\Phi_I)$	Isotropic fluid
Hydrodynamics decomposition	$\Phi = \Phi_I + \Phi_S + \Phi_R$	$\Phi = \Phi_I + \Phi_P$		N/A
Meshed free surface	$z = 0$	$z = 0$	$z = \eta_I(x,y,t)$	$z = \eta(x,y,t)$
Meshed body surface	$S(0)$		$S(t)$	
Hydrodynamic force computation	Sum of excitation and radiation forces	Integration of total pressure over the wetted body surface		Integration of total stress over the wetted body surface
Fluid vortices		NO		YES
Wave breaking		NO		YES
Drag force	NO, a Morison equation (as detailed in Section 3.1) can be added			YES
Surface piercing	NO		YES	YES
Computational speed	Extremely fast	Medium	Slow	Extremely slow

with large motion amplitudes, the potential flow methods should be used with care; whereas the RANS models, if proper convergence is reached, are capable of producing accurate results for all drop heights. In the Collaborative Computational Project in Wave Structure Interaction (CCP-WSI), numerical results are compared without prior access to experimental data in a blind test approach [653]. The project has been conducted via several steps with increasing complexity, and has considered interactions between focused waves and structures. It was concluded that Navier-Stokes solvers and hybrid methods were superior to linearised methods under these extreme wave conditions; however, heave motion was predicted reasonably well by all methods.

CHAPTER 3

Wave energy converter modelling, control, and power take-off design

Leandro S.P. da Silva[1], Boyin Ding[1], Bingyong Guo[2], Nataliia Y. Sergiienko[1]
[1]University of Adelaide, boyin.ding@adelaide.edu.au,
[2]Northwestern Polytechnical University, b.guo@nwpu.edu.cn

3.1 MODELLING OF WAVE ENERGY CONVERTER DYNAMICS

Currently, several WECs exist with different absorption mechanisms and subsystems. Hence, a general formulation to describe the dynamics of all possible devices is a nontrivial task. This section provides practical, condensed information, and derivations for the dynamics of oscillating bodies, as this category comprises most WECs, and relevant literature is recommended for more technical information. The approaches described here can be extended to the analysis of other WECs systems and examples of these are given throughout the book.

3.1.1 Time domain

The dynamics of WECs are generally characterised by a set of nonlinear equations from the hydrodynamic loads, mooring, structural behaviour, and PTO systems with their respective control strategy. Nonlinearities are relevant in WEC dynamics because their natural frequency is usually set within the sea spectrum range, which increases the energy transferred from waves to WECs, and leads to large displacements. Based on that, the use of time-domain models is a common practice in the wave energy field due to its capability to deal with nonlinearities. In such models, the system dynamics are deterministically solved in time via numerical integration using a specific time series of wave elevation/forces. Strictly speaking, several of the

DOI: 10.1201/9781003198956-3

aforementioned methods can be classified as time-domain models, such as linear/nonlinear potential flow models and CFD models. Hereafter, this subsection will focus mainly on time-domain models based on the Cummins equation [146], which relies on existing experiences from the offshore and ship industry. This type of model has been proved to be effective after extensive application to WECs. In addition, the foundation of the Cummins equation can be related straightforwardly to the frequency-domain model, presented in the following subsection, which is based on linear wave potential theory.

Let us consider general WECs dynamics, such as that shown in Eq. (2.46), in which the main sources of loads are:

$$\mathbf{M}\ddot{\mathbf{x}}(t) = \mathbf{F}_e + \mathbf{F}_r + \mathbf{F}_s + \mathbf{F}_v + \mathbf{F}_{add}, \qquad (3.1)$$

where the first four right-hand terms refer to the loads caused by wave-structure interactions: $\mathbf{F}_e$, $\mathbf{F}_r$, $\mathbf{F}_s$, and $\mathbf{F}_v$ refer to the wave excitation, wave radiation, hydrostatic, and viscous drag forces and moment terms respectively; the last term, $\mathbf{F}_{add}$, denotes additional forces, such the ones from the power take-off, mooring system, control strategy, and motion constraint forces.

For most WECs, the first-order wave excitations are sufficient to have reliable estimations of the body displacements because the natural frequency for the power absorption mode generally sits within the sea spectrum range. However, analysis using higher-order excitations is usually required during further stages of development. Assuming an irregular sea state, the wave excitation may be represented as a linear superposition of N sinusoidal load components, as:

$$\mathbf{F}_e(t) = \Re\left[\sum_{i=1}^{N} \eta_i(\omega_i)\mathbf{H}_{\mathbf{Fe},\mathbf{i}}(\omega_i)e^{(-i\omega_i t + \beta_{\eta,i})}\right], \qquad (3.2)$$

where the subscript i denotes the ith frequency component, η denotes the amplitude of wave elevation, which can be obtained from the sea spectrum as $\eta_i(\omega_i) = \sqrt{2S_\eta(\omega_i)\Delta\omega}$, as shown in Eqs. (2.36) to (2.38); $\mathbf{H}_{\mathbf{Fe},\mathbf{i}}(\omega_i)$ denotes the transfer function that relates the wave excitation loads and the wave elevation at frequency ω_i; $\beta_{\eta,i}$ is the wave surface phase angle, which is assumed to be uniformly distributed within $[0, 2\pi]$ radians. Note that for a regular sea state, the wave is composed of a single frequency.

The wave radiation loads may be represented by Cummins equation [146]:

$$\mathbf{F}_r(t) = -\mathbf{A}_\infty\ddot{\mathbf{x}}(t) - \int_{-\infty}^{t} \mathbf{K}_{rad}(t-\tau)\dot{\mathbf{x}}(t)\mathrm{d}\tau, \qquad (3.3)$$

where $\mathbf{A}_\infty = \lim_{\omega\to\infty} \mathbf{A}_\mathbf{m}(\omega)$ is the added mass matrix at infinite frequency, and $\mathbf{K}_{rad}$ denotes the radiation impulse response function matrix, also known as the memory function matrix. This term is related to the presence of the free surface, where the body motion generates propagating waves. The radiation load effects depend on any previous motion, which is shown in the integral limits. In practice, the contribution of the convolution integral is negligible after a relatively short time (approximately

20–80 s for some devices). Hence, a common practice in offshore engineering is to compute the convolution integral over a limited time when calculated by direct numerical integration. Several approximations have been proposed to circumvent the requirement to compute the convolution integral in Eq. (3.3), such as system identification based on frequency-domain, time-domain, and Prony's method [667, 629, 214]. Chapter 4 gives a more in-depth information regarding the radiation loads representation in the time domain.

The hydrostatic forces ($\mathbf{F}_s$) are originated by the static-pressure (still water) acting along the WEC's wet surface, and their influence is important to the dynamics and stability of the devices. As shown in Eq. (2.43), the hydrostatic force is expressed as a function of the displacement; therefore, it is common to represent this load in terms of hydrostatic stiffness and a constant force.

A common representation of the viscous drag loads can be formulated based on Morison's equation, which is characterised as a quadratic drag function. Initially, Morison's equation was applied to predict wave forces per unit of length, applied to a vertical pile. Since its first derivation, several modifications have been applied to the original formulation. In the wave energy field, the inclusion of viscous drag forces usually considers the relative velocity between the structure and the wave ($\dot{x}_{rel,i} = \dot{x}_i - u_i$) for translational modes of motion, and structure velocity for rotational modes. For example, the viscous drag component for the translational modes can be expressed as:

$$F_{v,j}(t) = -\frac{1}{2}\rho C_{D,j} S_{\perp,j} \dot{x}_{rel,j}(t)|\dot{x}_{rel,j}(t)|, \text{ for } j = [1,2,3] \qquad (3.4)$$

where $C_{D,j}$ denotes the viscous drag coefficient in the j-th mode, and $S_{\perp,j}$ is the cross-sectional area of the structure perpendicular to the j-th direction. An example for the rotational modes can be seen in Eq. (5.15). The drag coefficient is usually obtained via experimental results or CFD simulations; its magnitude is dependent on the geometry, Reynolds number, roughness number, and Keulegan-Carpenter number [309].

Additional forces act on the WEC dynamics, which are dependent on the characteristics of the devices. For example, some WECs might be connected to mooring systems composed of tethers connected to their PTO system, while others use catenary lines to restrain their motion [160]. The PTO system can be composed of hydraulic, pneumatic, mechanical, and electrical systems. Hence, different numerical models can be employed to describe additional loads, based on the characteristics of each device and assumptions. Although this topic is a broad field of knowledge, some extra information regarding control and PTO systems is provided in Section 3.3 to cover the main aspects of the wave energy conversion, and some are presented in subsequent chapters.

The main objective of a WEC and its PTO system is to extract power from ocean waves. In this regard, the mean mechanical power absorbed from the PTO system over a time T can be calculated as:

$$\overline{P}_{pto} = -\frac{1}{T}\int_0^T F_{pto}(t)\dot{x}(t)\mathrm{d}t, \qquad (3.5)$$

where F_{pto} is the PTO force, which was included as an additional force in Eq. (3.1). Assuming a mechanical PTO system, composed of a linear damper, the mean power absorbed is given by:

$$\overline{P}_{pto} = \frac{1}{T} \int_0^T B_{pto} \dot{x}^2 \mathrm{d}t, \tag{3.6}$$

where B_{pto} is the magnitude of the linear damping coefficient from the PTO system.

3.1.2 Frequency domain

Time-domain simulations are computationally expensive when compared with calculations in the frequency domain. Hence, they are generally employed after extensive investigations in the frequency domain, once the main features of the WEC have been defined, in order to demonstrate the validity of the device and control strategy. Therefore, the WEC dynamics during the operational regime and preliminary analysis are conducted in the frequency domain. In this regard, the wave-structure interaction is described based on the linear potential theory. Additional forces, such as the ones from PTO, mooring systems, and control, are linearised around an operating point by means of Taylor series up to first-order terms. Once the dynamics are described by a linear system, the principles of superposition and linear combination can be applied to the analysis of the WECs.

Considering harmonic excitation loads, the vector of response has also an harmonic motion that oscillates at same frequency and can be represented as:

$$\mathbf{x}(t) = \Re\{\mathbf{X}(\omega)e^{-i\omega t}\}, \tag{3.7}$$

where $\mathbf{X}$ is the vector of complex amplitude of the body displacement. Based on Eq. (3.7), the velocity and acceleration can be obtained as:

$$\dot{\mathbf{x}}(t) = \Re\{-i\omega\mathbf{X}(\omega)e^{-i\omega t}\}, \tag{3.8}$$

$$\ddot{\mathbf{x}}(t) = \Re\{-\omega^2\mathbf{X}(\omega)e^{-i\omega t}\}. \tag{3.9}$$

Based on the above, the WEC linear dynamics can be expressed in the frequency domain as:

$$\left[-\omega^2(\mathbf{M} + \mathbf{A_m}(\omega)) - i\omega(\mathbf{B} + \mathbf{B_{rad}}(\omega)) + \mathbf{K}\right]\mathbf{X} = \mathbf{F}_e(\omega), \tag{3.10}$$

where $\mathbf{M}$, $\mathbf{B}$, and $\mathbf{K}$ are the matrices of inertia, damping and stiffness, respectively; the hydrodynamic coefficients of added mass, $\mathbf{A_m}(\omega)$, radiation damping, $\mathbf{B_{rad}}(\omega)$, and excitation force, $\mathbf{F}_e(\omega)$, are usually obtained through BEM codes or analytically as per Section 2.2.2. In this formulation, additional forces from the PTO, mooring and control system are assumed to be functions of the displacements and velocities; hence, are described as a combination of linear stiffness and dampers that compose matrices $\mathbf{B}$ and $\mathbf{K}$. Note that some PTO systems present relevant inertial effects, e.g., translators, and flywheels [706], hence, inertia terms from the PTO appear in matrix $\mathbf{M}$. It is also important to highlight that some sources of nonlinearities vanish during the linearisation, such as the viscous drag loads under zero current speed.

Based on Eq. (3.10), the transfer function of the linear system dynamics can be established to relate the WEC response and excitation forces:

$$\mathbf{H_x}(\omega) = \left[-\omega^2(\mathbf{M} + \mathbf{A_m}(\omega)) - i\omega(\mathbf{B} + \mathbf{B_{rad}}(\omega)) + \mathbf{K}\right]^{-1}. \qquad (3.11)$$

In the offshore engineering field, the response is also usually expressed based on the response amplitude operator (RAO), which establishes a transfer function between the body response amplitude and the wave amplitude. Once the transfer function is derived, the dynamics can be obtained straightforwardly by solving the system of algebraic linear equations. For irregular sea states, the stochastic response can be expressed in terms of the power spectrum density (PSD) through the response spectrum matrix as:

$$\mathbf{S_x}(\omega) = \mathbf{H_x S_f H_x}^{\mathrm{T}*}, \qquad (3.12)$$

where $^{\mathrm{T}*}$ denotes the transpose conjugate of the matrix, and $\mathbf{S_f}$ is the force spectrum matrix given by:

$$\mathbf{S_f}(\omega) = \mathbf{H_{Fe} S_\eta H_{Fe}}^{\mathrm{T}*}, \qquad (3.13)$$

where $\mathbf{S_\eta}$ denotes the wave spectrum.

In Eq.(3.10), the damping matrix ($\mathbf{B}$) contains contributions from several sources of loads that can exist in the WEC's dynamics, such as those from the PTO system ($\mathbf{B_{pto}}$). In this regard, the mean power absorbed by a linear PTO system in an irregular sea state can be calculated in the frequency domain as:

$$\overline{\mathbf{P}}_{\mathbf{pto}} = \mathbf{B_{pto}} \int \omega^2 \mathbf{S_x} \mathrm{d}\omega. \qquad (3.14)$$

3.1.3 Statistical linearisation

Frequency-domain models have been used extensively for the initial stages of development due to their low computational cost compared with time-domain simulations. However, the application of such models is restricted to linear systems, which may lead to unreliable predictions of the system dynamics and the mean power absorbed. In this regard, and under certain assumptions, the linearisation in the frequency-domain models can be performed statistically for a more realistic prediction of the first-order motion of WECs, where the contribution of each source of nonlinearity is considered over the operating region. In such an approach, the wave excitation is described by a Gaussian distribution, as it is given by a linear transformation of the sea elevation, which is assumed Gaussian. Since the dynamics are approximated to a linear function, the WEC response is also assumed to be Gaussian. To date, the contributions of several nonlinear loads have been estimated using statistical linearisation for a variety of WECs [152, 247], such as oscillating wave surge converters [250, 778, 153], point absorbers [251, 147, 738, 150], and oscillating water columns [252, 737, 151].

Consider a general WEC described in the following form:

$$\mathbf{M\ddot{x}} + \mathbf{B\dot{x}} + \mathbf{Kx} + \mathbf{\Theta}(\mathbf{x}, \dot{\mathbf{x}}, \ddot{\mathbf{x}}) = \mathbf{F}_e(t) + \mathbf{F}_r, \qquad (3.15)$$

where Θ is a nonlinear vector, which is a function of the generalised coordinate vector and its first and second time derivative. Here, Θ comprises all sources of nonlinearities, such as those from viscous drag, PTO, and mooring systems. Based on Eq. (3.15), an equivalent linear system can be determined as:

$$(\mathbf{M} + \mathbf{M}_{eq})\ddot{\mathbf{x}} + (\mathbf{B} + \mathbf{B}_{eq})\dot{\mathbf{x}} + (\mathbf{K} + \mathbf{K}_{eq})\mathbf{x} = \mathbf{F_e}(t) + \mathbf{F}_r, \tag{3.16}$$

where the subscript $_{eq}$ denotes the equivalent linear terms, which are determined by minimising the difference between the nonlinear and equivalent linear systems, Eqs. (3.15) and (3.16) respectively, in a mean squared sense as:

$$\min \left\langle \varepsilon_{diff}^{T} \varepsilon_{diff} \right\rangle, \tag{3.17}$$

where $\langle \rangle$ denotes the mathematical expectation and the difference ε_{diff} is given by:

$$\varepsilon_{diff} = \Theta(\mathbf{x}, \dot{\mathbf{x}}, \ddot{\mathbf{x}}) - \mathbf{M}_{eq}\ddot{\mathbf{x}} - \mathbf{B}_{eq}\dot{\mathbf{x}} - \mathbf{K}_{eq}\mathbf{x}. \tag{3.18}$$

Under the assumption of a Gaussian response, it can be demonstrated that the minimisation condition is obtained systematically by setting the equivalent linear matrices as [674]:

$$M_{eq_{i,j}} = \left\langle \frac{\partial \Theta_i}{\partial \ddot{x}_j} \right\rangle, \tag{3.19}$$

$$B_{eq_{i,j}} = \left\langle \frac{\partial \Theta_i}{\partial \dot{x}_j} \right\rangle, \tag{3.20}$$

$$K_{eq_{i,j}} = \left\langle \frac{\partial \Theta_i}{\partial x_j} \right\rangle, \tag{3.21}$$

where the subscript $_{i,j}$ denotes the matrix index.

The equivalent random zero mean WEC steady-state response may be represented mathematically as:

$$\tilde{\mathbf{x}}(t) = \int_{-\infty}^{\infty} \mathbf{h_{x,eq}}(\tau)\tilde{\mathbf{f}_e}(t - \tau)\mathrm{d}\tau, \tag{3.22}$$

where $\mathbf{h_{x,eq}}$ is the equivalent linear impulse response function matrix for the body response, and $\tilde{\mathbf{f}_e}$ is a zero-mean, Gaussian, random excitation, which is given by the linear impulse response function of the incident wave elevation. Based on the above, an equivalent transfer function of the nonlinear system dynamics can be established to relate the WEC response and excitation forces, thus:

$$\mathbf{H_{x,eq}} = \left[-\omega^2(\mathbf{M} + \mathbf{A_m}(\omega) + \mathbf{M_{eq}}) - \mathrm{i}\omega(\mathbf{B_{rad}}(\omega) + \mathbf{B} + \mathbf{B_{eq}}) + (\mathbf{K} + \mathbf{K_{eq}}) \right]^{-1}. \tag{3.23}$$

Like Eq. (3.12), the stochastic response can be expressed in terms of PSD through the response spectrum matrix as:

$$\mathbf{S_x} = \mathbf{H_{x,eq}}\mathbf{S_f}\mathbf{H_{x,eq}}^{\mathrm{T}*}. \tag{3.24}$$

Based on Eq. (3.6), the mean power absorbed by a linear damper can be calculated as:

$$\overline{P}_{pto} = \left\langle B_{pto}\dot{x}^2 \right\rangle = B_{pto}\sigma_{\dot{x}}^2,$$ (3.25)

which is equivalent to the expression given in Eq. (3.14). Note that some PTO systems might be described by nonlinear forces as in Eq. (3.5). In this case, the power is given as:

$$\overline{P}_{pto} = \langle F_{pto}\dot{x} \rangle.$$ (3.26)

For nonlinear PTO systems, the mean power absorbed can be estimated using the equivalent linear terms as shown in [152]:

$$\overline{P}_{pto} = B_{eq,pto}\sigma_{\dot{x}}^2,$$ (3.27)

where $B_{eq,pto}$ is the equivalent damping coefficient from the nonlinear PTO.

3.1.3.1 *Asymmetric non-linearities and constant loads*

Note that in the case of asymmetric non-linearities or/and systems under constant loads, the response can be described in terms of mean values ($\mu_{\mathbf{x}}$) and random zero mean components ($\tilde{\mathbf{x}}$). For such conditions, Eq. (3.15) can be rewritten as:

$$\mathbf{M}\ddot{\tilde{\mathbf{x}}} + \mathbf{B}\dot{\tilde{\mathbf{x}}} + \mathbf{K}(\mu_{\mathbf{x}}+\tilde{\mathbf{x}}) + \Theta(\mu_{\mathbf{x}}+\tilde{\mathbf{x}},\dot{\tilde{\mathbf{x}}},\ddot{\tilde{\mathbf{x}}}) = \mu_{\mathbf{Fe}} + \tilde{\mathbf{F}}_{\mathbf{e}}(t) + \tilde{\mathbf{F}}_r,$$ (3.28)

where $\mu_{\mathbf{F}}$ denotes the vector of mean forces and moments. The mean response can be obtained taking the expectation of Eq. (3.28):

$$\mu_{\mathbf{x}} = \mathbf{K}^{-1}\left\langle \mu_{\mathbf{Fe}} - \Theta(\mu_{\mathbf{x}}+\tilde{\mathbf{x}},\dot{\tilde{\mathbf{x}}},\ddot{\tilde{\mathbf{x}}}) \right\rangle.$$ (3.29)

Subtracting Eq. (3.29) from Eq. (3.28), an equivalent form of Eq. (3.15) is obtained:

$$\mathbf{M}\ddot{\tilde{\mathbf{x}}} + \mathbf{B}\dot{\tilde{\mathbf{x}}} + \mathbf{K}\tilde{\mathbf{x}} + \mathbf{G}(\tilde{\mathbf{x}},\dot{\tilde{\mathbf{x}}},\ddot{\tilde{\mathbf{x}}}) = \tilde{\mathbf{F}}_{\mathbf{e}}(t) + \tilde{\mathbf{F}}_r,$$ (3.30)

where $\mathbf{G}(\tilde{\mathbf{x}},\dot{\tilde{\mathbf{x}}},\ddot{\tilde{\mathbf{x}}}) = \Theta(\mu_{\mathbf{x}}+\tilde{\mathbf{x}},\dot{\tilde{\mathbf{x}}},\ddot{\tilde{\mathbf{x}}}) - \left\langle \Theta(\mu_{\mathbf{x}}+\tilde{\mathbf{x}},\dot{\tilde{\mathbf{x}}},\ddot{\tilde{\mathbf{x}}}) \right\rangle.$

3.1.3.2 *Multivariate normal probability density function*

The mathematical expectations used to compute Eqs. (3.19) to (3.21) can be a function of several variables and their time derivatives. In this regard, multivariate normal probability density functions (MVNPDF) are required to estimate the equivalent linear terms. Considering the case of a nonlinearity function of the displacement, the MVNPDF to calculate the equivalent stiffness terms would have the following form:

$$f(\mathbf{x}) = \frac{1}{(2\pi)^{n/2}|\mathbf{V}|^{1/2}}\exp\left(-\frac{1}{2}(\tilde{\mathbf{x}}-\mu_{\mathbf{x}})^{\mathrm{T}}\mathbf{V}^{-1}(\tilde{\mathbf{x}}-\mu_{\mathbf{x}})\right),$$ (3.31)

where $\mathbf{V}$ denotes the covariance matrix of the coordinate $\mathbf{x}$, and $|\mathbf{V}|$ denotes the determinant. The symmetrical covariance matrix is given by:

$$\mathbf{V} = \begin{bmatrix} \sigma_{x_1}^2 & \sigma_{x_1 x_2} & \cdots & \sigma_{x_1 x_n} \\ & \sigma_{x_2}^2 & \cdots & \sigma_{x_2 x_n} \\ & & \ddots & \vdots \\ sym & & & \sigma_{x_n}^2 \end{bmatrix},$$ (3.32)

where the diagonal terms of the covariance matrix are obtained based on the auto-spectrum as:

$$\sigma_{x_k}^2 = \int_0^\infty S_{x_k x_k}(\omega)\mathrm{d}\omega, \tag{3.33}$$

and the non-diagonal terms of the covariance matrix $(j \neq k)$ are obtained based on the cross-spectrum as:

$$\sigma_{x_j x_k} = \int_0^\infty S_{x_j x_k}(\omega)\mathrm{d}\omega. \tag{3.34}$$

Some sources of nonlinearities may be written as a function of the coordinate **x** and their time derivatives. Considering the cases of nonlinearities with first time derivatives, the following relationships can be used to compute the covariance matrix and distributions:

$$\sigma_{\dot{x}_k}^2 = \int_0^\infty \omega^2 S_{x_k x_k}(\omega)\mathrm{d}\omega, \tag{3.35}$$

$$\sigma_{x_j \dot{x}_k} = \mathrm{i} \int_0^\infty \omega S_{x_j x_k}(\omega)\mathrm{d}\omega, \tag{3.36}$$

$$\sigma_{x_k \dot{x}_j} = -\mathrm{i} \int_0^\infty \omega S_{x_j x_k}(\omega)\mathrm{d}\omega, \tag{3.37}$$

$$\sigma_{x_k \dot{x}_k} = 0, \tag{3.38}$$

$$\sigma_{\dot{x}_j \dot{x}_k} = \int_0^\infty \omega^2 S_{x_j x_k}(\omega)\mathrm{d}\omega. \tag{3.39}$$

The same procedure can be extended for the nonlinearities functions of the coordinate, and their first and second-time derivatives. For nonlinearities dependent on a single variable, the Gaussian distribution can be used to calculate the mathematical expectation. Depending on the source of the nonlinearity, analytical solutions are available, as given in Table 3.1.

Table 3.1: Analytical solutions for the statistical linearisation coefficients of common nonlinearities for WECs.

$\Phi(x)$	$\left\langle \frac{d\Phi}{dx} \right\rangle$	$\langle \Phi(x) \rangle$
$\mathrm{sgn}(x)$	$\left(\frac{2}{\pi}\right)^{\frac{1}{2}} \sigma_x^{-1}$	0
$x\lvert x\rvert$	$\left(\frac{8}{\pi}\right)^{\frac{1}{2}} \sigma_x$	0
x^3	$3\sigma_x^2$	0

3.1.3.3 Iterative procedure

As observed in Eqs. (3.20) and (3.21), the mathematical expectations to calculate the equivalent linear terms require prior knowledge of the response distribution. The solution to this problem is usually not available analytically. Hence, the statistical linearisation is performed using an iterative procedure, wherein traditional frequency-domain models with Taylor linearisations are used as an initial guess. The iterative procedure runs until the error between the previous and the new steps is within an acceptable range.

3.1.4 Alternative numerical methods

Statistical linearisation provides reasonable estimates of the mean and mean-square responses of WECs, which are restricted to first-order dynamics. In addition, the equivalent linear solution leads to a Gaussian distribution, while the real response might deviate to a non-Gaussian one. As a result, equivalent linear responses might lose salient features of the actual nonlinear dynamics. It is well known that nonlinearities may lead to vibrations outside the excitation range frequency. Although WECs have a dominant mode that extracts energy, which is usually set to operate close to resonant conditions, higher-order responses can be relevant for other degrees of freedom (DoFs) and the response may be given by a non-Gaussian distribution.

The limitations of the statistical linearisation method can be circumvented by replacing nonlinear loads with an equivalent nonlinear one in a polynomial form, known as the statistical nonlinearisation method [182, 736, 776, 777]. This approach can be interpreted as an extension of statistical linearisation using high order Volterra series to describe the response of nonlinear systems [182]. Generally, polynomials up to the second or third-order are enough to describe nonlinear behaviour, known as the statistical quadratisation [182] and statistical cubicisation methods, respectively [776]. Like the SL (statistical linearization), the equivalent coefficients are obtained by minimising the difference between the nonlinear and the equivalent polynomial one in a mean square sense, while the probability density function is built using Gram-Charlier expansions of the Gaussian distribution [182]. Currently, the statistical nonlinearisation method has not been applied to WECs, however, additional examples applied to offshore systems can be found in [148, 444, 645]. Other examples of spectral methods can be applied to WECs; such as the harmonic balance method [540, 538, 735] and multiple timescale spectral analysis [171]. For a survey of the available methods, the reader is also encouraged to read references [674, 730, 728, 729].

3.1.5 Case study

This subsection provides a practical example of different methodologies being applied to estimate the response and power absorbed by a WEC. In this regard, a half-submerged spherical point absorber (PA) restrained to move only in the heave direction is chosen as a case study. The additional forces acting on the PA are from the mooring and PTO systems, which are represented by a linear stiffness and damper.

3.1.5.1 Time domain

As described in Eqs. (3.1) to (3.4), the governing equation of the PA heaving motion at the equilibrium position can be represented as:

$$(m + A_\infty)\ddot{x} + B_{pto}\dot{x} + K_{pto}x = -\int_{-\infty}^{t} \mathrm{k}_{rad}(t - \tau)\dot{x}(t)\mathrm{d}\tau + F_s + F_v + F_e(t), \quad (3.40)$$

where B_{pto} and K_{pto} are the damping and stiffness coefficients of the PTO system, which consists of a linear spring and a linear damper, and m is the mass of the point absorber.

For the following example it is important to notice that two nonlinear sources are included: viscous drag and nonlinear hydrostatic stiffness. The first nonlinearity is expressed as in Eq. (3.4); however, a simplification using only the body velocity is assumed here. Hence, the viscous drag is expressed as:

$$F_v = -\frac{1}{2}\rho C_D \pi r^2 \dot{x}|\dot{x}|, \tag{3.41}$$

where r denotes the radius of the sphere.

The second nonlinearity occurs due to the variation of the waterplane area with the depth, resulting in a nonlinear hydrostatic restoring force. For the spherical case, the variations of the immersed fluid volume around the centre of the sphere can be described as:

$$dV = \pi \left(r^2 - x^2\right) dx. \tag{3.42}$$

Hence, the nonlinear hydrostatic restoring force can be expressed as:

$$F_s = -\pi \rho g \left(xr^2 - \frac{x^3}{3}\right) = \underbrace{-\pi \rho g r^2}_{K_s} x + \frac{\pi \rho g}{3} x^3, \text{ with } -r \leq x \leq r, \tag{3.43}$$

where K_s is the linear hydrostatic stiffness of the buoy.

3.1.5.2 Frequency domain

In the frequency-domain model, the dynamics are linearised around the mean position. Hence, only the linear stiffness part is included and the viscous drag contribution is neglected. Based on that, the following transfer function is given to relate the buoy displacement and the wave excitation force:

$$\mathbf{H_x}(\omega) = \left[-\omega^2(m + A_m(\omega)) - i\omega(B_{pto} + B_{rad}(\omega)) + (K_{pto} + K_s)\right]^{-1}, \tag{3.44}$$

which is used to calculate the stochastic response, as shown in Eq. (3.12).

3.1.5.3 Statistical linearisation

Under the hypothesis of Gaussian excitations, statistical linearisation can be employed to estimate the contributions of nonlinear terms. In the following example, the viscous drag and nonlinear hydrostatic stiffness are included using equivalent linear terms, and the transfer function that relates the buoy displacement and wave excitation force is given by:

$$\mathbf{H_{x,eq}}(\omega) = \left[-\omega^2(m + A_m(\omega)) - i\omega(B_{pto} + B_{rad}(\omega) + B_{eq}) + (K_{pto} + K_s + K_{eq})\right]^{-1}. \tag{3.45}$$

If the PA oscillates within the range defined in Eq. (3.43), an analytical solution for the hydrostatic stiffness can be used. As a result, the equivalent coefficients are given by:

$$B_{eq} = \frac{1}{2}\rho C_D \pi r^2 \left(\frac{8}{\pi}\right)^{\frac{1}{2}} \sigma_{\dot{x}}, \tag{3.46}$$

$$K_{eq} = -\pi \rho g \sigma_x^2. \tag{3.47}$$

In the present example, no equivalent linear inertia term occurs because the nonlinearities are not dependent on the acceleration. However, nonlinearities as a function of the acceleration might be present in the numerical methods of other WEC mechanisms such as oscillating water columns, which are characterised by a variable mass system. Note that the relative velocity can be included in statistical linearisation of the viscous drag equation as shown in [149].

3.1.5.4 Results

The PA is subjected to a sea state described by the JONSWAP spectrum, with a significant wave height equal to 3 m and a wave peak period of 7 s, and a water depth is equal to 50 m. Table 3.2 presents the parameters used in this case study. Ten time-domain simulations were performed during 1,000 s, each using different sets of phase angles in the excitation force. The PA was set to operate under the maximum power configuration, which was optimised using the frequency-domain model.

Table 3.2: PA parameters used in the case study.

Property	Value	Unit
r	5	$[\mathrm{m}]$
m	2.68×10^5	$[\mathrm{kg}]$
K_{pto}	-450	$\left[\frac{\mathrm{kN}}{\mathrm{m}}\right]$
B_{pto}	100	$\left[\frac{\mathrm{kN.s}}{\mathrm{m}}\right]$
C_D	1	$[-]$

To compare the accuracy between methods, Figure 3.1a shows the PSD of the heave displacement using nonlinear time-domain simulations (TD), a statistical linearisation model (SL), and a frequency-domain model (FD). The SL model was able to estimate the effects of the nonlinearities accurately, which is captured using the nonlinear TD model. Depending on the source of nonlinearity, shifts in the peak response can be observed due to stiffness and inertial effects, while the magnitude is also affected by nonlinear terms described as a function of the velocities. For the following example, the main difference occurs due to the inclusion of viscous effects in the SL and TD approaches.

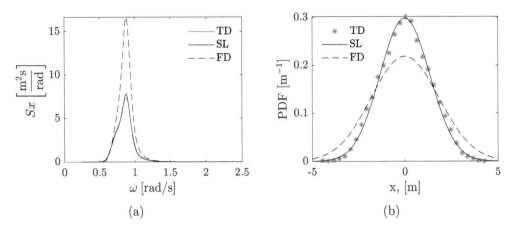

Figure 3.1: Comparison between the numerical methods: (a) PSD of the heave motion, (b) Probability distribution function of the heave motion.

To verify the applicability of the SL technique, Figure 3.1b shows a comparison of the probability distribution between the models. As mentioned previously, the SL model is built under the assumption of a Gaussian distribution, which is a valid approximation for the conditions investigated in this example. This led to the higher accuracy in the results predicted by the SL model, comparable to nonlinear TD simulations.

The mean power estimations using the different models and their associated simulation clock time are presented in Table 3.3. The power estimation using statistical linearisation obtained good agreement with nonlinear time-domain simulations, while the frequency-domain model was overestimated due to the viscous effects not being included. In terms of computational cost, the statistical linearisation model is one order of magnitude slower than the traditional frequency-domain model, and around three orders faster than the time domain, if multiple simulations with different sets of phase angles in the excitation force are considered.

Table 3.3: Power estimation and simulation wall-clock time comparison across models.

Model	Power estimated(kW)	Simulation wall-clock time(s)
Time domain	132	39.55 (each run)
Statistical linearisation	133	0.108
Frequency domain	255	0.005

3.2 PRINCIPLES AND BOUNDS OF WAVE POWER ABSORPTION

Understanding the principles and limits in WEC power absorption is crucial for efficient PTO system design. To avoid losing generality, this section treats the PTO system (a complex and multi-functional sub-system of WEC as detailed in

Section 3.3) as a generic feedback control system that generates a PTO control force modifying the power absorption performance of the WEC system.

3.2.1 Impedance matching

In general, a force-to-velocity model of a WEC in a single DoF in the frequency domain can be written as:

$$\frac{U(\omega)}{F_e(\omega)} = \frac{1}{Z_i(\omega) + Z_{pto}(\omega)}, \tag{3.48}$$

where U is the complex amplitude of the buoy velocity; F_e is the complex amplitude of the excitation force; $Z_i(\omega)$ is the frequency-dependent system intrinsic impedance (associated with wave-structure interaction dynamics); and $Z_{pto}(\omega)$ is the frequency-dependent PTO load impedance (associated with PTO system dynamics) that is introduced to modify the overall system behaviour via feedback control. For a half-submerged heaving point absorber, as shown in Figure 3.2, following Eq. (3.11), these two impedances are:

$$Z_i(\omega) = B_{rad}(\omega) - \mathrm{i}\left(\omega m + \omega A_m(\omega) - \frac{K_s}{\omega}\right), \tag{3.49}$$

$$Z_{pto}(\omega) = B_{pto} + \mathrm{i}\left(\frac{K_{pto}}{\omega}\right). \tag{3.50}$$

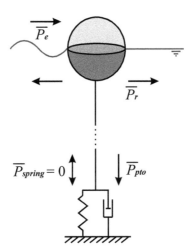

Figure 3.2: Average power flow in and out of a spring-damper controlled heaving point absorber wave energy converter under harmonic motion.

The following complex conjugate relationship between the system intrinsic impedance and the PTO load impedance was derived for maximising the power absorption of a WEC [227], often termed complex conjugate control or impedance matching in the literature:

$$Z_{pto}(\omega) = Z_i^*(\omega), \tag{3.51}$$

where ( )* denotes the complex conjugate. Substituting Eqs. (3.49) and (3.50) into Eq. (3.51), the optimal PTO parameters maximising power absorption can be obtained:

$$K_{pto} = \omega^2 (m + A_m(\omega)) - K_s, \tag{3.52}$$

and,

$$B_{pto} = B_{rad}(\omega). \tag{3.53}$$

Rearranging Eq. (3.52) specifies the WEC resonance condition, or phase optimality condition:

$$\omega_n = \sqrt{\frac{K_{pto} + K_s}{m + A_m(\omega_n)}}. \tag{3.54}$$

When the natural frequency of the controlled WEC system (as adjusted by the PTO spring) is equal to the incident wave frequency, the controlled WEC system operates in resonance with the wave excitation. This results in a complete cancellation of the reactive terms (or imaginary terms) in the $\frac{U(\omega)}{F_e(\omega)}$ relationship in Eq. (3.48), and consequently forces the buoy velocity in phase with the excitation force. Since the average power input for the WEC system arising from wave excitation, as shown in Figure 3.2, is given by:

$$\overline{P}_e = \frac{1}{2}\text{Re}\{F_e U^*\} = \frac{1}{2}|F_e|.|U|\cos(\gamma), \tag{3.55}$$

zero phase difference between the buoy velocity and the excitation force (i.e. $\gamma=0$) allows incident wave power to be input effectively into the WEC system. However, due to the existence of the PTO spring (for the tuning of the natural frequency of the WEC system), a bidirectional power flow occurs between the PTO spring and the buoy, which averages to zero (i.e., $\overline{P}_{spring} = 0$) under harmonic wave excitation. Therefore, the PTO control system including the function of the spring-like storage is often called reactive control because of the required reactive power flow between the PTO and the buoy.

Eq. (3.53) shows the optimal amplitude condition for the buoy velocity where the PTO damping is equal to the radiation damping. This condition is set to compromise the power outputs of the WEC system: the power returned back into the sea via radiation damping dissipation as shown in Figure 3.2:

$$\overline{P}_r = \frac{1}{2}B_{rad}|U|^2, \tag{3.56}$$

and the power absorbed by the PTO system via PTO damping dissipation, as shown in Figure 3.2:

$$\overline{P}_{pto} = \frac{1}{2}B_{pto}|U|^2. \tag{3.57}$$

A PTO damper, as a simplified analogy to an electricity generator, only dissipates power and thus any PTO control system modelled as a pure damper is often called a passive control due to the passive loading associated with the damping mechanism.

Substituting Eqs. (3.49), (3.50), and (3.51) into Eq. (3.48), the optimal buoy velocity (without considering the viscous drag) is obtained:

$$U_{opt}(\omega) = \frac{F_e(\omega)}{2B_{rad}(\omega)}. \tag{3.58}$$

Eq. (3.58) shows that the optimal buoy velocity is in phase with the excitation force and the optimal impedance of the controlled system becomes double that of the radiation damping, which indicates that half of the wave power input is radiated back to the sea and the other half is absorbed by the PTO damper. To gain a quantitative understanding on the power maximising theory under impedance matching, the averaged input power and averaged output power of a half-submerged heaving point absorber are plotted against the variations of the PTO spring and damper coefficients respectively, as shown in Figure 3.3. It can be seen that PTO absorbed mean power is given by the difference between the mean power input and the mean radiated power, regardless of the variations in K_{pto} and B_{pto}. The PTO absorbed mean power reaches a peak at the resonance condition by tuning K_{pto} in Figure 3.3a, while it reaches a peak at the optimal amplitude condition by tuning B_{pto} in Figure 3.3b, i.e., at the intersection between the $\overline{P}_r$ curve and the $\overline{P}_{pto}$ curve. It is also noticeable that with the increase of B_{pto}, the mean radiated power becomes negligible compared with the PTO absorbed mean power.

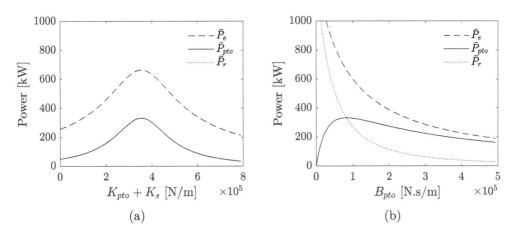

(a)   (b)

Figure 3.3: Mean power input arising from wave excitation and mean power output: radiated power and PTO absorbed power versus variations in (a) the PTO spring stiffness coefficient while the optimal amplitude condition is always achieved, (b) the PTO damping coefficient while phase optimality condition is always achieved. The wave conditions are described by a regular wave with 1 m amplitude and a period of 7 s.

3.2.2 Phi-method

Complex conjugate control or impedance matching has become a classic theory for the design of WEC PTO control systems, however, being a solution in the frequency domain, only linear terms were included in its derivations. For example, the nonlinear viscous drag loads were not included, and the control influence of its dynamics is not straightforwardly understood in the frequency domain, even when statistical linearisation is used, as it is estimated using iterative procedures. In these cases, control strategies were developed in the time domain for a better understanding of the impact of nonlinear dynamics on PTO control. The following time-domain approach is based on an algorithmic operator Phi-method [543] that determines the optimal PTO control force to maximise wave power absorption.

The WEC dynamics can be written as the following general expression in the time domain:

$$\mathrm{H}\left(t; x, \dot{x}, \ddot{x}, \ldots, x^{(n)}\right) = F_e(t) + F_{pto}(t), \tag{3.59}$$

where $x^{(n)} = \dfrac{\partial^n x}{\partial t^n}$ and n is the highest order of the time derivatives. H is a sufficiently smooth function of time derivatives describing the WEC dynamics in response to the excitation and PTO control forces, which are not necessarily linear.

The Phi-method derives the optimal control force as a function of the WEC time response from the analytical WEC dynamics H using the following expression [543]:

$$F_{pto}(t) = -\alpha + \int_0^t \dot{x}\frac{\partial}{\partial x}\mathrm{H}d\tau - \sum_{k=1}^{\infty}\left(-\frac{\partial}{\partial t}\right)^{k-1}\left(\dot{x}\frac{\partial}{\partial x^{(k)}}\mathrm{H}\right), \tag{3.60}$$

where α is a bias force equal to the mean wave force that prevents the WEC from drifting. Table 3.4 shows some typical dynamics terms and their corresponding optimal PTO control forces derived using Eq. (3.60).

Table 3.4: Typical dynamic terms and corresponding Phi-derived optimal control forces.

Dynamic terms	Analytical expression (H)	Phi-derived control force (F_{pto})
Spring	$a_0 x(t)$	$a_0 x(t)$
Damper	$a_1 \dot{x}(t)$	$-a_1 \dot{x}(t)$
Inertia	$a_2 \ddot{x}(t)$	$a_2 \ddot{x}(t)$
Viscous drag	$a_3 |\dot{x}(t)|\dot{x}(t)$	$-2a_3 |\dot{x}(t)|\dot{x}(t)$
Depth inertia	$m(x)\ddot{x}$	$m(x)\ddot{x} + \left(\dot{m}\dot{x} - \int_0^t \dot{m}\ddot{x}dt\right)$

Substituting the first four dynamic expressions in Table 3.4 and their corresponding Phi-derived control force into Eq. (3.59) and combining/cancelling common terms, the following observations can be made. The system intrinsic spring and inertia terms are respectively cancelled by the corresponding Phi-derived feedback control force, while the linear damper term is doubled by linear damper control. These follow impedance matching theory for the maximum power absorption of a WEC, in which the dynamics are assumed to be linear and under harmonic motion. On the other hand, the viscous drag term (the quadratic damper term) is tripled by the quadratic damper control according to the Phi-method, whose mathematical proof is shown in [308]. Table 3.4 also shows the Phi-derived optimal control force for the depth dependent inertia term, which readers can derive using the Phi-method as an exercise. Note that if m becomes a constant, then the optimal control force reduces to $m\ddot{z}$; identical to the solution for the standard inertia term.

3.2.3 Theoretical power limit

When the complex conjugate control in Eq. (3.51) is achieved by selecting a proper set of PTO coefficients, the amplitude of the optimal buoy velocity in Eq. (3.58) could be impractically large, especially when the wave period is relatively high and the radiation damping approaches zero. The optimal power capture may be overestimated and, hence, it is necessary to investigate the power limit with or without practical constraints. According to Eq. (3.58), the maximum absorbed power in a single DoF can be given as [227]

$$P_{max,j}(\omega) = \frac{|F_{e,j}(\omega)|^2}{8B_{rad,j}(\omega)}, \tag{3.61}$$

where $j = 1, 2, 3, \cdots 6$ represent the surge, sway, heave, roll, pitch, and yaw modes, respectively.

As the optimal velocity in Eq. (3.58) is achieved by design or control, the maximum absorbed power only depends on the amplitude of the excitation force and the radiation damping, as shown in Eq. (3.61). In this section, theoretical power absorption limits in two- and three-dimensional flows will be discussed, together with the influence of motion constraints on maximum power absorption.

3.2.3.1 *Power limits in two-dimensional flows*

The two-dimensional flow in the X-Z plane is considered, and the fluid motion is the same everywhere in the Y-direction. For WEC devices with a large dimension in the Y-axis (viewed as infinite in the Y-axis), the wave-structure interaction problem can be simplified as a two-dimensional flow problem. By assuming linear potential flow theory and a symmetrical body about the Z-axis, the amplitudes of the excitation forces in surge, heave or pitch modes in the two-dimensional flow, can be computed according to their radiation damping coefficients [338, 570], given as

$$|F_{e,j}(\omega)| = A \left(\frac{\rho g^2 D(kh)}{\omega} B_{rad,j} \right)^{\frac{1}{2}}, \tag{3.62}$$

where $D(kh)$ is the depth function defined in Eq. (2.26); $j = 1, 3$, and 5, represent the surge, heave, and pitch modes, respectively.

Substituting Eq. (3.62) into Eq. (3.61), the maximum power absorption can be rewritten as

$$P_{\text{max,j}}(\omega) = \frac{A^2 \rho g^2 D(kh)}{8\omega} = \frac{1}{2} \frac{\rho g^2 D(kh)}{4\omega} A^2 = \frac{J}{2}, \tag{3.63}$$

where J is defined in Eq. (2.25). This equation gives the theoretical wave-power absorption limit for symmetrical oscillating bodies in two-dimensional flow, first derived by Evans in 1976 [208].

In two-dimensional flow, the theoretical limit of the hydrodynamic efficiency for a transverse symmetry body in a single mode is 50%. For WECs operating in multiple modes, the theoretical limit of the hydrodynamic efficiency can reach 100% [208]. Alternatively, an asymmetric body, e.g., Salter's Duck [686], also has the possibility of achieving a hydrodynamic efficiency of 100%, even though it only oscillates in one mode.

3.2.3.2 Power limits in three-dimensional flows

In three-dimensional flows, the amplitudes of the excitation forces in surge or pitch modes, $j = 1, 5$, can be written as [570]

$$|F_{e,j}(\omega)| = A \left(\frac{4\rho g^2 D(kh)}{\omega k} B_{rad,j} \right)^{\frac{1}{2}}. \tag{3.64}$$

Substituting Eq. (3.64) into Eq. (3.61), the maximum power absorption can be rewritten as

$$P_{max,j}(\omega) = 2\frac{\rho g^2 D(kh) A^2}{4\omega k} = 2\frac{J}{k}, \tag{3.65}$$

where $k = \frac{\lambda}{2\pi}$ is the wave number. For a heaving axisymmetrical body, the amplitude of the excitation force is written as

$$|F_{e,3}(\omega)| = A \left(\frac{2\rho g^2 D(kh)}{\omega k} B_{rad,3} \right)^{\frac{1}{2}}. \tag{3.66}$$

Similarly, the maximum wave-power absorption is given as

$$P_{max,3}(\omega) = \frac{\rho g^2 D(kh) A^2}{4\omega k} = \frac{J}{k}. \tag{3.67}$$

To improve wave-power absorption, WECs can make further use of multiple modes. For instance, the theoretical wave-power absorption limit can reach $P_{\text{max}} = 3\frac{J}{k}$ for an oscillating body operating in surge-heave or surge-heave-pitch modes [571].

3.2.3.3 Power absorption with motion constraint

In practical engineering applications, the body motion cannot be infinite, but is constrained by the body dimension or PTO specifications. Here, the motion constraint of an oscillating body in a single mode is represented by the displacement stroke, defined as $x_{j,s}$. One practical method to determine this displacement stroke is the "maximum swept volume" method [99], which will be detailed in Section 5.2.2. In addition, $x_{j,s}$ can also be determined by the PTO design. By assuming harmonic wave and linear hydrodynamics, the velocity stroke can be expressed as $v_{j,s} = \omega x_{j,s}$.

When motion constraint is considered, the optimal velocity in Eq. (3.58) can be rewritten as

$$U(\omega)_{opt,j,s} = \frac{F_{e,j}(\omega)}{B_{pto,j,s}(\omega) + B_{rad,j}(\omega)}. \tag{3.68}$$

Substituting $|U(\omega)_{opt,j,s}| = \omega x_{j,s}$ into Eq. (3.68), the PTO damping coefficient can be determined as

$$B_{pto,j,s}(\omega) = \frac{|F_{e,j}(\omega)|}{\omega x_{j,s}} - B_{rad,j}(\omega). \tag{3.69}$$

Hence, the maximum power absorption with motion constraint is given as

$$P_{max,j,s}(\omega) = |F_{e,j}(\omega)|\omega x_{j,s} - \omega^2 x_{j,s}^2 B_{rad,j}(\omega). \tag{3.70}$$

One notable way to determine $x_{j,s}$ was developed by Budal in 1980 [99], the so-called Budal's upper bound in [228]. For a half-submerged heaving device, Budal's upper bound can be written as:

$$P_{B,3}(\omega) = c_0 \frac{VH}{T}, \tag{3.71}$$

where $c_0 = \pi \rho g / 4$; V, H and T represent the body volume, wave height, and period, respectively. An extension of the Budal's upper bound for multi-mode devices is discussed in Chapter 5. Consider both the theoretical limit and Budal's upper bound, a good example is shown in Figure 3.4. On the left side (the high frequency region), the power absorption is bounded by the theoretical limit represented by $P_A = c_\infty H^2 T^3$ with $c_\infty = (g/\pi)^3/128$, which is equivalent to Eq. (3.67). In the low frequency region (the right side), the power absorption is banded by Budal's upper bound in Eq. (3.71). The cross point of these two power bounds, $(T_{c,3}, P_{c,3})$, is given as

$$T_{c,3}(\omega) = \left(\frac{32\pi^4 V}{Hg^2}\right)^{1/4}, \tag{3.72}$$

$$P_{c,3}(\omega) = \frac{\rho g^{3/2}}{8}\left(\frac{V^3 H^5}{2}\right)^{1/4}. \tag{3.73}$$

In Figure 3.4, the power absorption by deploying passive control, reactive control, and latching control is well bounded by the theoretical limit and Budal's upper bound. Such a graphic plot also provides a basic guideline for WEC design, e.g., determining a proper PTO power rating or buoy volume. A good case study is given in Chapter 5, discussing the power absorption limits of a submerged multi-mode PA devices with or without displacement constraints.

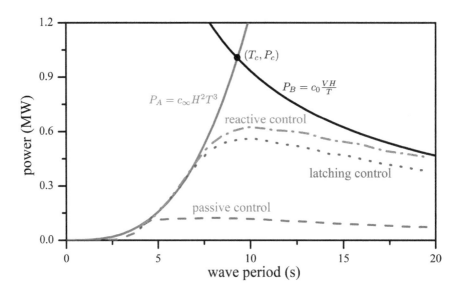

Figure 3.4: Power absorbed by a spherical heaving buoy in regular waves using passive, reactive and latching control strategies. This figure is adapted from [229]. The red and black solid curves are the radiation limit and the volume stroke limit, respectively. In this figure, the wave height is $H = 2.26$ m, and the radius of the sphere is $r = 5$ m.

3.3 CONTROL SYSTEM AND POWER TAKE-OFF DESIGN

The PTO system is the core of a WEC, and has multiple functions: 1) most obviously, it converts the mechanical power of the WEC mechanism into electricity; 2) it enhances the hydrodynamic efficiency of the WEC and thus maximises its power absorption in varying sea states; and 3) it ensures safe operation of the WEC in the harsh sea environment.

3.3.1 Power take-off control strategy classification

In practical applications, the motion of the wave energy device should be controlled to maximise the converted power subject to constraints imposed by the equipment and machinery, such as the maximum amplitude, velocity, load force, etc. [672]. Control is implemented by the PTO system, which exerts a load force on the oscillating body thereby tuning its dynamic response. The development of control strategies for the wave energy applications depends on parameters such as (i) the power take-off capabilities, (ii) availability and complexity of the system model, (iii) availability of information about any incident wave field, and (iv) number of controlled variables, etc. Based on these characteristics, control algorithms can be grouped as shown in Figure 3.5.

3.3.1.1 Power flow

The first classification relates to the ability of the PTO to provide either unidirectional or bi-directional power flow, as briefly introduced in Section 3.2.1:

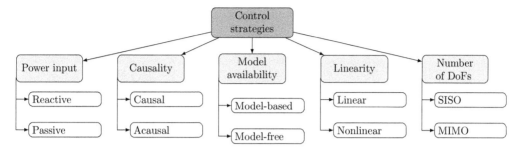

Figure 3.5: Classification of wave energy control strategies.

- reactive control $(F_{pto}(t) = B_{pto}\dot{x}(t) + K_{pto}x(t))$ implies that the machinery not only receives energy from the converter but also has to put some energy back during a part of a sinusoidal cycle (working partly as a motor) [227]. It requires a bi-directional power flow through the PTO and can be achieved by introducing energy storage into the system, e.g., through a series of accumulators for a hydraulic system or mechanical springs. Reactive control was first applied to the Edinburgh Duck (also known as Salter's Duck) [686], where the load force was proportional to the duck's velocity and position, imposing a complex-conjugate control law;

- passive control $(F_{pto}(t) = B_{pto}\dot{x}(t))$ means that the machinery is only required to extract power from the system, but not to return it [326], which is a reasonable assumption for some types of machineries, e.g., an electrical generator. Two variations of passive control exist: a phase control presented by latching and declutching [44, 49], and a resistive control where only a load resistance, e.g., a damper, is adjusted to the current sea state [186, 105].

In terms of the efficiency of these control strategies, it was demonstrated in [229] that reactive loading outperforms other controllers, as shown in Figure 3.4, and uses the full capacity of the oscillating body [229]. Both reactive and latching controls satisfy the phase optimality condition, or what is often called the resonance condition, while passive loading does not have this capability, which leads to a low efficiency in WECs when this strategy is applied. For more details on latching control refer to Section 6.4.1.2.

3.3.1.2 Causality

It is relatively easy to set the optimal phase and amplitude of the WEC oscillations when the incoming wave is sinusoidal with uniform frequency and height. However, real sea states have a stochastic nature and, in order to maximise the absorbed power by applying an optimal control force to the PTO, the wave excitation forces should be known in advance [672]. Thus, the wave energy control problem resulting from the optimality conditions is, in general, acausal (or non-causal) [225], relying on the future values of the wave excitation force. This future information can be gained by installing additional equipment and sensors at some distance from a converter in the direction of an incoming wave [225, 756] and/or using wave prediction algorithms (e.g., auto-regressive models [261, 263] or augmented Kalman filters [327]). As a result, the performance of acausal controllers is heavily dependent on real-time wave forecasting which deteriorates with any increase in the prediction horizon [264].

Other controllers that obviate uncertainties associated with wave predictions are called causal. For example, the works in [428, 696] suggested to include the knowledge about the spectrum of the operating sea state in the system model, forming a basis of the linear-quadratic-Gaussian control problem. Another method to achieve the causality of the system has been presented in [265], where a non-causal transfer function between the optimal velocity and the excitation force is replaced by the frequency-dependent coefficient of proportionality. Those controllers that tune their parameters to the peak (or energy) frequency of an incident wave [440, 20] also belong to the class of causal controllers. For example, the coefficients of the linear spring-damper controller can be tuned to the peak (or energy) frequency of the incident wave spectrum.

Comparing the performance of these two classes of controllers [326], the acausal strategy represented by model-predictive control (MPC) often demonstrates the best power generation efficiency, provided perfect wave force estimation is available. However, it comes at the cost of the power take-off complexity, which requires high peak-to-average power ratios, large energy storage, and high control forces [835]. As a result, similar to the techno-economic development of the wave energy devices, any implementation of advanced control strategies to increase power production must be subject to economic assessment, as discussed in Section 1.2.3.

3.3.1.3 *Model availability*

Low-cost models established based on linear potential theory, as discussed in earlier sections, are often necessary for WEC control implementation [672]. Optimisation-based control strategies such as MPC, dynamic programming (DP), and linear quadratic control (LQ) aim to maximise the performance objective (e.g., energy), subject to constraints, and their performances are heavily dependent on the fidelity of the WEC model used as a part of the optimisation algorithm [835]. On the other hand, whilst not directly applying the model in calculating the control force, the complex conjugate control (or spring-damper control) and latching control rely on the WEC model when determining the optimal control parameters. One category of control strategy that does not rely on the WEC model at all for maximising power absorption is termed a non-model based control. A typical example of this category is maximum power point tracking control (MPPT), which is widely used in maximising the energy conversion efficiency of wind turbines. However, it has been proven that MPPT does not work efficiently in a WEC system when compared with model-based control strategies, mainly due to the highly stochastic nature of ocean waves [117, 334].

3.3.1.4 *Linearity*

Most of the WEC control strategies are linear methods that are not capable of handling nonlinear dynamics such as viscous drag or other higher order nonlinear hydrodynamic effects. Therefore, these methods often require linearised models based on linear potential theory, although WECs do not always operate in the linear wave theory region as shown in Chapter 2. The Phi-method, as detailed in Section 3.2.2, belongs to the category of nonlinear control that derives the nonlinear optimal control force for the analytical nonlinear dynamics of the system. Thus, the control performance strongly depends on the fidelity of the nonlinear models in their analytical expressions. DP and shape-based (SB) control are also capable of handling nonlinear dynamics, however, their computational requirements are high and thus may not be suitable for real-time control implementation [835].

3.3.1.5 Multi-mode control

The vast majority of wave energy converters absorb power from only one hydrodynamic mode (heave or surge), so the development of control systems has also focused on those devices that operate at one degree of freedom, considering single-input-single-output (SISO) plants. Moreover, when developing new control principles and ideas, it is easier to work with systems that have relatively simple dynamics, where no attention should be paid to the coupling between the state variables. As a result, almost all effective controllers in the wave energy sector have been designed for single-DoF heaving devices [326, 835]. However, the number of converters that absorb power from several motion modes is increasing, leading to increased interest in developing controllers for multiple-input-multiple-output (MIMO) WECs, as discussed in more detail in Chapter 5.

3.3.2 Power take-off machinery

The objective of the PTO system machinery is to transform the mechanical power of the WEC actuator to useful electricity. The dynamic behaviour of the PTO machinery also dictates what control strategies can be applied to the WEC, which directly governs the power absorption capability of the WEC. Furthermore, taking techno-economic assessment into consideration, PTO machinery design is a very complex task, which plays an important role in raising the technology readiness level (TRL) and technological performance level of the WEC towards commercialisation. The main types of PTO machinery are shown in Figure 3.6 and their percentage of utilisation and their efficiencies are shown in Figure 3.7 and Table 3.5 respectively.

Table 3.5: Indicative efficiency of different PTO systems [765]

PTO system	Efficiency, %
Hydraulic	65
Pneumatic	55
Hydro	85
Direct mechanical drive	90
Direct electrical drive	95
Advanced electric materials	< 80

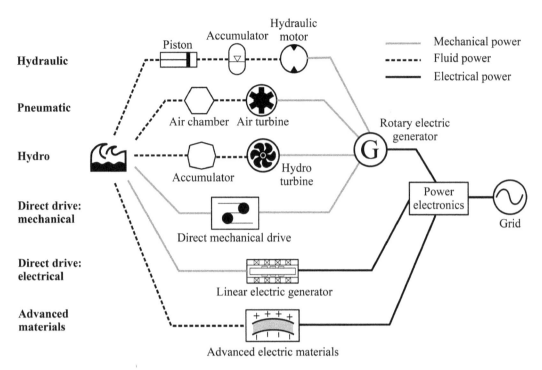

Figure 3.6: Power take-off principles utilised for the wave energy conversion, adapted from [73, 765].

3.3.2.1 Hydraulic power take-off machinery

Hydraulic PTO machinery utilises the cyclic motion of the hydraulic cylinder, driven by the buoy, to force pressurised fluid through a controlled manifold to the hydraulic motor, which is connected to the rotary electric generator. As hydraulic circuits are designed to operate with large forces at low speeds, they are well-suited for wave energy conversion purposes and make up the largest part of all PTO technologies [765], as shown in Fig. 3.7. Whilst being the most applicable for absorbing wave energy,

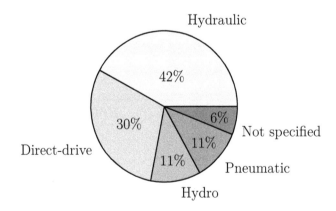

Figure 3.7: Breakdown of WECs depending on the utilised PTO principle [548].

hydraulic PTO machinery has relatively low power conversion efficiency (refer to Table 3.5), mainly because reactive control cannot be achieved due to the large time constant of the hydraulic PTO machinery dynamics. Figure 3.8 shows 3 CETO 5 WEC devices equipped with hydraulic PTOs, as developed by Carnegie Clean Energy prior to its patented CETO 6 technology, currently under development. The submerged buoys are moved by the ocean swell, driving pumps that pressurise seawater delivered ashore by a subsea pipeline. Once onshore, the high-pressure seawater is used to drive hydro-electric turbines, generating electricity.

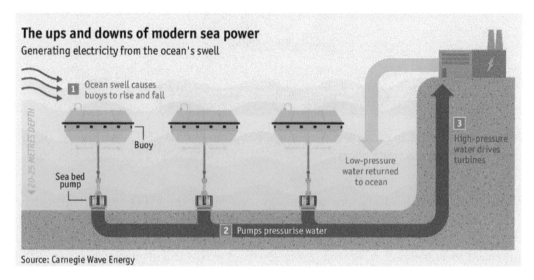

Figure 3.8: CETO 5 array supplying hydraulic power onshore, (image courtesy of Carnegie Clean Energy)

3.3.2.2 Pneumatic power take-off machinery

Pneumatic PTO machinery refers to those air-turbines that are driven by oscillating air pressure and directly coupled to a generator. This technology is utilised in OWCs and pressure differential devices, and has the advantage that no moving parts are under the water. Figure 3.9 shows Wave Swell's patented fixed-structure OWC technology, equipped with a unidirectional air-turbine that is simple, robust, and efficient. Conversely, a more complex turbine with pitching blades can be applied to absorb energy from bidirectional air flow, with the potential of achieving soft latching control.

3.3.2.3 Direct-drive power take-off machinery

Direct-drive PTO machinery refers to PTO systems where mechanical power from the buoy oscillations is directly transformed into electricity. While in a direct electrical drive system, the buoy is directly coupled to the moving part of the linear generator, which is demonstrated in Section 4.4.3. The direct mechanical drive is equipped with an additional mechanism to couple the buoy's motion to a rotary generator.

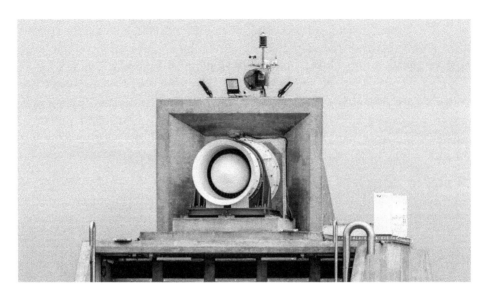

Figure 3.9: Oscillating Water Column applying pneumatic PTO machinery, (image courtesy of Wave Swell Energy).

Whilst being the least mature technology among all the PTO types, direct-drive PTO machinery is currently in the spot light of research and development, mainly due to its ability to generate any desired PTO control force and thus enable the potential to implement advanced power maximising control strategies such as MPC. Figure 3.10 shows the CorPower PA WEC, highlighting its PTO system. A composite buoy, interacting with wave motion, directly drives the generator via a mechanical drive-train located inside the buoy.

3.3.3 Performance measures of power take-off machinery

The performance measures specified in Section 1.2.3.1 are usually calculated using values of the mechanical power absorbed/captured by the WEC. As the power take-off machinery converts mechanical power to electrical power with associated losses, there is a need to quantify the efficiency of the entire power conversion chain, or each separate subsystem, including the power transmission (e.g., hydraulic, or mechanical drivetrain), the generator (rotary or linear), and the power converters (refer to Figure 3.6).

3.3.3.1 Power conversion efficiency (mechanical-to-electrical)

The power conversion efficiency (mechanical-to-electrical) quantifies how much of the mechanical power input to the PTO machinery is converted to useful electrical power output:

$$\eta_{\text{eff},\text{M}\to\text{E}} = \frac{\text{Electrical power output [W]}}{\text{Mechanical power input [W]}}. \tag{3.74}$$

Figure 3.10: Point absorber applying a direct mechanical drive PTO, (image courtesy of CorPower Ocean.)

Since the power conversion efficiency depends on the PTO input parameters (e.g., force and velocity, torque and angular speed, pressure difference and flow rate), it is usually represented as a matrix.

To assess the WEC's performance as a whole, particularly how much of the available wave energy is absorbed by the WEC, transferred to the generator and converted into electricity, the alternative power conversion efficiency (wave-to-electrical) coefficient can be used:

$$\eta_{\text{eff},\text{H}\rightarrow\text{E}} = \frac{\text{Electrical power output [W]}}{\text{Available hydrodynamic power [W]}}, \qquad (3.75)$$

where the available hydrodynamic power is calculated as (for a regular wave, heaving axisymmetric WEC) [228] as:

$$\text{Available hydrodynamic power [W]} = \overline{P}_{mech}^{\max} = \frac{|F_e(\omega)|^2}{8B_{rad}(\omega)}. \qquad (3.76)$$

3.3.3.2 *Maximum values, or peak values*

Maximum, or peak values of power, force, speed, and displacement are quantities that establish requirements for the PTO design (drivetrain, hydraulic circuit, electric generator, etc.). As the identification of the peak loads highly depends on the length of the numerical or physical experiments as well on the phasing of the incoming wave, it is more practical to report the "statistical peak" value that refers to a particular confidence level, 95%, 99%, or 99.9%, see Figure 3.11.

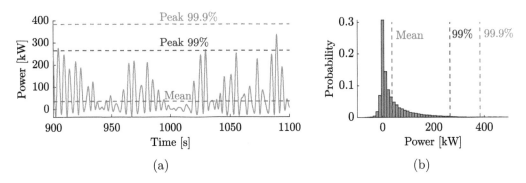

Figure 3.11: Demonstration of a "statistical" peak power calculation from the power time-series: (a) instantaneous power and (b) histogram.

3.3.3.3 Peak-to-average values

Peak-to-average values of power, control force, or speed are important parameters for the electrical part of any WEC. In particular, all electric generators are characterised by their nominal speed, torque, and current, where the generator operates at its maximum efficiency. Also, such generators are capable of handling some overload for a short period of time. Hence, the peak-to-average values guide whether the electrical system should be designed according to the mean power production, or to the peak power requirements. In addition, the peak-to-average values are highly dependent on the hydrodynamic control strategy used as demonstrated in [134]. If the signal has a zero mean, the peak-to-average value is calculated using its root-mean-square instead of the average value.

$$\text{Peak-to-average power} = \frac{\text{Peak power [W]}}{\text{Average absorbed power [W]}} = \frac{P^{\text{peak}}}{\overline{P}}. \tag{3.77}$$

$$\text{Peak-to-average force} = \frac{\text{Peak force [N]}}{\text{Force RMS [N]}} = \frac{F^{\text{peak}}}{F^{\text{RMS}}}. \tag{3.78}$$

3.3.3.4 Slew rate

The slew rate of force (or speed) is defined as the rate of change of the signal per unit time, and usually is expressed in units of [N/s] (or [m/s^2] for speed). This parameter defines how quickly the force should be changed by the mechanical or electrical part of the PTO machinery:

$$\text{Slew rate [N/s]} = \max \left| \frac{dF_{pto}(t)}{dt} \right|. \tag{3.79}$$

3.3.3.5 Damage equivalent loading

Damage equivalent loading (DEL) is a metric widely used in offshore applications to assess the capability of the structure to sustain the full spectrum of loads during its life [259, 129]. Recently, the DEL has been used in the design optimisation of wave energy converters [130, 272, 273], in particular assessing the DEL of PTO systems with joints (i.e., welding). The calculation of this metric spans multiple stages [130]:

- generate the PTO force time-series;
- convert force into stress;
- count the number of stress cycles;
- determine the cycles to failure according to S-N curves for a selected material and a joint type;
- estimate the accumulated damage.

3.3.3.6   *Flicker severity*

Once the wave energy converter is connected to the electricity grid, it can cause rapid fluctuations in the grid voltage, called flickers [58]. Thus, flicker evaluation determines the electricity quality in the power grid and is defined by the flicker severity measures. Short-term flicker severity is a measure that quantifies the visual severity of flicker over a 10 minute period, while long-term flicker severity is the cubic average of the short-term flicker over 2 hours.

3.3.4   Challenges in power take-off design

Various PTO mechanisms are illustrated in Section 3.3.2. These PTO mechanisms have been widely applied in industry, and are available on the market. It is commonly assumed that existing PTO systems can be used directly for WECs. However, the integration of existing commercial PTO systems to WECs is not that direct, mainly due to the harsh and time-varying deployment environment of WECs. Compared with conventional PTO systems, the design of WEC PTO systems is still challenging. The following problems have not been fully solved: (i) how to convert a slow motion of huge force or torque into electrical power? (ii) how to design a PTO mechanism operating in a wide range of varying sea states? and (iii) how to include wave-structure interaction (WSI) and control in the design and optimisation of PTO systems?

Ocean waves are typified as low frequency with huge hydrodynamic force or torque acting on WEC devices [184]. As a consequence, the WEC float's motion is slow, and the coupling force between the WEC float and the PTO system is large. Such a motion itself is naturally challenging for PTO design. Generally, PTO systems are optimised to operate efficiently at a high unidirectional speed, and such a slow WEC motion may result in a large physical dimension of the PTO generator. Hydraulic PTO systems show great advantage in WEC applications, as they can handle slow motion and large force simultaneously. However, their application in wave energy conversion is still constrained by some drawbacks, e.g., a relatively low efficiency [349], regular requirements for maintenance [64], and a high pollution risk [184]. Alternatively, mechanical motion rectifiers, e.g., gearboxes, are generally used to amplify the operating velocity for generators. The mechanical motion rectifiers increase the system complexity, and may weaken the reliability of WEC systems, e.g., including gearbox failure. To date, it is still challenging to design a PTO system to convert wave power into electrical power in a simple, efficient, and reliable manner.

Another obvious property of an ocean wave is its significant spatio-temporal variation. Even if the installation site is chosen properly, the wave power will still show high irregularity in wave frequency, amplitude and direction, and these changes across seconds and decades. Given that the lifespan, or design life, for a WEC is generally up to 20 years, the WEC may experience all kinds of sea states, from light to extreme. In general, a PTO system can only operate efficiently under its rated conditions, e.g., its rated speed, power, load, etc. Therefore, it is technically challenging to define the rated operation conditions for a PTO system. Even for an optimised PTO unit, its efficiency decreases dramatically when the load diverges from its rated value [217]. In addition, short-term variability induces a high peak-to-average power ratio, which may lead to overrated conditions. Meanwhile, extreme waves may cause severe damage to WEC devices, as PTO units are prone to being damaged in extreme waves. Therefore, PTO systems are required to operate efficiently in a wide range of sea states, to run shortly at an overrated power, and to survive reliably in extreme waves.

For PTO operating efficiently in a wide range of sea stages, control can play an important role to extend the WEC's bandwidth. Properly designed power maximising control may cause more strict requirements for PTO design specifications, e.g., the maximum slew rate, force/torque, power, etc. On the other hand, control always tends to exaggerate the WEC motion and the PTO motion as a consequence. Such an exaggeration in WEC and PTO motion may induce rich and complex nonlinear dynamics, including mechanical friction [443, 399], hysteresis effect [349], dead-zone, end-stop saturation, load effect [217], and so forth. To considering both the wave-structure and structure-PTO interactions, a high-fidelity, computationally efficient wave-to-wire (W2W) modelling method is required [623], leading to a co-design concept. However, such a model is of high system complexity, and it is challenging to optimise the system holistically.

In summary, it is technically challenging to design an efficient, reliable and cost-effective PTO system that can operate in a wide range of sea states, run at overrated power levels, execute various control strategies, require limited maintenance and survive in extreme waves. Although some novel PTO technologies, e.g., the dielectric elastomer generator [550] and the triboelectric-electromagnetic hybrid nanogenerator [237], show some advantages in wave energy conversion, their TRLs are low and more sea trials are required for validation and verification.

3.3.5 Challenges in wave energy converter control

Over the past 20 years, WEC control has progressed significantly, with emerging algorithms to tackle various technical challenges. However, it is still challenging to develop a "perfect" control system considering PTO implementation, mainly because of the time-varying sea states. To date, there still exist some challenges in WEC control, including (i) non-causality in control, (ii) constraint handling, (iii) control robustness, (iv) real-time implementation etc.

As mentioned in Section 3.3.1, WEC modelling and control problems are well known as non-causal. Such a non-causality falls into two key parts, (i) modelling non-causality in the physical process from the wave elevation to the wave excitation force and (ii) control non-causality in generating velocity reference from the excitation force [265]. Modelling the non-causality problem can be tackled by upstream wave measurement or wave prediction [312, 3, 315, 619]. For non-causality in control, future knowledge of the excitation force is required [670]. In addition, MPC naturally requires further prediction of wave excitation force [213], which asks for an even longer prediction window and may increase the prediction error. Although the non-causality in WEC modelling and control can be handled well either by excitation force prediction [619] or causal control approaches [696], the system complexity may increase dramatically.

Another challenge is to limit WEC motion by control according to varying sea states and WEC physical constraints. As mentioned in Section 3.2.3.3, motion constraints, i.e., displacement or/and velocity, can significantly influence WEC power capture, both in regular and irregular waves. In addition, a practical PTO unit, working as an actuator for control implementation, has some physical constraints, e.g., the maximum force/torque, maximum instantaneous power, and velocity/displacement limit. Considering both constraints in WEC motion and PTO force, the existence of an optimal control solution is not fully understood [57]. Given that sea states vary all the time, from moderate to extreme conditions, a supervisory control, to switch the WEC system between operational and survival modes according to sea sates, and a fault tolerant control, to improve system reliability, are needed. However, these problems are seldom investigated [319].

Since WEC hydrodynamics are governed by the Navier-Stokes equations, as discussed in Section 2.1.1.1, WEC models to represent wave-structure interactions are not straightforward for control design. Parametrised time-domain models, e.g., transfer functions or state-space models, are preferred for control. Modelling error and uncertainty, induced by parametrisation or system identification, are inevitable. In addition, WEC hydrodynamic properties also drift over time, given that a life cycle is generally up to 20 years. Thus, control strategies are required to be robust and insensitive to modelling errors, sensor faults, external disturbances, and estimation/prediction errors of the excitation force. Model sensitivities for various WEC control architectures have been investigated both analytically and numerically in [673], aiming to address the robustness problem in WEC control. Alternatively, model-free control can be useful in WEC control to make WEC performance more immune to model uncertainty, external perturbation and unmodelled dynamics. However, some model free control methods inherently contain a large number of optimisation or training iterations [18, 17, 94, 19], resulting in a challenge in real-time online implementation.

A PTO unit has its own dynamics, but is generally simplified as an ideal mass-spring-damper system for theoretical study. To model WEC dynamics and evaluate their performance in a more realistic manner, non-linear wave-WEC interactions and a non-ideal PTO model should be considered with control in loop, to derive a high-fidelity wave-to-wire model. However, such a wave-to-wire model is systematically complex, and complexity reduction is crucial for real-time control implementation [624]. For real-time implementation, another aspect is to avoid (i) excitation force estimation (current value), (ii) excitation force prediction (future value), and (iii) online optimisation [670].

In general, power maximising control always trends to exaggerate the WEC's motion, and, hence, amplifies the nonlinearity in wave-WEC interactions. Such nonlinear hydrodynamic-control interactions should be considered both in WEC modelling and control [287]. Meanwhile, a practical PTO unit is non-ideal and has its own limits. Therefore, a co-design framework is required for WEC modelling, performance assessment, optimisation, control, and so forth [319, 706].

III

Wave Energy Converters

CHAPTER 4

Point absorber wave energy converters

Bingyong Guo

Northwestern Polytechnical University, China, b.guo@nwpu.edu.cn

4.1 INTRODUCTION

This chapter is dedicated to point absorber wave energy converters, which belong to the oscillating body WECs introduced in Chapter 1. The concept of a point absorber (PA) was first defined in 1975 [100], referring to a wave energy converter with a relatively small dimension with respect to the prevailing wavelength. Nowadays, some PAs' dimensions are ever increasing; however, their working principles are the same as for the original PA concepts, so-called quasi-PAs [698]. The PA definition in [100] does not depict the installation environment, operating modes, and geometry design of a PA device. Thus, PAs can be either floating or submerged, may use either single or multiple bodies to interact with waves, and may operate in either single or multiple modes.

As discussed in Chapter 1, the PA concept is one of the simplest and the most promising concepts, having attracted intensive global research interest. This chapter aims to introduce floating PA devices, while submerged PAs will be fully discussed in Chapter 5. The rest of this chapter is organised as follows: Section 4.2 introduces the working principles of various one- and two-body PAs; Section 4.3 depicts the design, testing, operation, and development details of some typical PA prototypes; some generic time-domain modelling approaches for a floating one-body PA in heave are discussed in Section 4.4 along with detailed case studies; and Section 4.5 summarises the research and development perspectives of floating PAs.

4.2 WORKING PRINCIPLES

The PA concept has been investigated intensively, and various PAs have been proposed and tested. Even though the design of PAs varies from case to case, PAs can be classified according to their manner of installation, geometry designs, and operating

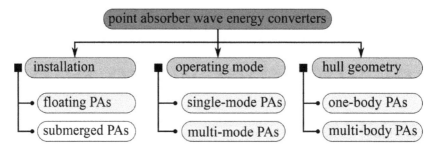

Figure 4.1: Classification of point absorber wave energy converters according to the manner of installation, operating modes, and hull geometries.

modes, as shown in Figure 4.1. With regards to the manner of installation, PAs can be classified as floating and submerged devices. The former type is addressed in this chapter, while the latter will be detailed in Chapter 5.

According to their operating modes, PAs can also be divided into two subgenera, single- and multi-mode PAs. If there is no motion constraint applied, PAs generally operate in all six modes, i.e., surge, sway, heave, pitch, roll, and yaw. However, some dominant oscillating modes exist for some PAs, e.g., the Seabased devices mainly operating in heave [649], which relies significantly on PAs' geometric design. PAs can also be classified as one- and multi-body systems, according to their geometries. As hull geometry has a significant influence on PAs' dynamics and performance, this chapter introduces the working principles of some typical one- and two-body PAs. Although some PAs consist of more than three bodies, few of them are of high TRL and, hence, only one- and two-body PA devices are addressed in this chapter.

Three typical PA concepts are shown in figure 4.2, illustrating the working principles of a one-body PA, a self-reacting two-body PA, and a self-containing PA, in Figure 4.2(a)–(c), respectively. Their working principles are detailed in the following subsections.

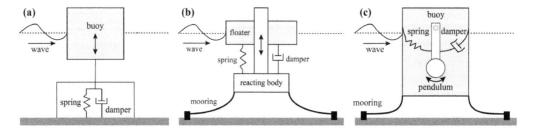

Figure 4.2: Working principles of (a) a heaving one-body PA, (b) a self-reacting two-body PA, and (c) a self-containing two-body PA. The PTO mechanism is simplified as, and represented by, a spring-damper system.

4.2.1 One-body point absorbers

A one-body PA utilises a floating or submerged body to interact with incident waves, and the body's motion is generally referred to an absolute reference point, e.g., the sea bed. As shown in Figure 4.2(a), a floating PA device usually comprises a symmetrical structure and a PTO system (simplified as a spring-damper system). The PTO system can be anchored on the seabed, e.g., the Seabased device [115]. A taut cable is used to connect the floating structure to the PTO system. The PTO unit can also be fixed to an offshore platform via a rotational arm, e.g., the WaveStar device [822]. In engineering practice, the PTO unit can be a hydraulic, mechanical or an all-electric system (see Section 3.3.2). Excited by incident waves, the floating structure oscillates and drives the PTO system to generate electricity. Some typical prototypes of one-body PAs, including the Seabased [115], the CorPower [141], the LifeSaver [722] and the WaveStar [822] devices, are discussed in Section 4.3.1, with design details.

The dynamics of one-body PAs are typified by low-pass characteristics in the frequency domain, and the natural frequency and RAO of a PA depend significantly on the geometric shape of the body. One-body PAs mainly operate in the heave mode and are largely insensitive to wave direction, due to their small dimensions and symmetrical geometries. When resonance occurs, PAs can achieve maximum power output; however, the resonance bandwidth is narrow. In addition, waves are highly irregular in period, height and direction, and power maximising control strategies can play an important role in tuning the device's natural frequency towards wave frequency accordingly [672, 671], especially when the device's natural frequency is distant from the prevailing frequency of ocean waves.

4.2.2 Two-body point absorbers

A two-body PA is characterised by its self-referenced property and the relative motion between the two bodies drives the PTO system to generate electricity. Since a two-body PA is not attached to an absolute reference point, a mooring system is indispensable, and it is possible to achieve deep-water installation. Two-body PAs can be further classified as the self-reacting PAs, as shown in Figure 4.2(b), and the self-contained PAs, as shown in Figure 4.2(c).

The self-acting PA shown in Figure 4.2(b) consists of a floater oscillating with waves and a reacting body providing reference. A PTO mechanism couples these two bodies and utilises their relative motion to generate electricity. Such a PTO system can be a hydraulic, mechanical or an all-electrical unit. Some of the well-known devices are the OPT PowerBuoy device in the USA [190] and the Wavebob prototype in Ireland [553], which are detailed in Section 4.3.2. For these self-reacting two-body PAs, both the floating and the reacting bodies interact with incident waves.

Although a self-contained PA is still self-reacting, only the hull structure interacts with incident waves, as shown in Figure 4.2(c). The inner body is enclosed by the hull and linked with it via a PTO system. When incident waves excite the hull, the inner body oscillates accordingly, and their relative motion is used by the PTO system to produce electricity. The operating modes can be heave, pitch or/and roll. Among the various self-contained two-body PAs, the most notable are the vertical pendulum SEAREV device [45] and the horizontal pendulum Penguin device [829], which are discussed further in Section 4.3.2.

4.3 EXISTING PROTOTYPES

This section introduces some typical prototypes of floating point absorbers, with the one- and two-body prototypes detailed in Sections 4.3.1 and 4.3.2, respectively.

4.3.1 One-body point absorber prototypes

Among various one-body floating PAs, the notable ones are shown in Figure 4.3, including the Seabased, the LifeSaver, the CorPower, and the WaveStar devices, detailed below.

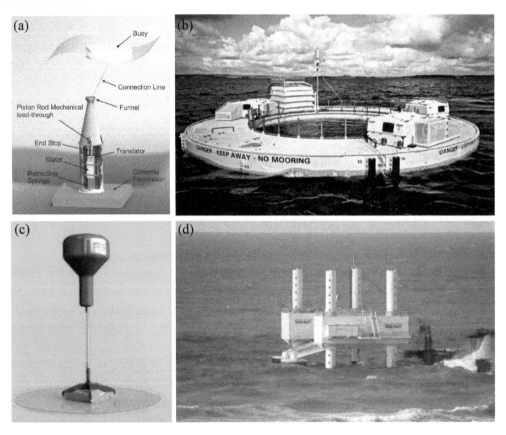

Figure 4.3: Typical prototypes of one-body floating point absorbers: (a) the Seabased device [115], (b) the LifeSaver device [306], (c) the CorPower device [786], and (d) the WaveStar device [306].

4.3.1.1 Seabased device

One of the most notable floating one-body PA is the Seabased device, as shown in Figure 4.3(a). A floating buoy is used to interact with surface waves and its motion is coupled with a seabed-mounted linear generator via a line or rope, in order to move the translator of a linear generator to generate electricity [204]. Although the Seabased device oscillates in six modes, its dynamics are dominated by the heave mode, and a simplified mathematical model in heave alone can give a good representation of the device's performance [649]. The technical details are given below.

- The original design of the Seabased device uses a semi-submerged cylindrical buoy to capture wave power. At full scale, the cylinder has a diameter of 3 m, a height of 0.8 m, and a draft of 0.4 m. A permanent magnate linear generator is used as the PTO system, with a rated power of 10 kW at a nominal speed of 0.67 m/s [819, 205]. A torus buoy was developed in 2009 as an alternative to the cylindrical buoy [437].

- The design of the Seabased device dates back to 2002, at the Swedish Centre for Renewable Electric Energy Conversion at Uppsala University. In 2006, a full-scale prototype was tested off the Swedish west coast in the proximity of Lysekil (also referred to as the Lysekil Project), where the water depth is about 25 m[819]. A 3 km long cable was used to connect this prototype to an onshore measuring station. By comparing the experimental data with numerical results based on potential wave theory, it was concluded that potential wave theory is well-suited for modelling of the Seabased device in normal operation and for the design of future Seabased WECs [205]. In 2009, another three prototypes were deployed and the electricity generated by the four devices was connected to a substation [82]. This sea trial concluded that the generator load played a significant role in power optimisation, indicating the importance of passive control.

- The Lysekil Project also conducted some unique research activities on the environmental impact of WEC installation and operation [435]. Even though the corrected data were insufficient, it can be concluded that WECs have limited negative impact on the environment. By 2017, more than 10 Seabased devices had been deployed in Norway, and the installation experience was summarised in [115]. According to the sea trial described in [115], the Seabased technology has demonstrated its ability to operate using a small array in a real environment, satisfying the definition of TRL 9 in [357].

4.3.1.2 LifeSaver device

The Fred Olsen company has developed the LifeSaver device, shown in Figure 4.3(b). The design details are given below.

- The LifeSaver device is a circular PA with an external diameter of 16 m, an inner diameter of 10 m, a height of 1 m, and a freeboard of 0.5 m [723, 724]. The circular buoy consists of three segments, on which three PTO units are installed. Meanwhile, the PTO units are anchored to the sea floor through winch lines, with a total rated power of 30 kW, to form an all-electrical PTO design [724].

- In 2007 and 2008, a small proof-of-concept prototype, named B33, was operated outside Akland, Norway. Based on the B33 experience, the second generation

B22 was operated outside Risor, Norway in 2008 and 2009. In 2012 and 2013, a full-scale LifeSaver device was installed at the test site, FaBTest, located in Falmouth Bay, just outside Cornwall [722]. In October 2018, a LifeSaver device was deployed at the US Navy's Wave Energy Test Site in Hawaii to provide electricity for several sensing systems and an autonomous underwater vehicle. At full power, the instrumentation and recharging system consumed 600 W, which was relatively high in oceanographic terms. As a first of its kind installation, this system demonstrated the transformative potential of wave energy to power oceanographic instrumentation and extend the endurance of autonomous underwater vehicles [383].

- For the LifeSaver device, the mooring lines are uniquely integrated as part of the winch and rope PTO units. Thus, the PTO units can only generate electricity during the upwards motion of the float. Such a unidirectional production pattern had led to a very high peak-to-average power ratio (up to 60) [723]. During the 108-day sea trial in Hawaii, the winch and rope systems experienced a couple of failures, but no downtime on power production was caused, thanks to the redundancy design of the three PTO units [383].

4.3.1.3 CorPower device

The CorPower device uses a heaving buoy on the water surface to capture energy from ocean waves, as shown in Figure 4.3(c). The buoy is connected to the seabed using a tensioned mooring line. Inside the buoy, a gearbox is used to transfer reciprocating motion to spinning motion and, thus, a conventional rotating generator is used for producing electricity. Design details of the CorPower device are given below.

- The CorPower Ocean company developed a 1/2 scale prototype, called C3, which has a diameter of 4.3 m, a height of 10 m, and a power rating of 25 kW. The CorPower Ocean company is manufacturing its first full-scale commercial device for a sea trial at EMEC. The full-scale CorPower device has a diameter of 9 m, a height of 18 m, a weight of 60 tonnes, and a nominal power rating of 300 kW [141].

- The development of the CorPower device was initiated in 2009. In 2013, a 1/16 scaled down model was tested in a wave tank, with specific focus on its patented WaveSpring technology [330]. In 2018, the CorPower device, C3, was tested at EMEC with a grid connection, after an accumulated investment of 6.5 M€ [141]. In 2020, the Swedish company secured 9 M€ equity funding towards its first commercial pilot array at EMEC, which will commence between 2022 and 2023.

- The WaveSpring component was invented at NTNU (Norwegian University of Science and Technology), which uses passive, pneumatic machinery to provide negative stiffness. As heaving PAs often have a larger hydrostatic stiffness than required, the WaveSpring inherently reduces the total stiffness, alleviates the phase shift between the wave excoriation force and the PA's velocity, and provides phase control [330]. Tank tests revealed that the CorPower device, integrated with the WaveSpring technology, is able to absorb three times more power in realistic sea conditions than its counterpart without the WaveSpring technology, and can simultaneously reduce the maximum mooring line tension. In addition, the CorPower device was used as a case study to assess the techno-economic performance of WEC farms at 20 MW installation capacity, aiming to compare several techno-economic metrics for LCOE reduction [163].

4.3.1.4 WaveStar device

The WaveStar device was developed and tested in Denmark [420], as shown in Figure 4.3(d). Two PAs are attached to a jacked-up frame or structure via hydraulic PTO units for harvesting wave power. The details of this device are given below.

- A full scale WaveStar platform should have 40 PAs to form an WEC array. Each PA is in the shape of a semi-sphere with a diameter of 5 m, and attached to the jacked-up platform via a steel arm of 10 m. Incident waves excite the semi-sphere buoy to oscillate in heave and pitch modes, and these motions feed in to a hydraulic PTO unit, with a power rating of 55 kW, to generate electricity [499].

- In 2004, a 1/40 rig with 40 floats was tested at Aalborg University in Denmark. In 2006, a 1/10 scale model with 40 floats was installed in the sea at Nissum Bredning in Denmark. The experimental tests of the 1/40 and the 1/10 scale WaveStar prototypes showed an energy conversion efficiency up to 40–60% [420]. In 2009, a full-scale prototype with two floats was installed at a water depth of 6 m at Hanstholm, Denmark. The device was connected to the utility grid and supplied electricity to the public [499].

- Sea trials have demonstrated that the WaveStar device has good survivability, high reliability, and very limited maintenance requirements [499]. In storms, the WaveStar device can pull the buoys up to the frame to enter survival mode, thanks to the jacked up platform. Meanwhile, the jacked-up structure only allows the device to be deployed in shallow water. In addition, the WaveSpring technology has also demonstrated its effectiveness in improving the response properties of the WaveStar device, in both numerical and laboratory testing in 2018 [766].

4.3.2 Two-body point absorber prototypes

Some typical prototypes of two-body floating PAs are shown in Figure 4.4, including the OPT PowerBuoy, the Wavebob, the SEAREV, and the Penguin devices, which are detailed below.

4.3.2.1 OPT PowerBuoy device

The PowerBuoy device has been developed by the company called Ocean Power Technologies (OPT), and consists of a torus float and a spar with a heavy plate at

Figure 4.4: Typical prototypes of two-body floating point absorbers: (a) the Power-Buoy device [857], (b) the Wavebob device [754, 753], (c) the SEAREV device [139], and (d) the Penguin device [306].

its bottom, as shown in Figure 4.4(a). The float and the spar are coupled via a PTO system, and their relative motion is used by the PTO to generate electricity. The float heaves up and down, while the spar remains stationary due to the heavy plate at the bottom. The PTO is situated inside the spar and uses a ball-screw or rack-pinion mechanism to transfer the bidirectional linear motion to the uni-directional rotation for a conventional generator [191, 597]. The design details are given below.

- The first commercial model of the PowerBuoy is the PB3 device, which can act as an uninterruptible power supply (UPS) for offshore applications. The float has a diameter of 3 m, and the spar has a diameter of 1 m, a height of 13.3 m with a draft of 9.28 m. The total weight of the device is about 8.3 tonnes. The typical average power is 8.4 kWh per day [597]. The PB3 can operate in water depth ranging from 20 m to 3,000 m, and the maintenance intervals by design are every 3 years [597].

- The PowerBuoy concept dates back to 1997. In 2010 and 2011, the PB150 prototype, with a power rating of 150 kW, was tested at EMEC. In 2011, the Autonomous PowerBuoy prototype was deployed off the coast of New Jersey for a 3-month sea trial [597]. In 2015, the PowerBuoy prototype, PB-3-50-A1, was tested off the coat of New Jersey, which produced 1.3 MWh energy during a

108-day operation [87]. Since 2015, the OPT company pivoted to commercialising the PowerBuoy device to provide autonomous, zero-carbon power and data solutions for the increasing electrical needs of offshore industries, scientific research and territorial security, resulting in the company's first fully commercial product the PB3 PowerBuoy [597].

- The PowerBuoy PB150 prioritised system survivability at an early stage of its development. Hundred-year storms and extreme wave conditions were used to design the structure to improve its survival loads. As a consequence, the device survived Hurricane Irene in 2011 [87, 597].

4.3.2.2 Wavebob device

The Wavebob device is an axisymmetrical, self-reacting point absorber that primarily operates in the heave mode, as shown in Figure 4.4(b). It consists of two concentric floating buoys, known as a torus and a float-neck-tank (FNT). The torus is essentially a wave follower over the range of wave frequency of interest for power conversion, while the FNT has a much lower natural frequency and acts as a reference for the torus [475]. The torus and the FNT are coupled via a hydraulic PTO unit or a direct-drive linear generator, using their relative motion to generate electricity. The development details are given below.

- A full-scale torus has a diameter of 17.6 m, a draft of 4.86 m and a freeboard of 3.0 m. The FNT has a draft of 57 m [474].

- The company, Wavebob, commenced in 1997 and had its first 1/100 scale model tested in 1999. A 1/25 scale model was tank tested between 2001 and 2003, and a 1/4 scale prototype was deployed at the Galway Bay Ocean Energy Test Site twice, in 2006 and 2007, becoming the first sea-tested device in Ireland to produce electricity from waves. During these sea trials, some technical problems associated with PTO, end-stop, etc., were identified. As a result, more analysis and further tests were conducted, with a 1/19 scale model tested at MARIN in 2011 and 2012, with a direct-drive linear generator [474, 753]. In addition, a 1/17 scale model, equipped a PTO, was tested at the Ecole Centrale de Nantes in 2010, and the nonlinear phenomenon of parametric resonance in roll and pitch was observed and recorded [754, 753].

- Wavebob went into liquidation in 2013 due to cashflow problems. The development trajectory of the Wavebob device strongly suggests the necessity to keep design decisions simple to improve survivability and reliability in extreme waves.

4.3.2.3 SEAREV device

The SEAREV concept was first proposed in 2002 at the Ecole Centrale De Nantes. The device consists of a closed floating hull in which a heavy pendulum oscillates. The controlled relative motion of the pendulum is used to produce electricity, as shown in

Figure 4.4(c). The PTO system can be either a direct-drive generator with a super capacitor as energy storage or a hydraulic PTO unit. The development trajectory of the SEAREV device is detailed as follows.

- At full scale, the SEAREV device was designed as 26.64 m in length, 13.25 m in width, 20.7 m in height, and 14.6 m in draft [43]. The nominal power can range up to 400 kW [384].

- From 2002 to 2014, three key generations of the SEAREV prototype were developed. From 2002 to 2006, the SEAREV G1 prototype was developed to validate the concept and mathematical modelling approach. Shape optimisation was conducted intensively at this stage [45]. In 2006, tank testing was conducted and parametric resonance was recorded [187]. From 2007 to 2009, the second generation device, G21, was developed, with specific foci on modelling [384], control development and further refinement of the WEC design. From 2007 to 2009, the G3 prototype was developed to optimise the shape further and to assess the economic performance of a SEAREV farm. In 2013, a company named Oceanwing was founded to demonstrate the SEAREV technology using sea trials. Based on the SEAREV concept, a wave farm economic model was developed, which projected that the LCOE of the SEAREV farm is about 400 €/MWh at an installation capacity of 20 MW, which is not competitive in the utility market.

- The development trajectory of the SEAREV project showed that shape optimisation can improve energy capture performance significantly. However, an optimal shape may be difficult to manufacture, consequently resulting in a high CAPEX, and, consequently, a high LCOE. In addition, power maximising control can play an important role in WEC applications, as it improved the average power of the G21 prototype from 84 kW to 150 kW, while the CAPEX only increased 300 k€ for the control components.

4.3.2.4 Penguin device

The two-body Penguin PA device was developed by a Finnish company called Wello Oy. Wello's Penguin device is characterised by its simple structure and survivability capacity. As shown in Figure 4.4(d), the Penguin device consists of an inner mass rotating around a shaft like a horizontal pendulum. The hull oscillates in pitch and roll modes under the excitation of ocean waves, and the eccentric pendulum spins. Such a relative motion directly drives a rotational electric generator to generate electricity. The development details are given below.

- At full scale, the size of the device ranges from 30 m to 56 m with its nominal power rated from 0.5 MW to 1 MW. The LCOE is estimated to be around 0.06–0.32 €/kWh [829]. As the dimension is relatively large with respect to the prevailing wavelength, the Penguin device is more like a "quasi-PA" than a PA. These design parameters, or economic estimates, rely significantly on the wave conditions of installation sites.

- In 2010, the first full scale Penguin device was deployed at EMEC, and tested from 2011 to 2014. During the sea trial, several storms occurred. The device survived successfully in extreme waves of 18 m [829], showing high survivability. Another Penguin prototype, with a width of 30 m, a height of 9 m, a draft of 7 m and a power rating of 1 MW, was also tested at EMEC from 2017 to 2019. The average power ranged from 160 kW to 180 kW, and the peak power was up to 700 kW [764]. In 2019, Wello Oy completed the construction of a 44-m Penguin WEC.

- The Penguin device has demonstrated its high survivability and low LCOE in real sea environment over a long period of time. The simple design is one main reason for these favourable characteristics. In addition, an environmental impact study of the Penguin device was conducted by measuring its acoustic noise [65]. The mean value of ambient noise was 112 dB re 1 μPa, and the source sound pressure level of the Penguin device was measured at 140.5 dB re μPa. The noise frequency was lower than 300 Hz [65].

4.4   CASE STUDY – A HEAVING POINT ABSORBER

In general, the WEC modelling problem mainly focuses on wave-structure interactions (WSI), for which several modelling methods have been discussed in Chapter 2, including analytical, BEM, CFD and experimental methods. However, the hydrodynamic models derived from these methods cannot be used straightforwardly for WEC control design, geometric optimisation, and performance assessment. Hence, simple, parametrised models are preferred, which can be classified into frequency-domain (FD), time-domain (TD), and spectral-domain (SD) models. FD and SD modelling methods have been detailed in Chapter 3, and TD modelling approaches are detailed in this section. Compared with FD and SD modelling methods, TD modelling approaches are feasible for dealing with various nonlinearities, to depict WECs' transient and extreme responses, and to implement real-time control algorithms.

In this section, a cylindrical one-body PA, as shown in Figure 4.5, is taken as an example to demonstrate some basic time-domain modelling approaches for WECs. As shown in Figure 4.5(a), the semi-submerged buoy has a diameter of 0.3 m, a height of 0.56 m, and a draft of 0.28 m. The design sea state was characterised by a significant wave height of 4.3 m, a peak frequency of 0.6 rad/s and a prevailing wavelength of 130 m. Thus, the original buoy diameter was set as 15 m, to satisfy the PA definition. To conduct tank testing, the scale ratio is selected as 1/50 [310], according to the Froude scaling law described in Section 2.4.4. An experimental rig was developed and tested in a wave tank for model validating, as shown in Figure 4.5(b). In this section, the buoy was constrained in heave mode only.

In this case study, hybrid modelling methods, referred to as methods that augment the linear potential flow-based Cummins' equation with some critical nonlinear terms to derive parametrised WEC models are demonstrated. Based on the input-output relation, TD models can be classified as force-to-motion (F2M), wave-to-motion (W2M), and wave-to-wire (W2W) models, suiting various application scenarios.

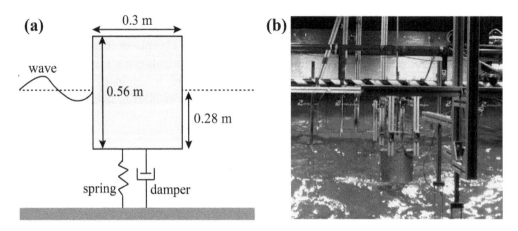

Figure 4.5: A 1/50 scale heaving PA is taken as an example to demonstrate modelling approaches for PAs. The buoy is a semi-submerged cylinder, with a schematic plot shown in (a) and an experimental rig under tank testing shown in (b).

For simplicity, a one-body floating PA, operating in heave mode only, is investigated in this section, and the modelling methods of a multi-mode PA will be discussed in Chapter 5. A couple of time-domain modelling approaches for the heaving PA are introduced in this section, to derive F2M, W2M, and W2W models in Sections 4.4.1–4.4.3, respectively.

4.4.1 Force-to-motion modelling

F2M modelling focuses on the radiation problem of WECs, together with the inertial and restoring forces. A F2M model can be derived analytically based on Cummins' equation [676, 313, 214], or directly identified from numerical or experimental data by externally forcing the body to oscillate in still water [159, 291]. F2M models utilise a known excitation force as system input, and the output signal is usually the displacement or/and velocity of the buoy, creating so-called F2M models. Section 4.4.1 handles the radiation force approximation problem from the BEM results, to derive a state-space model to approximate the radiation convolution term in Cummins' equation. Then, linear and nonlinear F2M models are derived in consequence.

4.4.1.1 Radiation force approximation

The dynamics of the heaving buoy in Figure 4.5 are governed by Newton's second law [227], given as

$$M\ddot{z}(t) = F_e(t) + F_r(t) + F_s(t) + F_f(t) + F_v(t) + F_{pto}(t) + F_{add}(t), \qquad (4.1)$$

where $F_e(t)$, $F_r(t)$, and $F_s(t)$ are the excitation, radiation, and hydrostatic forces. $F_f(t)$ and $F_v(t)$ represent mechanical friction and viscous force, respectively. $F_{pto}(t)$ is the power take-off (PTO) or control force, while $F_{add}(t)$ represents some additional force, e.g., the mooring force. M is the buoy mass; $z(t)$ is the PA's displacement in heave mode, for which the positive direction is defined as upward.

For the vertical cylinder in Figure 4.5, the hydrostatic force is proportional to the heaving displacement, given as

$$F_s(t) = -\rho g \pi r^2 z(t) = -K_s z(t), \tag{4.2}$$

where ρ, g are the water density and gravity constant, respectively. r and $K_s = \rho g \pi r^2$ represent the buoy radius and hydrostatic stiffness, respectively. The excitation force $F_e(t)$ is viewed as a known system input here, and its modelling is discussed in Section 4.4.2.

According to Cummins' equation [146], the radiation force in the time domain is written as

$$F_r(t) = -A_\infty \ddot{z}(t) - k_{rad}(t) * \dot{z}(t), \tag{4.3}$$

where A_∞ is the added mass at infinite frequency and $k_{rad}(t)$ is the impulse response function (IRF) also called the kernel function. The symbol $*$ represents the convolution operation.

Alternatively, the radiation force can be expressed in the frequency domain, as

$$F_r(i\omega) = [\omega^2 A_m(\omega) - j\omega B_{rad}(\omega)] Z(i\omega), \tag{4.4}$$

where ω is the angular frequency. $F_r(i\omega), A_m(\omega), B_{rad}(\omega),$ and $Z(i\omega)$ are the frequency-domain representations of the radiation force, the added mass, the radiation damping and the heaving displacement, respectively. Thus, the radiation frequency response function (FRF) can be written as

$$K_r(i\omega) = \int_0^\infty k_{rad}(\tau) e^{-i\omega\tau} d\tau. \tag{4.5}$$

The relationship between the time- and frequency-domain coefficients is derived by Ogilvie [590], so referred to as the Ogilvie relation, given as

$$A_m(\omega) = A_\infty - \frac{1}{\omega} \int_0^\infty k_{rad}(t) \sin(\omega t) \, dt, \tag{4.6}$$

$$B_{rad}(\omega) = \int_0^\infty k_{rad}(t) \cos(\omega t) \, dt. \tag{4.7}$$

The hydrodynamic coefficients are computed by solving a boundary value problem in the BEM code NEMOH, as shown in Figure 4.6. NEMOH is an open-source BEM package for WEC modelling, which provides a useful alternative to commercial software packages, like WAMIT and Ansys AQWA [621]. A variety of BEM packages for WEC modelling are compared in Table 2.1.

The convolution operation in Eq. (4.3) is not convenient and straightforward for buoy hydrodynamic analysis and control design. Hence, it is important to approximate the convolution term with a model of a finite order number. The causality of the radiation process is proved in [826]. Thus the radiation force can be approximated by a finite order system with constant parameters via frequency- or time-domain system identification approaches. These approaches are investigated in [752, 675, 794, 158, 422]. Thus these radiation force approximation approaches are mature, and this section gives a brief description below.

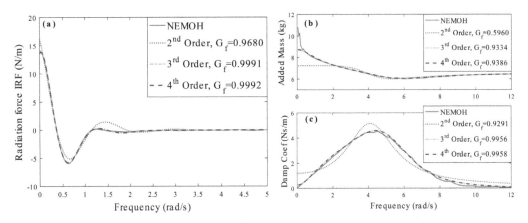

Figure 4.6: Comparison of the radiation IRF, added mass and damping coefficient between NEMOH results (solid curves) and approximated results (dahsed curves) in (a)–(c), respectively. For the approximated results, the identified system order numerbers are 2, 3, and 4, respectively. The goodness of fit is defined in Eq. (4.11).

As a brief introduction to the convolution approximation procedure, the radiation force convolution term is defined as a subsystem $f'_r(t)$, written as:

$$F'_r(t) = k_{rad}(t) * \dot{z}(t) = \int_0^t k_{rad}(t - \tau)\dot{z}(\tau)d\tau. \tag{4.8}$$

The IRF $k_{rad}(t)$ in Eq. (4.8) is gained from NEMOH simulation. The realisation theory is applied on the IRF to get a linear state-space model directly, using the MATLAB function *imp2ss* [426] from the *Robust Control Toolbox*. Thus, the convolution term can be approximated as

$$\dot{x}_r(t) = A_r x_r(t) + B_{rad}\dot{z}(t), \tag{4.9}$$
$$F'_r(t) = C_r x_r(t), \tag{4.10}$$

where $x_r(t) \in \mathbb{R}^{n_r \times 1}$ is the state vector for the identified system and n_r is the system order. $A_r \in \mathbb{R}^{n_r \times n_r}$, $B_{rad} \in \mathbb{R}^{n_r \times 1}$, $C_r \in \mathbb{R}^{1 \times n_r}$ are the system matrices.

The order of the initially identified system is quite high and determined by the IRF data. Model reduction is required and can be achieved by the square-root balanced model reduction method with the MATLAB function *balmer* [685]. The appropriate order is determined by the goodness of fit function G_f, defined by the normalised mean square error, given as

$$G_f = 1 - \left\| \frac{X - \widetilde{X}}{X - \overline{X}} \right\|_2^2, \tag{4.11}$$

where X is the original data (it can be the added mass, radiation damping or the IRF) and $\widetilde{X}$ represents the approximation of X. $\|X\|_2$ is the two-norm operation of X and $\overline{X}$ is the mean value of X. The goodness of fit tends to 1 for perfect fitting and tends to $-\infty$ for the worst fitting. In Figure 4.6, the identified results for $n_r = 2$,

$n_r = 3$, and $n_r = 4$ are compared with the NEMOH results. For $n_r = 3$, the values of the goodness of fit are 0.9991, 0.9334, and 0.9956 for the radiation IRF, added mass and radiation damping, respectively. Thus, $n_r = 3$ is selected for this case study.

Finite order approximations of the radiation force in the time-domain are proposed and applied for offshore structure motion prediction by [752, 628, 858, 422, 158, 675, 278]. Alternatively, the radiation convolution term can also be approximated by a transfer function or a state-space model using system identification technology in the frequency domain. In the frequency domain, the MATLAB functions *invfreqs* and *tf2ss* can be applied to the FRF $K_{rad}(i\omega)$ in Eq. (4.5) to obtain a transfer function or a state-space model with a finite order number. In addition, there are plenty of FD system identification techniques to approximate the convolution term. These are investigated in [794, 675, 752, 629, 630]. It is worth noting that the moment-matching algorithm in [214] can exactly match the frequency response of the original system at chosen frequencies and retain specific physical properties, e.g., passivity.

4.4.1.2 Linear force-to-motion modelling

To derive a linear F2M model, the nonlinear forces F_f, F_v, and F_{add} are neglected. As the wave excitation force is viewed as a known system input, and buoy displacement is set as the system output, a linear F2M model can be written as

$$x_{f2m} = [z \quad \dot{z} \quad x_r]^T, \tag{4.12}$$
$$\dot{x}_{f2m} = A_{f2m}x_{f2m} + B_{f2m}F_e + B_{f2m}F_{pto}, \tag{4.13}$$
$$y_{f2m} = C_{f2m}x_{f2m}, \tag{4.14}$$

with

$$A_{f2m} = \begin{bmatrix} 0 & 1 & 0 \\ -\frac{K_s}{M_t} & 0 & -\frac{C_r}{M_t} \\ 0 & B_r & A_r \end{bmatrix}, \tag{4.15}$$

$$B_{f2m} = \begin{bmatrix} 0 & -\frac{1}{M_t} & 0 \end{bmatrix}^T, \tag{4.16}$$

$$C_{f2m} = \begin{bmatrix} 1 & 0 & 0 \end{bmatrix}, \tag{4.17}$$

where $M_t = M + A_\infty$ represents the total mass. $x_{f2m} \in \mathbb{R}^{(n_r+2)\times 1}$ is the F2M state vector. $A_{f2m} \in \mathbb{R}^{(n_r+2)\times(n_r+2)}$, $B_{f2m} \in \mathbb{R}^{(n_r+2)\times 1}$, and $C_{f2m} \in \mathbb{R}^{1\times(n_r+2)}$ are the system matrices. This linear F2M model is convenient and straightforward for numerical analysis of PA hydrodynamics and for the development of a control system. If the initial condition $x(0)$, the PTO/control force f_{pto} and the excitation force f_e are applied to excite the linear model at time $t = 0$ s, the response can be written as

$$z = C_{f2m}e^{A_{f2m}t}x_{f2m}(0) + \int_0^t C_{f2m}e^{A_{f2m}(t-\tau)}B_{f2m}[F_e(\tau) + F_{pto}(\tau)]d\tau. \tag{4.18}$$

The linear system response can be separated into three parts, as

- *Free-decay test*: if $F_e = 0$, $F_{pto} = 0$, and $x_{f2m}(0) \neq 0$, the system response is defined as the *free-decay test*, written as

$$z = C_{f2m} e^{A_{f2m} t} x_{f2m}(0). \qquad (4.19)$$

- *Forced-motion test*: if $F_e = 0$, $x_{f2m}(0) = 0$, and $F_{pto} \neq 0$, the system response is identified as the *forced-motion test*, written as

$$z = \int_0^t C_{f2m} e^{A_{f2m}(t-\tau)} B_{f2m} F_{pto}(\tau) d\tau. \qquad (4.20)$$

- *Free-motion test*: if $F_{pto} = 0$, $x_{f2m}(0) = 0$, and $F_e \neq 0$, the system response is identified as the *free-motion test*, written as

$$z = \int_0^t C_{f2m} e^{A_{f2m}(t-\tau)} B_{f2m} F_e(\tau) d\tau. \qquad (4.21)$$

The free-decay test and forced-motion test are conducted both in numerical and physical tank testing, for identifying or validating linear F2M models. One main drawback of this linear F2M model is that the omission of nonlinear effects leads to overestimation of buoy motion, especially when the incoming wave frequency is close to the PA natural frequency. This work places special emphasis on the nonlinear viscous and friction phenomena; these nonlinear effects are discussed in further detail, below.

4.4.1.3 Nonlinear force-to-motion modelling

In practice, the viscous force due to fluid viscosity and the mechanical friction due to relative motion cannot be neglected for the buoy in Figure 4.5. In this section, the viscous force is modelled as the drag term in the Morison equation [551], while the friction is modelled as a combination of the Stribeck, the Coulomb, and the damping terms [784]. The summation of the viscous and the friction forces is defined as a lumped nonlinear force, and is the focus of this section, leading to a more practical nonlinear model for the 1/50 buoy in Figure 4.5.

As suggested by [192, 672], the viscous force f_v follows the drag force in the Morison equation [551], given as

$$F_v = -0.5 \rho C_D \pi r^2 (\dot{z} - u) |\dot{z} - u|, \qquad (4.22)$$

where C_D is the viscous coefficient; $u(t)$ is the vertical velocity of water particles around the buoy. When the wave height is small and the buoy motion is exaggerated by power maximising control, $u(t) = 0$ can be used. C_D is a function of the Reynolds number, the Keulegan-Carpenter number (K_c), and the roughness number [309]. As suggested in [309], the empirical value of C_D varies from 0.6 to 1.2. For a small K_c value ($K_c \approx 3.67$ for the 1/50 scale PA rig in Figure 4.5), the appropriate range of

C_D from 0.8 to 1 is commonly acceptable [692] and $C_D = 0.93$ is selected in this case study on a semi-empirical basis.

Several mechanical friction models are reviewed in [33], from which the Tustin friction model is selected in this case study, as it fits the experimental results well. The Tustin model is expressed as a combination of the Stribeck, the Coulomb, and the damping friction terms in [33, 784] (see Figure 4.7), and the components are given as

$$f_c = -s_v F_c, \tag{4.23}$$
$$f_s = -s_v F_s e^{-C_s|\dot{z}|}, \tag{4.24}$$
$$f_d = -s_v C_f \dot{z}, \tag{4.25}$$

where $s_v = sgn(\dot{z})$ is the sign of the buoy velocity; f_c is the Coulomb friction with its coefficient F_c; f_s is the Stribeck friction with its coefficient F_s and shape factor C_s; f_d is the damping friction with its coefficient C_f. The negative symbol means that the friction force always impedes the PA velocity.

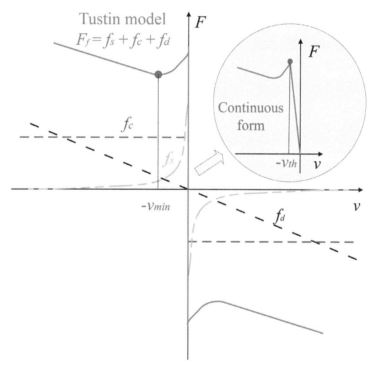

Figure 4.7: The Tustin friction model and its components.

Therefore, the Tustin model can be expressed as

$$F_f = -s_v(F_c + F_s e^{-C_s|\dot{z}|} + C_f|\dot{z}|). \tag{4.26}$$

The Stribeck shape factor can be determined by the intersection point of the Stribeck friction curve and the damping friction line, given as

$$C_s = \frac{1}{V_{min}} \ln \frac{C_f V_{min}}{F_s}, \tag{4.27}$$

where V_{min} is the velocity related to the intersection point (also the point associated with the minimal friction in Eq. (4.26)).

A key characteristic of this model is that the friction is discontinuous at the zero-velocity point. The discontinuity may cause difficulties for numerical modelling. In addition, the mechanical friction can be estimated from the velocity measurement made during wave tank tests. Measurement noise is unavoidable and has a significant influence on the friction modelling, especially when the velocity is close to zero. Therefore, a velocity threshold V_{th} is applied to the Tustin model in this work to improve its continuity within the zero-velocity region. The continuous formulation of the Tustin model can be rewritten as

$$F_f = \begin{cases} -s_v(F_c + F_s e^{-C_s|\dot{z}|} + C_f|\dot{z}|), & |\dot{z}| \geq V_{th}; \\ -s_v(F_c + F_s e^{-C_s V_{th}} + C_f V_{th})\frac{|\dot{z}|}{V_{th}}, & |\dot{z}| < V_{th}. \end{cases} \quad (4.28)$$

V_{th} is always set with a very small value. For the continuous form, the Stribeck shape factor can be rewritten as

$$C_s = \frac{1}{V_{min} - V_{th}} \ln \frac{C_f(V_{min} - \frac{V_{th}F_c}{F_s + F_c})}{F_s}. \quad (4.29)$$

In the friction model in Eq. (4.28), there are five unknown parameters (F_s, F_c, C_f, V_{th}, and V_{min}) to be determined experimentally. The method to determine these parameters is detailed in [313, 310, 378], and the experimentally estimated values are listed in Table 4.1.

Table 4.1: Parameters of the viscous and friction models in Eqs. (4.22) and (4.28), respectively.

Parameter	Unit	Value	Parameter	Unit	Value
F_s	N	5.0065	f_d	N·s·m^{-1}	2.7200
F_c	N	3.1160	v_{th}	m·s^{-1}	0.038
C_d	/	0.9300	v_{min}	m·s^{-1}	0.1056

During wave tank tests, the viscous, and friction forces were lumped together and could not be decoupled from each other. Therefore, a lumped nonlinear force F_{ln} is defined as a summation of the viscous force F_v and the friction F_f, given as

$$F_{ln} = F_v + F_f. \quad (4.30)$$

If the lumped nonlinear force is considered, Eq. (4.1) is rewritten as:

$$M\ddot{z} = F_e + F_r + F_s + F_{pto} + F_{ln}. \quad (4.31)$$

Therefore, the 1/50 scale PA hydrodynamics can be expressed as a nonlinear F2M model in the state-space formulation, as

$$\dot{x}_{f2m} = A_{f2m}x_{f2m} + B_{f2m}F_e + B_{f2m}F_{pto} + B_{f2m}F_{ln} \quad (4.32)$$

$$y_{f2m} = C_{f2m}x_{f2m}. \quad (4.33)$$

The system matrices are given in Eqs. (4.15)–(4.17).

4.4.1.4 *Results and discussion*

In this case study, the linear and nonlinear F2M models are validated against tank tests, and the setups for the tank tests are shown in Figure 4.8. The wave tank was 13 m in length, 6 m in width, 2 m in height, and 0.9 m in water depth. On the left side, there were 8 pistons, which acted as a wave-maker and were capable of generating both regular and irregular waves. The 1/50 scaled down buoy was installed at the centre of the wave tank. Five wave gauges (WGs) were mounted to measure the water elevation in real-time, and five pressure sensors (marked PS1 to PS5) were installed at the buoy bottom to measure the dynamic pressure acting on the hull. To investigate the buoy motion, a linear variable displacement transducer (LVDT) was used to record the buoy displacement and a three-axis accelerometer (Acc) was mounted at the top of the buoy to measure its heaving acceleration. All the sensing signals were collected via a 16-bit data acquisition system (USB-6210) and the data were collected by a self-assembled graphic user interface in LABVIEW. The sampling frequency was set at 100 Hz.

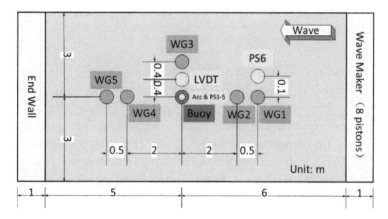

Figure 4.8: Setup for tank testing.

Intensive free-decay tests described in Eq. (4.19) were conducted, by pushing the buoy down to a non-zero initial position. The buoy was held stable for a short term and then released. The data from the LVDT, Acc and PS1–PS5 were recorded. The free-decay test with an initial position of −0.2 m was selected as a typical example to validate the linear and nonlinear F2M models, and the experimental data were compared with the numerical results in Figure 4.9(a). All data were normalised to the initial position of −0.2 m. The natural periods (frequencies) for the experimental rig, the linear and nonlinear F2M models were 1.233 s (5.0959 rad/s), 1.215 s (5.1713 rad/s), and 1.230 s (5.1083 rad/s), respectively. In terms of the normalised displacement in Figure 4.9(a), the nonlinear model results only slightly differed from their experimental counterparts. For the numerical results of the nonlinear F2M model, its goodness of fit, G_f, defined in Eq. (4.11), was as high as 0.9794. Thus, the proposed nonlinear F2M modelling approach can provide a more accurate model than its linear counterparts.

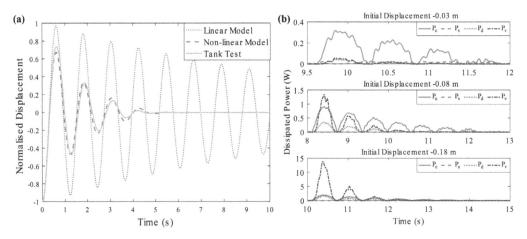

Figure 4.9: Comparison of normalised displacement between experimental and numerical free-decay tests in (a), and comparison of power dissipating effects between various nonlinear factors in (b). P_c, P_s, P_d, and P_v represent the power dissipated by the Coulomb friction, the Stribeck friction, the damping friction, and the viscous force, respectively.

Within the first period in figure 4.9(a), the buoy velocity was high. Thus, the viscous force predominated the decay process, dissipated a large amount of power, and consequently damped the displacement quickly. During the second and third periods, the Coulomb and the damping friction components became the major factors to consume power and impede the buoy. When the velocity became small in the fourth period, the Coulomb and the Stribeck friction components drove the buoy back to its equilibrium point. However, the power dissipating effects of these nonlinear factors depended significantly on the buoy's velocity range (changed in tank tests by adjusting the initial position). For the free-decay tests with initial positions of –0.03 m, –0.08 m, and –0.18 m, the dissipated power by the aforementioned nonlinear factors is compared in Figure 4.9(b), within which P_c, P_s, P_d, and P_v represent the power dissipated by the Coulomb friction, the Stribeck friction, the damping friction, and the viscous force, respectively. It can be seen from Figure 4.9(b) that: (i) for the free-decay test with an initial position of –0.03 m, the Coulomb friction force dissipated most of power; (ii) for the free-decay test with an initial position of –0.08 m, the Coulomb and the fluid viscous forces consumed the main part of power; and (iii) for the free-decay test with an initial position of –0.18 m, the fluid viscous force used most of power.

Intensive free-decay tests were conducted with initial displacements ranging from –0.02 m to –0.2 m. The simulation results of the nonlinear F2M model were compared with the experimental data and shown in Table 4.2. The simulation results of the nonlinear model fitted well with the experimental data, with a goodness of fit higher than 0.95. Therefore, it can be concluded that the nonlinear model can represent the 1/50 scale PA hydrodynamics for a wide range of the free-decay tests.

Table 4.2: Goodness of fit of the nonlinear F2M model results with respect to the experimental data.

Initial position	Goodness of fit	Initial position	Goodness of fit
−0.02 m	0.9582	−0.12 m	0.9576
−0.04 m	0.9536	−0.14 m	0.9755
−0.06 m	0.9578	−0.16 m	0.9810
−0.08 m	0.9808	−0.18 m	0.9875
−0.10 m	0.9808	−0.20 m	0.9724

4.4.2 Wave-to-motion modelling

For W2M modelling, not only is the radiation problem included, but also the excitation process, dealing with the incident and diffraction problems, is considered. The physical process from wave elevation to excitation force is well known as non-causal [227]. In practical WEC implementation, the excitation force *cannot be measured directly* [575], since it cannot be decoupled from the other hydrodynamic forces. However, knowledge of the excitation force is required for generating optimal velocity reference signals for some power maximising control strategies. Consequently, there are some papers dedicated to modelling the excitation force from incident waves [312, 3, 315], which is one of the foci in this section. Alternatively, the non-causal problem can be handled implicitly by estimating the excitation force via unknown input observers [6, 315, 618, 619], or by identifying W2M models directly from CFD or experimental data [290, 159, 291].

Section 4.4.2.1 deals with the excitation force approximation problem from BEM results, to derive a state-space model to approximate the excitation force from incident waves. Then, the wave-to-excitation-force (W2EF) model is integrated with the linear and nonlinear F2M models, to form linear and nonlinear W2M models, detailed in Section 4.4.2.2. The numerical and experimental results are compared and discussed in Section 4.4.2.3.

4.4.2.1 Excitation force approximation

Based on linear potential flow theory, the amplitude of the excitation force is related to the radiation damping, given by the Haskind's relation [570]. For a regular wave of a given angular frequency, the excitation force $F_e(t)$ can be analytically given as

$$F_e(t) = \frac{H}{2} \left(\frac{2\rho g^3 B_{rad}(\omega)}{\omega^3} \right)^{1/2} \cos(\omega t). \tag{4.34}$$

Although Eq. (4.34) is derived based on regular waves, it can be modified for irregular waves. Based on the superposition principle, the analytical form in

Eq. (4.34) can be extended to a spectral form [51], given as

$$F_e(t) = \Re\left[\sum_j \sqrt{2S_{\hat{\eta}}(\omega_j)\Delta\omega}\, H_{fe}(i\omega_j)e^{i(\omega_j t + \phi_j)}\right], \tag{4.35}$$

where $\Re$ is the real operator. $S_{\hat{\eta}}(\omega)$ and $H_{fe}(i\omega)$ represent the wave spectrum and the excitation force FRF, respectively. ω_j is a set of the angular frequencies with the subscript j with an interval of $\Delta\omega$, and ϕ_j is a set of random phases. These analytical representations in Eqs. (4.34) and (4.35) are widely applied to assess the performance of various WECs. However, the modelled excitation force according to Eqs. (4.34)–(4.35) fails to reflect the varying real-time sea waves, making it inapplicable for real-time reference generating.

As the excitation force is physically induced by the incident wave, it is possible to approximate the excitation force direct from wave elevation. The FRF of the W2EF process can be obtained by BEM packages, and the excitation force FRF $H_{fe}(i\omega)$ for the 1/50 PA rig is obtained for NEMOH and shown in Figure 4.10(a), in terms of the amplitude response $|H_{fe}(i\omega)|$ and the phase responses $\angle H_{fe}(i\omega)$. In the frequency domain, the relationship between the wave elevation and the excitation force can be written as

$$F_e(i\omega) = H_{fe}(i\omega)A_{\hat{\eta}}(i\omega), \tag{4.36}$$

where $A_{\hat{\eta}}(i\omega)$ is the frequency-domain representation of the wave elevation $\eta(t)$. The excitation force IRF and FRF are a Fourier transform pair. Thus, the excitation force IRF $k_e(t)$ can be computed from its FRF, given as

$$k_e(t) = \frac{1}{2\pi}\int_{-\infty}^{\infty} H_e(i\omega)e^{i\omega t}d\omega. \tag{4.37}$$

For the 1/50 scale PA rig in Figure 4.5, the excitation force IRF is shown in Figure 4.10(b). Therefore, the excitation force in the time domain can be written as

$$F_e(t) = k_e(t) * \eta(t) = \int_{-\infty}^{t} k_e(t-\tau)\eta(\tau)d\tau. \tag{4.38}$$

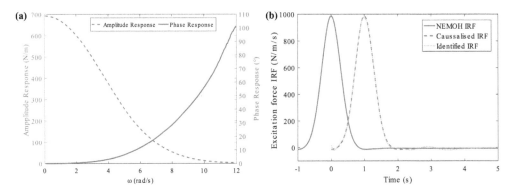

Figure 4.10: Frequency response function of the wave-to-excitation-force process in (a) and its impulse response functions in (b).

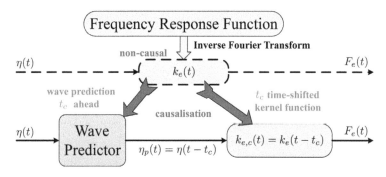

Figure 4.11: Methodology for excitation force approximation.

Although the convolution operation can be numerically solved directly (with wave prediction), a parametrised model is preferred for real-time control realisation and performance assessment. Similar to the radiation approximation method in Section 4.4.1.1, the convolution term in Eq. (4.38) can also be approximated by a state-space model, which is the focus of this section.

It is clear that the W2EF process is non-causal since $k_e(t) \neq 0$ N/m/s for $t < 0$ s, to see the blue curve in Figure 4.10(b). Physically, such a system is unimplementable. In addition, the $t < 0$ s part is as large as the $t > 0$ s part. Therefore, neglecting the non-causality of $k_e(t)$ will in general lead to significant errors in the excitation force modelling. Therefore, IRF causalisation is required before implementing system identification techniques.

A time-shift technique is applied to causalise the non-causal IRF $k_e(t)$ to a causal form $k_{e,c}(t)$ with the causalisation time t_c. As shown in Figure 4.11, the causalised system with wave prediction can provide the same approximation of the non-causal system, since

$$\begin{aligned} F_e(t) &= k_e(t) * \eta(t) & (4.39) \\ &= k_{e,c}(t) * \eta_p(t), & (4.40) \end{aligned}$$

where

$$\begin{aligned} k_{e,c}(t) &= k_e(t - t_c), & (4.41) \\ \eta_p(t) &= \eta(t + t_c). & (4.42) \end{aligned}$$

$\eta_p(t)$ represents the predicted wave with a time window of t_c, which is indispensable for the implementation of Eq. (4.40). It is noteworthy that direct solving the convolution operation in Eq. (4.38) also requires such a wave prediction. The expression in Eq. (4.40) can be realised in practice and gives the same output of the non-causal system, assisted by wave prediction. The causalised system in Eq. (4.40) is achieved with a time-shifted IRF $k_{e,c}(t)$ and wave prediction $\eta_p(t)$. The wave prediction horizon is the same as the causalisation time t_c. The W2EF modelling procedure is summarised as follows.

- Obtain the excitation force FRF via BEM packages, e.g., WAMIT, AQWA, and NEMOH.

- Compute the excitation force IRF from its FRF according to Eq. (4.37).

- Causalise the excitation force IRF by selecting an appropriate causalisation time t_c.

- Identify the causalised W2EF process in Eq. (4.40), for which the method is the same as the radiation approximation problem in Section 4.4.1.1.

- Use auto-regressive (AR), auto-regressive-moving-average (ARMA), or fast Fourier transform (FFT) approaches [262] to predict the wave elevation $\eta_p(t)$ as the system input signal .

For the causalised system in Eq. (4.40), the convolution operation can be approximated by a finite order system, and the method is the same as the radiation approximation in Section 4.4.1.1. The MATLAB functions *imp2ss* and *balmr* can be applied to deduce a state-space model of the W2EF process in Eq. (4.40), given as

$$\dot{x}_e(t) = A_e x_e(t) + B_e \eta_p(t), \tag{4.43}$$
$$F_e(t) = C_e x_e(t), \tag{4.44}$$

where $x_e(t) \in \mathbb{R}^{n_e}$ is the state vector for the identified system. $A_e \in \mathbb{R}^{n_e \times n_e}$, $B_e \in \mathbb{R}^{n_e \times 1}$ and $C_e \in \mathbb{R}^{1 \times n_e}$ are the system matrices.

The causalisation time t_c and system order number n_e are selected by trial and error via evaluating the truncation error and the goodness of fit defined in [315]. In this study, $t_c = 1$ s and $n_e = 6$ are used, and the identified excitation IRF is compared with its counterpart in Figure 4.10(b), with a truncation error less than 0.0104 and a goodness of fit of 0.9993. In addition, the W2EF model in Eqs. (4.43)–(4.44) is validated by tank testing, detailed in Section 4.4.2.3.

4.4.2.2 Wave-to-motion modelling

By integrating the W2EF model in Eqs. (4.43)–(4.44) into the nonlinear F2M model in Eqs. (4.32)–(4.33), a nonlinear W2M model is derived, written as

$$x_{w2m} = [x_{f2m} \quad x_e]^T, \tag{4.45}$$
$$\dot{x}_{w2m} = A_{w2m} x_{w2m} + B_{w2m} \eta_p + D_{w2m}(F_{ln} + F_{pto}), \tag{4.46}$$
$$y_{w2m} = C_{w2m} x_{w2m}, \tag{4.47}$$

with

$$A_{w2m} = \begin{bmatrix} A_{f2m} & B_{f2m} C_e \\ 0 & A_e \end{bmatrix}, \tag{4.48}$$

$$B_{w2m} = \begin{bmatrix} 0 & B_e \end{bmatrix}^T, \tag{4.49}$$

$$D_{w2m} = \begin{bmatrix} B_{f2m} & 0 \end{bmatrix}^T, \tag{4.50}$$

$$C_{w2m} = \begin{bmatrix} C_{f2m} & 0 \end{bmatrix}. \tag{4.51}$$

where $x_{w2m} \in \mathbb{R}^{(n_e+n_r+2)\times 1}$ is the state vector of the W2M system. $A_{w2m} \in \mathbb{R}^{(n_e+n_r+2)\times(n_e+n_r+2)}$, $B_{w2m} \in \mathbb{R}^{(n_e+n_r+2)\times 1}$, $D_{w2m} \in \mathbb{R}^{(n_e+n_r+2)\times 1}$, and $C_{w2m} \in \mathbb{R}^{1\times(n_e+n_r+2)}$ are the system matrices.

For some application scenarios, the nonlinear friction and viscous effects can be neglected. For example, the WaveStar device utilises a rounded base to minimise the viscous effect, and its mechanical friction is relatively small with respect to the hydrodynamic forces. Under these conditions, the nonlinear W2M model in Eqs. (4.46)–(4.47) degenerates to a linear W2M model, expressed as

$$\dot{x}_{w2m} = A_{w2m}x_{w2m} + B_{w2m}\eta_p + D_{w2m}F_{pto}, \tag{4.52}$$

$$y_{w2m} = C_{w2m}x_{w2m}. \tag{4.53}$$

These linear and nonlinear W2M models can be directly used for WEC modelling, control design, shape optimisation and performance assessment, especially when wave elevation is measured or known.

4.4.2.3 Results and discussion

The setup for tank testing is the same as the radiation problem, which has been detailed in Section 4.4.1.4. To validate the W2EF modelling methods in Section 4.4.2.1, nine excitation tests were conducted by fixing the buoy to the tank gantry. Even though the buoy was excited by the incident waves, it could not oscillate and radiate waves outwards. At the bottom of the buoy, five pressure sensors were mounted to measure the dynamic pressure acting on the hull. Thus, the measured wave excitation force in heave can be approximated by

$$F_{e,m} = \iint p\,dS = \pi r^2 \bar{p}, \tag{4.54}$$

where $\bar{p}(t)$ represents the average measured pressure of the five pressure sensors.

The wave conditions were configured as wave height $H = 0.08$ m (4 m for the full scale case) and wave frequencies $\omega = 2.512 : 0.628 : 7.536$ rad/s (0.0355 : 0.0888 : 1.0658 rad/s for the full scale case). For harmonic waves, precise wave prediction with $t_c = 1$ s is easy to achieve. Therefore, the W2EF modelling approach always provides accurate approximations of the excitation force, both in terms of the amplitude response and the phase response, as shown in Figure 4.12(a)–(b), respectively.

For the harmonic wave with frequency $\omega = 4.396$ rad/s (0.6217 rad/s for the full scale case), the typical time series of the measured and modelled excitation forces are compared and shown in Figure 4.12(c). The modelled excitation force, according to the W2EF model in Eqs. (4.43)–(4.44), shows a high agreement with the experimental data, which validates the W2EF modelling method under regular wave conditions. In addition, irregular waves, described by the Bretschneider spectrum, and defined in Eq. (2.36), were also applied as test wave conditions to validate the W2EF model, and one typical example is shown in Figure 4.12(d). In Figure 4.12(d), the Bretschneider spectrum is characterised by a peak frequency of $\omega_p = 3.768$ rad/s (0.5329 rad/s for the full scale case) and a significant wave height of $H_s = 0.11$ m (5.5 m for the full scale case). The modelled excitation force via the W2EF modelling approach

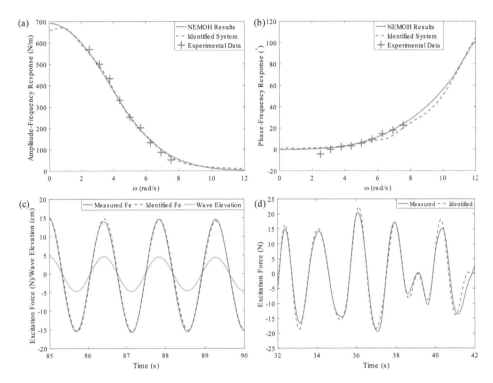

Figure 4.12: Comparison of the excitation force FRFs between BEM, numerical and experimental results in terms of the amplitude response in (a) and the phase response in (b). Typical time traces of measured and identified excitation forces are compared under regular and irregular waves and shown in (c) and (d), respectively.

shows a good accordance with the experimental data most of the time. It only differs slightly from the measured excitation force when the wave elevation is small. Thus, the W2EF modelling method demonstrates its capability and validity in estimating the excitation force from regular or irregular incident waves.

Releasing the buoy to move freely, the free-motion test defined in Eq. (4.21) was also conducted to verify the linear W2M model in Eqs. (4.52)–(4.53), and the nonlinear W2M models in Eqs. (4.46)–(4.47). During free-motion tests, the PTO force was configured as zero. As the buoy oscillated freely, Eq. (4.54) cannot be used to represent the measured excitation force for this set of tests, as the radiated pressure cannot be decoupled from the incident pressure. A series of free-motion tests were conducted for validating the linear and nonlinear W2M models. The numerical results of the linear and nonlinear W2M models were compared with the experimental data in Figure 4.13.

In Fig. 4.13(a), the incident wave conditions were configured with a wave height of $H = 0.08$ m (4 m for the full scale case) and wave frequencies ranging across in $\omega \in [2.1991, 7.2257]$ rad/s ([0.3110, 1.0219] rad/s for the full scale case). For the free-motion tests with a low wave frequency (2.5133–4.3782 rad/s) or a high wave frequency (above 5.6549 rad/s), both linear and nonlinear W2M models fitted the

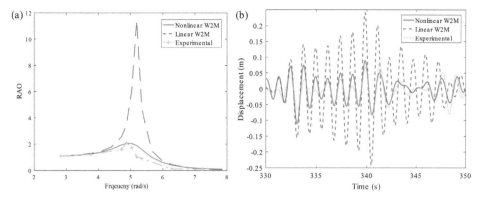

Figure 4.13: Comparison of the RAOs between the linear and nonlinear W2M models, and the free-motion tests in (a), and comparison of the buoy displacements between the linear and nonlinear W2M models, and a free-motion test in irregular waves in (b).

experimental RAO well. However, within the resonance frequency region (4.3782–5.6549 rad/s), only the nonlinear W2M model worked. At resonance frequency, the linear M2M model gave an unrealistic RAO peak value of 11.77, where the RAOs for the nonlinear W2M model and the free-motion tests were 2.13 and 2.10, respectively.

In Figure 4.13(b), the irregular wave, characterised by the Bretschneider spectrum wave with a peak frequency of $\omega_p = 3.7699$ rad/s (0.5331 rad/s for the full scale case) and a significant height of $H_s = 0.11$ m (5.5 m for the full scale case), was used to excite the buoy to oscillate freely. The displacements of the linear and nonlinear W2M models are compared with their experimental counterpart. The linear W2M model significantly overestimated the buoy displacement, whereas the nonlinear W2M model gave a relatively close approximation of the displacement response with respect to the experimental data.

By comparing the numerical results of the linear and nonlinear W2M models with the experimental data in Figure 4.13, it can be concluded that the nonlinear W2M modelling approach, considering radiation approximation, excitation approximation, nonlinear friction, and viscous forces, is capable of depicting the WSI process of the 1/50 scaled buoy under a wide range of wave conditions. This comparison further addresses the importance of nonlinear friction and viscous force in the modelling of the 1/50 buoy, especially when the body motion is exaggerated by resonance.

4.4.3 Wave-to-wire modelling

W2W modelling includes both the wave-structure and the structure-PTO interactions, and can depict all the power conversion trains from wave power to mechanical power, and then to electrical power. Thus, W2W models are capable of including more realistic nonlinear W2M and non-ideal PTO models. In the literature, a number of W2W models for various WEC devices are summarised in [622, 815]. Some W2W models are of high complexity and can be simplified according to the modelling purpose [624].

Section 4.4.3.1 aims to derive a mathematical model for a permanent magnate linear generator (PMLG) as a direct-drive PTO unit for the 1/50 scaled down PA rig. Then, the derived PMLG model is integrated into the linear and nonlinear W2M models to form nonlinear W2W models, detailed in Section 4.4.3.2. Based on the derived W2W models, a tracking control approach is demonstrated in Section 4.4.3.3, with some results and discussion in Section 4.4.3.4.

4.4.3.1 Modelling of a permanent magnet linear generator

A variety of PTO systems, including hydraulic, pneumatic, and direct-drive mechanisms, are discussed and compared in Section 3.2.4. Among these PTO mechanisms, the direct-drive PMLG is selected here as the PTO unit to convert the buoy motion into electricity, as permanent magnets can provide a low cost solution to achieve high force/power density. However, WEC motion is typically slow, leading to a large generator size. Thus, it is still challenging to reduce the PMLG size.

The PMLG is a special topological structure of a conventional permanent magnet synchronous machine (PMSM). As shown in Figure 4.14, a PMSM can be modified into a PMLG within two steps. The first step is to cut the PMSM through its radial direction and lie it on a horizontal plane. In this case, the PMSM is transformed to a flat linear generator. The second step is to rotate the flat linear generator around the longitudinal direction of the iron core. What emerges from this is a tubular PMLG. From this viewpoint, most design and optimisation techniques developed for PMSMs are portable in terms of the PMLG design.

As described in Figure 4.14, the PMLG is a special topological structure of conventional PMSMs. Hence, the PMLG mathematical model is the same as conventional PMSMs, given in the *dq-axis*, as

$$U_d = -R_s I_d - L_d \dot{I}_d + \omega_e I_q L_q, \tag{4.55}$$

$$U_q = -R_s I_q - L_q \dot{I}_q - \omega_e (I_d L_d - \phi_{pm}), \tag{4.56}$$

where U_d, U_q, and I_d, I_q are the voltages and currents for the *d-axis* and *q-axis*, respectively. ϕ_{pm} is the flux linkage constant and ω_e is the electrical angular velocity. R_s, L_d, and L_q are the synchronous resistance, and the armature inductance in the *d-axis* and *q-axis*, respectively. The model described by Eqs. (4.55)–(4.56) is for a generator rather than a motor. The difference lies in the definition of the directions of the *dq-axis* currents.

The PMLG's translator is directly and rigidly coupled with the heaving buoy, while the stator is mounted to an absolute reference point. Thus, the vertical buoy motion can directly drive the PMLG to produce electricity. In turn, the PMLG feeds a PTO force back to the mechanical system. Therefore, the electrical angular velocity and PTO force [135, 169] can be expressed as

$$\omega_e = \frac{\pi}{\tau_p} \dot{z}, \tag{4.57}$$

$$F_{pto} = \frac{3\pi}{2\tau_p} [I_q (I_d L_d - \phi_{pm}) - I_d I_q L_q]. \tag{4.58}$$

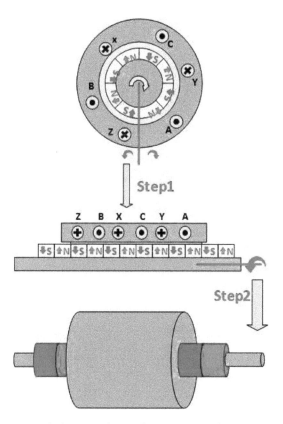

Figure 4.14: Comparison of the topological structures between a permanent magnet synchronous machine and a permanent magnet linear generator.

In general, the *d-axis* current of a PMLG is controlled to achieve $I_{d,ref} = 0$ A, to maximise the force-to-current ratio. Thus, the PTO force can be rewritten as

$$F_{pto} = -\frac{3\pi}{2\tau_p}\phi_{pm}I_q. \tag{4.59}$$

The instantaneous power captured by the generator can be written as

$$P_g = -F_{pto}\dot{z} = -\frac{3\pi}{2\tau_p}\phi_{pm}I_q\dot{z}. \tag{4.60}$$

It is worth noting that the copper loss by R_s is not excluded in Eq. (4.60). Thus, to exclude the copper loss, the useful power generated by the PMLG can be defined as

$$P_{g,u} = U_dI_d + U_qI_q. \tag{4.61}$$

For convenience, the following constants, i.e., the back-electromotive-force (back-EMF) constant K_e, the thrust force coefficient K_f and the permanent magnet 'wave number' K_{pm}, are defined, respectively, as

$$K_e = \frac{\pi}{\tau_p}\phi_{pm}, \tag{4.62}$$

$$K_f = \frac{3\pi}{2\tau_p}\phi_{pm}, \tag{4.63}$$

$$K_{pm} = \frac{\pi}{\tau_p}. \tag{4.64}$$

To represent the PMLG model by ordinary difference equations, Eqs. (4.55)–(4.56) can be rewritten as

$$\dot{I}_d = -\frac{R_s}{L_d}I_d - \frac{1}{L_d}U_d + \frac{\pi L_q}{\tau L_d}\dot{z}I_q, \tag{4.65}$$

$$\dot{I}_q = +\frac{\pi\phi_{pm}}{\tau L_q}v - \frac{R_s}{L_q}I_q - \frac{1}{L_q}U_q - \frac{\pi L_d}{\tau L_q}\dot{z}I_d. \tag{4.66}$$

Substituting Eqs. (4.62) and (4.64) into Eqs. (4.65)–(4.66) gives

$$\dot{I}_d = -\frac{R_s}{L_d}I_d - \frac{1}{L_d}U_d + \frac{L_q}{L_d}K_{pm}\dot{z}I_q, \tag{4.67}$$

$$\dot{I}_q = +\frac{1}{L_q}K_e v - \frac{R_s}{L_q}I_q - \frac{1}{L_q}U_q - \frac{L_d}{L_q}K_{pm}\dot{z}I_d. \tag{4.68}$$

For the 1/50 scaled down PA in Figure 4.5, a specific PMLG, with a power rating of 9 W at a rated speed of 0.24 m/s, was designed and optimised in MAXWELL [310]. The PMLG parameters associated with the model in Eqs (4.67)–(4.68) are given in Table 4.3.

Table 4.3: The PMLG electromagnetic parameters.

Contents	Units	Values
Force Constant (K_f)	N $\cdot$A	54.4019
Back EMF Constant (K_e)	V $\cdot$ s $\cdot$ m $^{-1}$	36.2679
Pole Pitch (τ)	mm	18
Flux Linkage Constant (ϕ_{pm})	Wb $\cdot$ m	0.2078
d-axis Inductance (L_d)	mH	32.78
q-axis Inductance (L_q)	mH	35.47
Synchronous Resistance (R_s)	Ω	1.04

4.4.3.2 Wave-to-wire modelling

To couple the PMLG model in Eqs. (4.67)–(4.68) with the nonlinear W2M model in Eqs. (4.46)–(4.47), a nonlinear W2W model is derived, written as

$$x_{w2w} = \begin{bmatrix} I_d & I_q & x_{w2m} \end{bmatrix}^T, \tag{4.69}$$

$$\dot{x}_{w2w} = A_{w2w}x_{w2w} + B_{w2w,c}U + B_{w2w,uc}\eta_p + F_{nl}(t, x_{w2w}), \tag{4.70}$$

$$y_{w2w} = C_{w2w}x_{w2w}, \tag{4.71}$$

with

$$u = \begin{bmatrix} U_d & U_q \end{bmatrix}^T, \tag{4.72}$$

$$A_{w2w} = \begin{bmatrix} -\frac{R_s}{L_d} & 0 & 0 \\ 0 & -\frac{R_s}{L_q} & A_{v,w2w} \\ 0 & K_f D_{w2m} & A_{w2m} \end{bmatrix}, \tag{4.73}$$

$$B_{w2w,c} = \begin{bmatrix} -\frac{1}{L_d} & 0 \\ 0 & -\frac{1}{L_q} \\ 0 & 0 \end{bmatrix}, \tag{4.74}$$

$$B_{w2w,uc} = \begin{bmatrix} 0 & 0 & B_{w2m} \end{bmatrix}^T, \tag{4.75}$$

$$F_{nl}(t, x_{w2w}) = \begin{bmatrix} \frac{L_q}{L_d} K_{pm} \dot{z} I_q \\ -\frac{L_d}{L_q} K_{pm} \dot{z} I_d \\ D_{w2m}F_{ln} + \frac{3}{2}K_{pm}(L_d - L_q)D_{w2m}I_d I_q \end{bmatrix}, \tag{4.76}$$

$$C_{w2w} = \begin{bmatrix} I_4 & 0_{4\times(n_r+n_e)} \end{bmatrix}, \tag{4.77}$$

$$A_{v,w2w} = \begin{bmatrix} 0 & \frac{K_e}{L_q} & 0_{4\times(n_r+n_e)} \end{bmatrix}. \tag{4.78}$$

It is worth noting that the nonlinearities in Eq. (4.76) have two origins, including the WSI nonlinearity represented by the lumped nonlinear force F_{ln}, and the structure-PTO interaction nonlinearity represented by the quadratic terms $\dot{z}I_q$, $\dot{z}I_d$, and $I_d I_q$. Thus, neglecting the lumped nonlinear force, i.e., $F_{ln} = 0$, can result in a linear F2M or W2M model, but it cannot lead to a linear W2W model. This nonlinear W2W model looks very complex; however, its concept is simple. It is a combination of the nonlinear F2M model in Eqs. (4.32)–(4.33) derived in Section 4.4.1, the W2EF model in Eqs. (4.43)–(4.44) derived in Section 4.4.2, and the PMLG model in Eqs (4.67)–(4.68) derived in this section. Thus, this nonlinear W2W model is capable of investigating the influence of hydrodynamic, mechanic, and electric nonlinearities in WEC dynamics, control design, and performance assessment.

4.4.3.3 Tracking control based on a wave-to-wire model

This section exemplifies a reference tracking control strategy based on the derived nonlinear W2W model, with the control structure given in Fig. 4.15. In this figure, a three-level hierarchical control strategy is used for power maximising of the 1/50 scaled down PA rig in Figure 4.5.

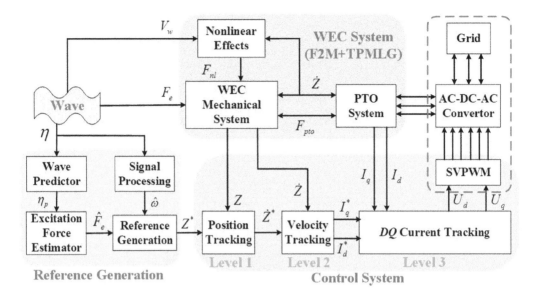

Figure 4.15: A hierarchical control structure based on the wave-to-wire model.

In Figure 4.15, there are three main parts, i.e., WEC modelling, reference generation, and control. For the WEC modelling part, the nonlinear W2W model in Eqs. (4.70)–(4.71) can be used directly. For the reference generation, the F2M models, rather than the W2W model, should be used, as the optimal velocity depends on the excitation force and system's damping coefficient, as shown in Eq. (3.58). The PTO unit works as an actuator to track the optimal velocity for power maximising. However, the displacement tracking is set as the first tracking level in Figure 4.15, in order to constrain the buoy's displacement.

For the control part in Figure 4.15, a three-level hierarchical control structure is applied. This displacement reference feeds into the first level controller, which deals with the displacement tracking and hence generates the velocity reference for the second level controller. The second level controller forces the buoy velocity to track its reference for power maximisation and generates the *dq-axis* current references for the third level controller. The third level controller only deals with the voltage regulation in the *dq-axis*. One potential implementation of U_d and U_q regulation can be achieved by controlling the AC/DC/AC converters/inverters. However, this part of the work is beyond the scope of this chapter, and readers are referred to [193, 436, 81].

Assuming a properly designed control system can ensure resonance, the optimal velocity is proportional to the excitation force, as shown in Eq. (3.49). Therefore, a real parameter is defined as the *force-to-velocity ratio* (FVR) [311], given as

$$R_{fv} = \frac{F_e}{\dot{z}}. \tag{4.79}$$

For the linear F2M model, the radiation damping is the only dissipating factor. According to Eq. (3.49), the optimal FVR should be

$$R_{fv} = 2B_{rad}. \tag{4.80}$$

In extreme waves, the buoy displacement may exceed its physical constraints. By limiting the buoy displacement within $\pm L_z$, this FVR should satisfy

$$R_{fv} \geq \frac{|F_e|}{|\dot{z}|} \geq \frac{|F_e|}{\omega L_z}. \tag{4.81}$$

Summarising, the FVR for linear F2M can be expressed as

$$R_{fv,l} = max\left(\frac{|F_e|}{\omega L_z}, 2B_{rad}\right). \tag{4.82}$$

Thus, the displacement reference can be written as

$$z_{ref,l} = \frac{F_e}{\omega R_{fv,l}} e^{-j\frac{\pi}{2}}. \tag{4.83}$$

For the nonlinear F2M model, radiation, friction and viscosity phenomena dissipate energy and thus the FVR is modified as

$$R_{fv,nl} = max\left(\frac{|F_e|}{\omega L_z}, 2B_{nl}\right), \tag{4.84}$$

where $B_{nl} = B_{rad} + B_f + B_v$ represents the total linearised damping of the radiation, friction (B_f), and viscosity (B_v) factors. Thus, the displacement reference can be rewritten as

$$z_{ref,nl} = \frac{F_e}{\omega R_{fv,nl}} e^{-j\frac{\pi}{2}}. \tag{4.85}$$

As illustrated in Figure 4.15, signal processing is required to estimate the incoming wave excitation characteristic parameters. These estimates are used to: (i) update the radiation damping coefficient, (ii) compute the optimal FVR, and (iii) generate reference signals. Instantaneous frequency and amplitude estimates can be obtained by using Teager's energy operation (TEO) [494]. Compared with estimation via either the fast Fourier transform or the Hilbert transform approaches, the TEO method offers a good accuracy with rapid computation.

For the current tracking loop, the reference of the *d-axis* current is set to $I_{d,ref} = 0$ A, (i) to maximise the force-to-current ratio, and (ii) to compensate the nonlinear terms $I_q I_d$ in Eqs. (4.58) and (4.76). As illustrated in Eqs. (4.65)–(4.68), the nonlinear terms $\frac{L_q}{L_d} K_{pm} \dot{z} I_q$ and $\frac{L_d}{L_q} K_{pm} \dot{z} I_d$ can significantly influence the W2W modelling. These nonlinear terms can be compensated via *dq-axis* decoupling technology, as shown in Figure 4.16, as has been widely used in wind turbine control [617, 694].

The *dq-axis* PMLG model in Eqs. (4.67)–(4.68) can be rewritten as their Laplace transform, as

$$I_d = \frac{-U_d + \omega_e I_q L_q}{R_s + L_d s}, \tag{4.86}$$

$$I_q = \frac{-U_q - \omega_e I_d L_d + \omega_e \phi_{pm}}{R_s + L_d s}. \tag{4.87}$$

Thus, the I_d and I_q decoupling can be achieved as illustrated in Figure 4.16. Such a *dq-axis* coupling method requires real-time measurements of the ω_e, I_d, and I_q. In addition, ω_e can also be derived from the buoy's velocity, as $\omega_e = K_{pm} \dot{z}$ holds for rigid structure-PTO connections.

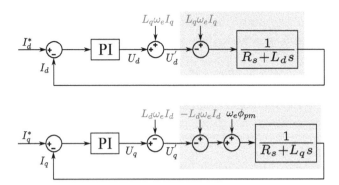

Figure 4.16: Decoupling of dq-$axis$ currents.

4.4.3.4 *Results and discussion*

Based on the derived nonlinear W2W model in Section 4.4.3.2 and the proposed tracking control structure in Section 4.4.3.3, this section aims to demonstrate the utility of the W2W model for control development by numerical simulation. Figure 4.17 shows the performance of W2W-based tracking control under irregular wave conditions.

For an irregular wave characterised by the Bretschneider spectrum of $H_s = 0.06$ m (3 m for the full scale case) and $\omega_p = 5.0265$ rad/s (0.7109 rad/s for the full scale case), the simulation results of the 1/50 PA displacement and accumulated energy conversion are given in Figure 4.17(a). It is clear that the linear model (neglecting the lumped nonlinear force $F_{ln} = 0$) overestimates the buoy motion as well as the converted energy under this wave condition. The wave is small and, hence, the wave excitation force is also small. In this case, the nonlinear friction impedes the buoy motion and thus the overall efficiency is low for the nonlinear model. This shows a good correspondence with the power dissipation results of the -0.03 m free-decay test in Figure 4.9(b).

For the irregular wave $H_s = 0.25$ m (7.5 m for the full scale case) and $\omega_p = 2.5133$ rad/s (0.3554 rad/s for the full scale case), the simulation results of the PA displacement and captured energy are given in Figure 4.17(b). This wave condition is more severe than for the design modal spectrum and, hence, the overall efficiencies cannot reach very high values. Under this wave condition, the nonlinear viscous force dissipates more energy than the friction force. From the viewpoint of PA displacement, end protection is required for the linear model. However, the simulated displacement response of the nonlinear model seldom reaches its displacement constraints. For regular waves, similar conclusions have been drawn in [314].

It is worth noting that the legends *Linear* and *NLM* in Figure 4.17 indicate $F_{ln} = 0$ and $F_{ln} \neq 0$, respectively, for reference generation and W2W modelling. The W2W model is always nonlinear, as the structure-PTO interaction has the quadratic terms $I_d I_q$, $\dot{z} I_d$, and $\dot{z} I_q$. The dq-$axis$ coupling in Figure 4.16 can only be used for control design, rather than W2W modelling.

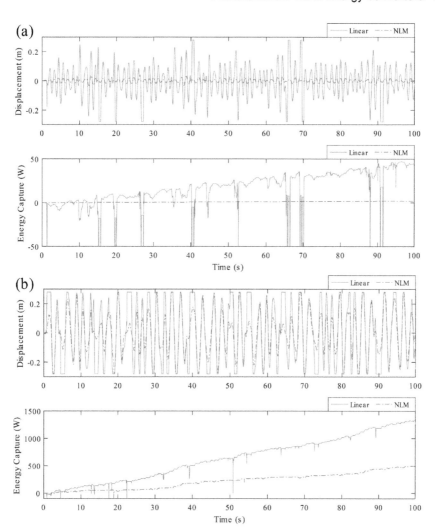

Figure 4.17: W2W tracking control performance based on linear/nonlinear W2W models under irregular wave conditions of $H_s = 0.06$ m and $\omega_p = 5.0265$ rad/s in (a), and $H_s = 0.25$ m, and $\omega_p = 2.5133$ rad/s in (b). The legends *Linear* and *NLM* indicate $F_{ln} = 0$ and $F_{ln} \neq 0$, respectively, for reference generation and W2W modelling.

The simulation conditions for Figure 4.17 are given below: (i) For the linear F2M model, the damping coefficient is optimised at the natural frequency 5.1522 rad/s. Thus, $B_{rad} = 4.55$ Ns/m is selected, according to Figure 4.13(a). (ii) For the nonlinear F2M model, $B_{nl} = 21.32$ Ns/m is optimised at the frequency 4.9009 rad/s according to Figure 4.13(a). (iii) Three proportional integral (PI) controllers are applied here to achieve the position, velocity, and current tracking with parameters $P_z = 128$, $I_z = 20$ for position tracking, $P_v = 220$, $I_v = 42$ for velocity tracking, and $P_i = 300$, $I_i = 75$ for *dq-axis* current tracking. These parameters are selected by trial and error. (iv) The irregular wave data were obtained from tank testing.

4.5 SUMMARY

This chapter aims to give an overview of floating point absorber wave energy converters, with specific foci on the working principles and typical prototypes of one- and two-body PAs. One-body PAs are characterised by low-pass characteristics, while two-body PAs perform as band-pass filters. Most PAs are oscillatory and, hence, end-stop protection is required. Some two-body PAs, e.g., the SEAREV and Penguin devices introduced in Section 4.3.2, utilise rotational pitch or/and roll modes for energy harvesting, and the need for end-stop protection is avoided. For both one-body and two-body PAs, their energy capture efficiencies have a narrow bandwidth and hence power maximisation control is essential to achieve a high energy conversion efficiency under varying sea states.

A cylindrical one-body PA is selected as a typical case study in this chapter to demonstrate some basic and generic time-domain modelling approaches. The approach to approximate the radiation convolution term is detailed in Sections 4.4.1.1. Then, linear and nonlinear F2M models are derived, and validated by tank testing, as detailed in Sections 4.4.1. Both numerical and experimental results emphasise the importance of considering some critical nonlinear factors in WEC modelling. However, the *critical* nonlinear factors should be considered on a case-by-case basis [620]. In this case study, only the nonlinear friction and viscous forces are considered. As the cylindrical buoy has a sharp edge at the bottom, viscous effect is severe and can be minimised by shape optimisation, e.g., using a rounded base [379]. In addition, the scale number of the rig is as small as 1/50. For such a small scale device, the mechanical friction is relatively large, with respect to the hydrodynamic forces. In general, other nonlinearities, e.g., the nonlinear hydrostatic force [539], nonlinear radiation force [539], nonlinear Froude-Krylov force [287], nonlinear end-stop force [316, 318, 317] etc., may have significant influence on WEC dynamics and performance, mainly depending on WECs' geometries, control algorithms, and wave conditions.

The approach to approximate the excitation force from wave elevation is exemplified in Section 4.4.2.1. Integrating the W2EF model into the derived linear and nonlinear F2M models results in linear and nonlinear W2M models, which are also validated by tank testing in Section 4.4.2. Both the numerical and experimental results of the free-motion tests, once again, emphasise the importance of the nonlinear friction and viscous forces. The W2EF modelling approach can be directly and straightforwardly applied for WEC modelling and control design. However, this approach requires wave measurement, which may be not applicable for some offshore applications. In this case, one promising approach is to estimate the excitation force by designing unknown input observers [6, 315, 618, 352].

A PMLG model is also derived in this chapter, and integrated into the derived W2M models to form W2W models. As the PMLG is rigidly and directly connected with the buoy, the structure-PTO interaction is always nonlinear, as the nonlinear terms of $\dot{z}I_q$, $\dot{z}I_d$, and I_dI_q never vanish in the W2W models. For W2W-based control design, these nonlinearities can be mitigated by $I_d = 0$ control and $dq - axis$ decoupling. However, the numerical simulation results still suggest inclusion of the nonlinear friction and viscous nonlinearities in W2W modelling and reference generation. In this chapter, the most classic PI control strategy is applied for position, velocity, and current tracking, to demonstrate the utility of the derived W2W model. However, more advanced control strategies should be applied for WEC power maximising.

CHAPTER 5

Multi-mode wave energy converters

Nataliia Y. Sergiienko, Boyin Ding

University of Adelaide, nataliia.sergiienko@adelaide.edu.au

5.1 INTRODUCTION

This chapter is dedicated to multi-mode wave energy converters, a sub-class of oscillating-type WECs. The term *mode*, in the context of this chapter, refers to the hydrodynamic mode involved in the power absorption process, i.e., surge, heave, and pitch. It is necessary to distinguish clearly between multi-mode converters and converters with multiple degrees of freedom, as these terms can be used interchangeably in the literature. Some wave energy converters, consisting of several oscillating bodies, have more than one degree of freedom, or rigid body mode, but the wave power is predominantly absorbed from only one hydrodynamic mode, for example, heave (refer to Chapters 4 and 6).

The concept of multi-mode WECs came about when it was proved that the maximum amount of absorbed wave power is not solely related to the geometry of the converter, but mainly to its oscillation mode. Theoretically, it is possible to triple the power absorption potential of a heaving axisymmetric body if other modes of oscillation (surge or pitch) are involved in power generation. The number of multi-mode prototypes is relatively low compared with single-mode WEC types, and the most interesting concepts amongst them are demonstrated in this chapter.

5.2 WORKING PRINCIPLE

The power absorption mechanism and power absorption limits of multi-mode WECs are explained in this section.

DOI: 10.1201/9781003198956-5

5.2.1   Power absorption by a body oscillating in several modes of motion

Any kind of body placed in water, whether it is a ship, offshore platform, or wave energy converter, interacts with incident waves: some parts of the wave field are reflected from the structure, while other parts continue propagating forward. If the body is not stationary and experiences an oscillatory motion, it generates (radiates) waves that spread out from the body along the water surface. It is interesting to note that the radiated wave does not depend on the size of the oscillating body, but is a function of the oscillation mode and amplitude of the oscillations that determine how large a generated wave is [227]. Destructive interference between incident and radiated waves leads to the wave energy absorption phenomenon: in order to absorb a wave (power), a "counter-wave" should be generated by the wave energy device to interfere with and cancel an incident wave [227, 329]. This principle is clearly demonstrated in Figure 5.1, which shows 100% power absorption by an infinite array (perpendicular to the figure) of evenly-spaced, small floating bodies or an elongated body that can oscillate both vertically and horizontally. As a result, almost all the energy contained in a wave can be absorbed by an oscillating terminator-type device.

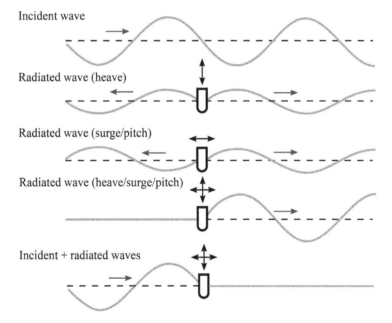

Figure 5.1: Principle of wave power absorption in a two-dimensional case, adapted from [227].

As shown in Figure 5.1, the pattern of the radiated wave depends on the degree of freedom of the oscillating system (for the definition of motion modes adopted in the marine industry, refer to Figure 2.5 in Chapter 2). Thus, heave is considered as a symmetric, or source-type, mode, as an axisymmetric body oscillating in heave radiates circular waves (Figure 5.2a), while motion in surge or pitch generates dipole waves (Figure 5.2b) [227]. This difference in the radiated wave patterns is a key factor that defines the power absorption limits of oscillating wave energy converters [227, 329].

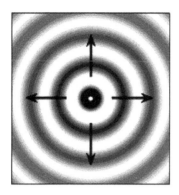

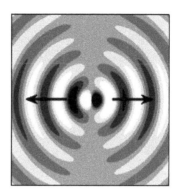

(a) Monopole or circular waves (heave).

(b) Dipole waves (surge or pitch).

Figure 5.2: Types of waves radiated by an axisymmetric body in different modes, adapted from [329].

As stated in Chapter 3, an infinitely long oscillating terminator-type device (or an array of point absorbers) can harvest 100% of the incident wave power, when it oscillates in heave and surge simultaneously [227]. However, for an axisymmetric body, the amount of energy that can be removed from a sinusoidal wave is limited by the radiation properties of the system [98, 208, 571]:

$$P_{\max} = \alpha \frac{J}{k} = \alpha \frac{\lambda}{2\pi} J, \tag{5.1}$$

where $\alpha = 1$ for oscillations in heave and $\alpha = 2$ for surge or pitch, J is the wave energy transport per unit frontage of the incident wave (defined in Eqs. (2.24) and (2.25)), λ is the wavelength, and k is the wavenumber.

As a result, the efficiency of an oscillating WEC is maximised when surge, heave, and pitch modes are involved in power production, reaching the level of $P = (3\lambda/2\pi)J$. In this case, 1/3 of the total power will be absorbed from oscillations in heave, while the remaining 2/3 will be captured from surge and/or pitch. As both surge and pitch are dipole-type modes that radiate waves of the same pattern, only one of the two can be considered for power absorption purposes, as it is not possible to improve the efficiency of the system, including the second dipole mode [227].

5.2.2 Practical limits to the power absorption per mode

5.2.2.1 Heaving mode

The limit presented in Eq. (5.1) shows how much energy can be removed from the ocean wave, regardless of the WEC size or motion amplitude. However, any converter is designed to operate within a constrained working envelope that imposes an additional upper bound on the amount of absorbed power [98, 229]. As a result, the power absorption of a <u>floating</u> *heaving* buoy has two boundaries:

(i) a high-frequency limit P_A defined by the body's ability to radiate waves (from Eq. (3.67) assuming deep water conditions $\omega^2 = kg$) [98, 571, 227]:

$$P_A = \frac{J}{k} = \frac{\rho g^2 A^2}{4\omega k} = \frac{\rho g^3 \left(\frac{H}{2}\right)^2}{4\omega^3} = c_\infty T^3 H^2, \qquad (5.2)$$

where $c_\infty = \rho(g/\pi)^3/128$, $H = 2A$ is the wave height, and $T = 2\pi/\omega$ is the wave period;

(ii) a low-frequency limit P_B defined by the maximum swept volume of the body, which applies when the velocity of the converter is smaller than the optimal value due to physical constraints [99, 229]:

$$P_{B,f} = \frac{1}{2}|\hat{F}_{e,3}\hat{u}_3| = \frac{\rho g \omega V A}{4} = \frac{c_0 V H}{T}, \qquad (5.3)$$

where $c_0 = (\pi/4)\rho g$ and the subscript f corresponds to the floating case.

These boundaries have been derived for floating bodies that move in heave only regardless of shape. In general, the P_A-limit depends only on the mode of motion and has the same expression for submerged and floating bodies. In terms of the P_B curve, the power absorption limit of a fully submerged converter is strongly dependent on shape and should be derived for each case under consideration independently. Thus, for a spherical body with its centre placed d_s below the water surface, the P_B-limit can be expressed as [700]:

$$P_{B,s} = 4\pi^3 \rho e^{-kd_s} s_{3,\max} \frac{V H}{T^3}, \qquad (5.4)$$

where the subscript s corresponds to the submerged case and $s_{3,\max}$ is the maximum displacement of the sphere in heave.

A comparison between power limits for floating and submerged spherical heaving bodies with different submergence depths (see Figure 5.3) is demonstrated in Figure 5.4. All the spheres have the same physical volume of 524 m^3 (radius is $a = 5$ m) and the motion amplitudes are constrained by $s_{3,\max} = 0.67a = 3.3$ m. Regular waves of height $H = 2$ m are considered. The most important difference between the power absorption of floating and submerged heaving systems is that the latter has a faster decay rate at a low frequency or long wave period range. Comparing Eqs. (5.3) and (5.4), it is obvious that $P_{B,f} = \mathcal{O}(T^{-1})$, while $P_{B,s} = \mathcal{O}(T^{-3})$, which leads to a decrease in power absorption at longer wavelengths. Moreover, due to the fact that the hydrodynamic pressure on the body surface decays exponentially with depth, the presence of $\exp(-kd_s)$ in Eq. (5.4) shows a reduction in power for deeper submergences. Consequently, a sphere submerged to $2a = 10$ m extracts less power than that submerged to $1.2a = 6$ m.

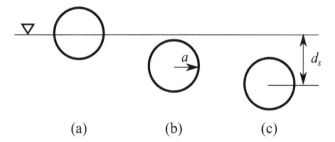

Figure 5.3: Schematic representation of the floating and submerged spheres of radius $a = 5$ m: (a) $d_s = 0$, (b) $d_s = 1.2a = 6$ m and (c) $d_s = 2a = 10$ m.

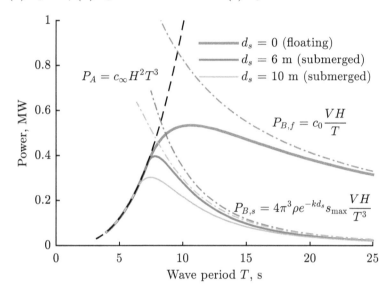

Figure 5.4: Power absorbed by the floating and submerged spheres in regular waves vs. wave period. Sphere radius is $a = 5$ m, displacement in heave is constrained to $0.6a$, wave height is $H = 2$ m.

5.2.2.2 Surging and pitching modes

According to Eq. (5.1) the P_A-bound of the surging body is twice as high as that of a converter that moves in heave only [227]. The low frequency limits, P_B, are also different for these motion modes where Eq. (5.4) describes heave oscillation, while an expression for the surging floating sphere takes the form [700]:

$$P_{B,f}^{surge} = 2\pi^3 \rho s_{1,\max} \frac{VH}{T^3}. \tag{5.5}$$

Analysing Eqs. (5.3) and (5.5), it is clear, that the P_B-bound for the surging body is $\mathcal{O}(T^{-3})$, while for the heaving body this bound has a smaller decay rate and is $\mathcal{O}(T^{-1})$. These results are very similar to the comparison of floating and submerged heaving bodies, meaning that a surging floating sphere is a poorer power absorber at long wavelengths (low frequency range) than the same body that oscillates in heave.

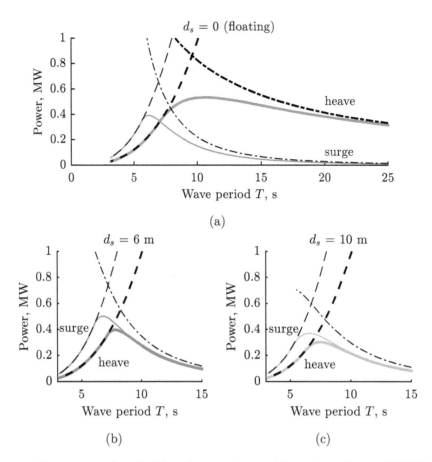

Figure 5.5: The power absorbed by the surging and heaving spherical WECs of 5 m radius with different submergence depth: (a) floating $d_s = 0$ and submerged (b) $d_s = 6$ m, (c) $d_s = 10$ m. Motion amplitudes in heave and surge are constrained by $s_{1,\max} = s_{3,\max} = 0.67a = 3.3$ m. Wave height is set to $H = 2$ m.

Comparing motion modes of the floating sphere, the following features should be outlined:

- $P_A^{surge} = 2P_A^{heave}$;

- $P_{B,f}^{surge} = \mathcal{O}(T^{-3})$, while $P_{B,f}^{heave} = \mathcal{O}(T^{-1})$.

Unlike floating converters, fully submerged buoys have almost the same power efficiency from oscillations in heave or surge. Thus, the P_B-bounds for the surging and heaving submerged spheres have the same expressions described by Eqs. (5.4) and (5.5), as shown in [328, 700]. Hence, for the fully submerged, spherical WEC:

- $P_A^{surge} = 2P_A^{heave}$,

- $P_{B,s}^{surge} = P_{B,s}^{heave} = \mathcal{O}(T^{-3})$.

The difference in power efficiency between surging and heaving spheres is demonstrated in Figure 5.5. The sphere radius is $a = 5$ m: the motion in surge (mode 1) and heave (mode 3) is constrained by $s_{1,max} = s_{3,max} = 0.67a = 3.3$ m; the submergence depths are $d_s = 0$ m, $1.2a = 6$ m and $2a = 10$ m, and the wave height is taken as $H = 2$ m. It can be seen that at longer wave periods the heave motion is dominant for floating converters, showing that the power contribution from the surge mode may be marginal for floating systems. In contrast, a submerged sphere that oscillates in surge is more efficient across the entire frequency range. Therefore, the power efficiency of the submerged system may increase two to three times due to the additional controllable degree of freedom. Also, the ratio between the power levels from surge and heave does not change with the submergence depth, as shown in Figures. 5.5b and 5.5c. It should be noted that the surging floating sphere utilises only half of its volume to couple with the fluid (at nominal depth), while for a submerged sphere the total volume is involved in power absorption. This explains why the power level of a surging floating sphere is lower than that of a fully submerged sphere.

5.3 EXISTING PROTOTYPES

Based on the analysis presented in Section 5.2.2, it could be concluded that employment of several motion modes in power generation could be more advantageous for *fully submerged* converters, while any benefit for their *floating* counterparts would be marginal. This might explain the design solutions of some of the multi-mode wave energy converters presented in this chapter and shown in Figure 5.6.

5.3.1 Bristol cylinder, UK

One of the first developments of multi-mode wave energy converters relates to the Bristol cylinder invented by Evans et al., [212] from the University of Bristol, UK. This is a fully submerged long circular cylinder with its main axis parallel to the incident wave (refer to Figure 5.6a), which uses the principle described in [789, 589] to achieve a 100% power absorption. When the cylinder axis moves in a circle, undergoing heave and surge oscillations with a phase difference of $\pi/2$ radians, radiated waves generated above the cylinder on the water surface propagate away from the oscillating body in only one direction. This combination of surge and heave motion is used to cancel the transmitted wave behind the cylinder, achieving full absorption of the incoming wave. The designed size of the cylinder is 75–100 m in length and 12–15 m in diameter, and the power is captured by means of six hydraulic pistons located along the device, three on each side as shown in Figure 5.6a. Although the efficiency of this WEC can reach 65% in irregular waves, the initial design had a high estimated cost for the produced power due to the high cost of installation and power take-off [658]. As a result, it has not been tested at full scale and has not been commercialised. In order to offer a more affordable solution for the power take-off, an alternative version of the Bristol cylinder was developed in [145], where the power is captured from surge motion alone (similar to OWSCs) and the PTO system is placed inside the cylinder

(a) Bristol cylinder [658].

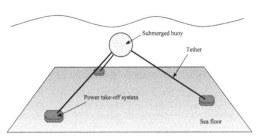

(b) Artistic impression of the Srokosz three-tether sphere.

(c) WaveSub (concept 1) by Marine Power Systems, UK, (image courtesy of Marine Power Systems).

(d) WaveSub (concept 2) by Marine Power Systems, UK, (image courtesy of Marine Power Systems).

(e) The Triton WEC by Oscilla Power™, US, (image courtesy of Oscilla Power).

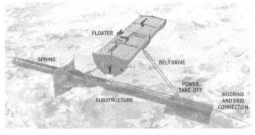

(f) The NEMOS wave energy converter, Germany, (image courtesy of NEMOS GmbH).

(g) The CETO 6 multi-moored system by Carnegie Clean Energy, Australia, (image courtesy of Carnegie Clean Energy).

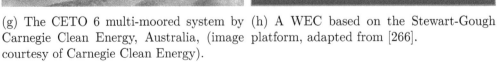

(h) A WEC based on the Stewart-Gough platform, adapted from [266].

Figure 5.6: Examples of wave energy converters that capture power in several modes of motion.

itself. Despite the difficulties associated with practical implementation of the Bristol cylinder, its initial design served as a striking example of a very efficient terminator device (similar to the Edinburgh Duck [686]).

5.3.2   Multi-tether spherical buoy

In contrast with the Bristol cylinder, Srokosz [741] suggested applying the concept of multi-mode motion to the point absorbing device. The wave energy converter consists of a spherical fully submerged buoy which is connected to three inclined tethers enabling power to be captured from all translational degrees of freedom (see Figure 5.6b). It is assumed that each tether is connected to the spring-loaded cable drum attached to the power generator that converts the tether motion into electricity, similar to a direct mechanical drive mechanism. The power output of the system has been analysed for three different arrangements of tethers: (i) one tether (tether angle from the vertical is $0°$), (ii) three tethers (inclined by $45°$), and (iii) a tether angle of $60°$. As a result, the three-tether configuration of the power take-off provides an increase in power absorption of two to three times in regular waves when compared with a single-tether mooring. Even though the analysis is dimensionless and purely theoretical, it demonstrates that this system could be an effective absorber of wave power, as its capture width in regular waves can be greater than the diameter of the sphere, leading to a hydrodynamic efficiency higher than 100% [741].

5.3.3   WaveSub by Marine Power Systems, UK

The concept of a spherical device with a multi-mooring system has been utilised by Marine Power Systems, UK, in collaboration with the University of Bath, UK, [833, 113] to develop the first prototype of their WaveSub WEC. The device consists of two main parts: a submerged inertia platform (reaction barge) and a subsurface spherical buoy (see Figure 5.6c). The buoy is connected to the hydraulic power take-off system located on the platform through the four flexible tethers. The inertia platform is kept in place by means of four taut mooring lines attached to the seabed. The main power absorption principle of the WaveSub is to follow the circular orbit of the water particles while capturing energy from the combination of heave and surge motion modes. The system adapts the concept of varying geometry WECs as the submergence depth of the buoy and of the barge can either be tuned to maximise power production, or detuned to avoid high wave loads under storm conditions. In 2019, the device undertook 1/4 scale testing at the marine test site FaBTest in Cornwall, UK [497].

The next generation of the WaveSub, shown in Figure 5.6d, has a cylindrical buoy instead of the originally-proposed spherical buoy, resembling the Bristol cylinder concept. The WEC developer claims that one such float can capture over 1.5 MW of power, and it is intended to have three operating floats on a platform. Moreover, Marine Power Systems plan to combine their WEC technology with a floating offshore wind turbine to achieve a higher rated power of 15 MW [497].

5.3.4   Triton WEC by Oscilla Power™, US

A similar approach to the two-body system with multiple mooring lines has been adapted to design the Triton WEC by Oscilla Power™, US [555]. In contrast with the WaveSub, the Triton WEC employs an asymmetric buoy floating on the water's surface and a ring-shaped submerged reaction plate with a U-shaped (elliptical ogive) cross-section (see Figure 5.6e). The power is captured from the heave, surge, sway, pitch, and roll motion of the floating body that reacts against a relatively stable submerged platform by means of three flexible tether lines [678]. The power take-off is placed inside the float and designed as a direct mechanical drive system for each tether separately. The Triton WEC took four place (out of 92 initial submissions) in the Wave Energy Prize [790] competition arranged by the US Department of Energy in 2015/2016. Based on the results from the 1/20 scale testing [791], this WEC has an efficiency of 23.5% in irregular sea states, which is slightly higher than the average efficiency of just heaving devices.

5.3.5   The NEMOS wave energy converter by NEMOS GmbH, Germany

Another commercial prototype of the multi-mode system is the NEMOS wave energy converter developed by NEMOS GmbH, Germany [562]. The elongated floating body (8×2 m) is attached to a 16 m long substructure located at the sea bottom by two tethers as shown in Figure 5.6f. Wave-induced motion of the buoy transmits mechanical energy to a generator by a spring-loaded belt drive. One of the tethers can behave as an actuator to control the motion of the buoy and achieve the maximum power output. The orientation of this terminator device can be adjusted according to the direction of the wave propagation. Moreover, this WEC has the ability to submerge the floater in order to protect the system in heavy storms. According to the 1/5 scale testing [615], the hydrodynamic efficiency of NEMOS can reach up to 60% in natural sea conditions.

5.3.6   The CETO system by Carnegie Clean Energy, Australia

The CETO system, shown in Figure 5.6g, utilises an idea originally proposed by Srokosz [741], where a fully submerged buoy is connected to three tethers/mooring lines to absorb power from surge, heave, and pitch modes. The CETO 6 multi-moored unit evolved from a heaving CETO 5 prototype that could only absorb power from the vertical motion in waves. The CETO 5 system had a hydraulic power take-off unit, while, in the CETO 6 unit, each individual mooring line is connected to the direct mechanical drive PTO located inside the buoy hull. The shape of the buoy has been optimised such that the majority of the power is absorbed from the heave mode, minimising the LCOE value [633].

5.3.7   Wave energy converter based on a Stewart-Gough platform, Mexico

Several research groups [467, 801, 267, 268] have adapted the Stewart-Gough platform for wave energy conversion purposes. This parallel mechanism with six actuators allows full spatial control (6-DoF) of the system. The WEC depicted in Figure 5.6h

[801, 267, 266] consists of three rigidly interconnected floating buoys that are attached to the lower submerged platform by means of six cables (legs). Each leg is equipped with an electric generator to control the motion of the floats and generate electricity. The lower submerged platform works as a reaction plate and is connected to the seabed via a system of slack mooring lines. The project shown in Figure 5.6h is under development by the Mexican Centre for Research and Innovation on Marine Energy, and the scaled physical model (1:20) has already been laboratory tested in a wave tank.

5.3.8   Overview of existing multi-mode prototypes

It is worth noting that all designs of the multi-mode WECs presented in this section employ inclined degrees of freedom to provide the required controllability of the converter in both heave and surge modes. The only difference is in the number of tethers utilised to control the buoy and extract the wave power:

(i) two tethers inclined at 45° are usually used to demonstrate a possible mooring configuration of the 2D-system, such the Bristol cylinder, where the actual number of required PTO units is strongly correlated with the length to diameter ratio of the terminator device;

(ii) three is the minimum number of tethers that allows full controllability of the 3D-WEC in all translational degrees of freedom (Srokosz' sphere or NEMOS) or, in planar case, control of heave, surge, and pitch;

(iii) the four-tether solution used in the WaveSub WEC introduces an over-constrained control problem if 3-DoF control is the target, or an under-constrained control problem if 6-DoF control is required;

(iv) the six-tether solution is the only mechanism which provides controllability for all 6 rigid-body DoFs. However, such control comes at significantly increased capital cost and, therefore, is unlikely to be commercialised.

The number of tethers, and consequently the number of power take-off machineries, is directly related to the capital and operational expenditures of the wave energy converter. Therefore, according to the law of diminishing returns, there should be an optimal configuration, after which an additional tether and the costs associated with it are no longer justified by the increased power production of the multi-mode converter.

5.4   CASE STUDY – SUBMERGED THREE-TETHER SYSTEM

A generic three-tether WEC has been chosen as a case study to demonstrate the modelling approach, power absorption potential, and design of control systems that can be applied to any multi-mode wave energy converter. Also, the difference in performance measures between multi-mode and just heaving WEC (Figure 5.7) is shown throughout this section. The geometric parameters of the system are specified in Table 5.1.

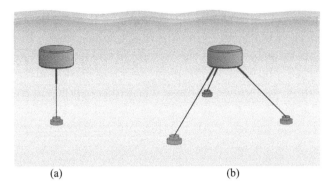

Figure 5.7: Power take-off configurations for the submerged WECs: (a) a generic heaving buoy connected to one tether and (b) a three-tether system.

Table 5.1: Parameters of the WEC.

Parameter	Value
Shape	vertical cylinder
Radius of the cylinder, a	5.5 m
Height of the cylinder, h_c	5.5 m
Water depth	50 m
Submersion (top of the buoy)[*]	3.75 m
Submergence depth, d_s (centre of the buoy)	6.50 m
Volume[*], V	524 m^3
Surface area[*]	380 m^2
Mass of the buoy, m_b	268 t
Displaced mass of fluid[*], m_w	537 t
Stroke length, $\Delta\ell_{max} - \Delta\ell_{min}$	6 ($\pm$3) m
Tether inclination angle from the vertical, α_t	44°
Initial tether length[*], ℓ_0	56.6 m
Pretension force in each tether[*]	1.2MN

[*] not independent parameters.

5.4.1 Dynamic modelling

The wave-to-wire model employed in this study is based on linear wave theory, assuming small motion amplitudes of the buoy, as compared with the length of the mooring lines (tethers). The only second-order hydrodynamic effect included in the model is a viscous drag force which is proportional to the square of the body velocity relative to the fluid.

5.4.1.1 Kinematics

A schematic of a three-tether WEC is shown on Figure 5.8. The spatial arrangement of all tethers is defined by $\mathbf{s}_i$:

$$\mathbf{s}_i = \mathbf{r} + \mathbf{R}\mathbf{n}_i - \mathbf{d}_i, \quad i = 1 \ldots 3, \tag{5.6}$$

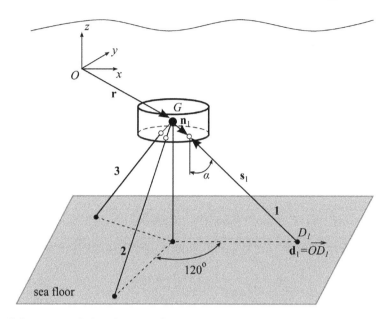

Figure 5.8: Schematic of the three-tether wave energy converter. The wave propagates along the x-axis parallel to the first tether.

where $\mathbf{r}$ is the position vector of a buoy in the reference coordinate frame $Oxyz$, $\mathbf{R}$ is the rotation matrix of the buoy with respect to the reference frame, $\mathbf{n}_i$ denotes the position vector of the anchor point of the i-th tether on the hull relative to the buoy centre of mass G, and $\mathbf{d}_i$ is the position vector of the anchor point of the i-th tether on the sea floor in the $Oxyz$ coordinate frame. The instantaneous tether length is:

$$\ell_i = \|\mathbf{s}_i\| = \sqrt{\mathbf{s}_i^\mathsf{T}\mathbf{s}_i}, \quad i = 1\ldots 3, \tag{5.7}$$

and the change in length of the i-th tether is $\Delta\ell_i = \ell_i - \ell_0$.

Mapping from the buoy velocities in a Cartesian coordinate frame to the rate of change of the tether length is provided by the inverse kinematic Jacobian:

$$\dot{\mathbf{q}} = \mathbf{J}^{-1}\dot{\mathbf{x}}, \tag{5.8}$$

where $\mathbf{x} = [\mathbf{r} \quad \boldsymbol{\theta}]^\mathsf{T}$ is the buoy location (pose) vector with three translational and three rotational motions, $\mathbf{q} = [\Delta\ell_1 \quad \Delta\ell_2 \quad \Delta\ell_3]^\mathsf{T}$ is a vector of three tether length variables and the inverse kinematic Jacobian $\mathbf{J}^{-1}$ can be obtained as:

$$\mathbf{J}^{-1} = \begin{pmatrix} \mathbf{e}_{s1}^\mathsf{T} & (\mathbf{R}\mathbf{n}_1 \times \mathbf{e}_{s1})^\mathsf{T} \\ \mathbf{e}_{s2}^\mathsf{T} & (\mathbf{R}\mathbf{n}_2 \times \mathbf{e}_{s2})^\mathsf{T} \\ \mathbf{e}_{s3}^\mathsf{T} & (\mathbf{R}\mathbf{n}_3 \times \mathbf{e}_{s3})^\mathsf{T} \end{pmatrix}, \tag{5.9}$$

where $\mathbf{e}_{si}$ is a unit-vector along the tether i, so $\mathbf{e}_{si} = \dfrac{\mathbf{s}_i}{\|\mathbf{s}_i\|}$.

5.4.1.2 Forces

The generalised forces that act on the body are expressed in the Cartesian coordinate frame $Oxyz$ as vectors of six elements including horizontal and vertical forces and rotational moments.

Hydrostatic force. As the buoy is fully submerged, the generalised hydrostatic force is $\mathbf{F}_{buoy} = [0 \quad 0 \quad (m_w - m_b)g \quad 0 \quad 0 \quad 0]^\mathsf{T}$, where $m_w = \rho V$ is the mass of the displaced water, ρ is the water density, V is the displaced volume of the buoy, and m_b is the mass of the WEC.

Power take-off forces. The behaviour of the linear PTO system is modelled as [50]:

$$F_{pto,i} = (C_{pto} - B_{pto}\Delta\dot{\ell}_i - K_{pto}\Delta\ell_i), \quad i = 1\ldots 3, \tag{5.10}$$

where K_{pto} and B_{pto} are the PTO stiffness and damping coefficients (control parameters) respectively, and

$$C_{pto} = -\frac{(m_w - m_b)g}{3\cos\alpha_t} \tag{5.11}$$

is the force that counteracts the hydrostatic force in an undisturbed position, and α_t is the tether angle relative to the vertical. It is assumed that all PTO machineries have the same stiffness and damping coefficients.

End stop forces. As each PTO system has a limited stroke, an additional force from the hard stop mechanism is exerted on the body to constrain its motion. As a result, the hard stop system is modelled by the repulsive energy potential [50]:

$$\begin{aligned} F_{es,i} = &-K_{es,\min}(\Delta\ell_i - \Delta\ell_{\min})u(\Delta\ell_{\min} - \Delta\ell_i) \\ &-K_{es,\max}(\Delta\ell_i - \Delta\ell_{\max})u(\Delta\ell_i - \Delta\ell_{\max}), \end{aligned} \tag{5.12}$$

$i = 1\ldots 3$, $u(\cdot)$ is the Heaviside step function, $K_{es,\min}$, and $K_{es,\max}$ are the end stop spring coefficients, $\Delta\ell_{\min}$ and $\Delta\ell_{\max}$ are the stroke limits relative to the nominal position.

Tension in tether. A linear superposition of the PTO and hard stop forces governs the tension in each tether. In addition, the tethers should always be under tension in order to transmit forces to the machinery:

$$F_{t,i} = \min(0, F_{pto,i} + F_{es,i}), \quad i = 1\ldots 3. \tag{5.13}$$

Noting that $F_{pto,i}$, $F_{es,i}$, and $F_{t,i}$ are the forces that act along the mooring line i, the generalised tether force in the Cartesian coordinate frame is:

$$\mathbf{F}_{tens} = \mathbf{J}^{-\mathsf{T}}\mathbf{F}_t, \tag{5.14}$$

where $\mathbf{F}_t = [F_{t,1} \quad F_{t,2} \quad F_{t,3}]^\mathsf{T}$ and $\mathbf{J}^{-\mathsf{T}}$ is the transposed inverse kinematic Jacobian.

Viscous damping forces. The viscous damping force is modelled according to the Morison equation [551, 50]:

$$\mathbf{F}_v = \begin{pmatrix} -\frac{1}{2}\rho\mathbf{C}_d\mathbf{A}_d\|\mathbf{V}_b - \mathbf{V}_f\|(\mathbf{V}_b - \mathbf{V}_f) \\ -\frac{1}{2}\rho\mathbf{b}_Q D^4 D\|\dot{\boldsymbol{\theta}}\|\dot{\boldsymbol{\theta}} \end{pmatrix}, \tag{5.15}$$

where $\mathbf{V}_b = \dot{\mathbf{r}}$ is the buoy velocity, $\mathbf{V}_f$ is the fluid particle velocity at the position of the centre of mass of the buoy, and

$$\mathbf{C}_d = \begin{pmatrix} C_x & 0 & 0 \\ 0 & C_y & 0 \\ 0 & 0 & C_z \end{pmatrix}, \quad \mathbf{A}_d = \begin{pmatrix} A_x & 0 & 0 \\ 0 & A_y & 0 \\ 0 & 0 & A_z \end{pmatrix} \tag{5.16}$$

are matrices of drag coefficients and the cross-section areas of the buoy perpendicular to the direction of motion, respectively,

$$\mathbf{b}_Q = \begin{pmatrix} b_{yz} & 0 & 0 \\ 0 & b_{xz} & 0 \\ 0 & 0 & b_{xy} \end{pmatrix} \tag{5.17}$$

is the matrix of quadratic angular damping coefficients, D is the buoy diameter, and $\dot{\boldsymbol{\theta}}$ is the vector of angular velocities of the buoy.

5.4.1.3 Time-domain model

The buoy motion in the time domain can be described using the Cummins equation [146], including the wave excitation and radiation forces and other forces specified in Section 5.4.1.2:

$$(\mathbf{M} + \mathbf{A}_\infty)\ddot{\mathbf{x}} = \mathbf{F}_e - \int_{-\infty}^t \mathbf{K}_{rad}(t-\tau)\dot{\mathbf{x}}(\tau)d\tau + \mathbf{F}_{buoy} + \mathbf{F}_v + \mathbf{F}_{tens}, \tag{5.18}$$

where $\mathbf{M}$ is the mass matrix, $\mathbf{A}_\infty$ is the matrix with infinite-frequency added mass coefficients, $\mathbf{F}_e$ is the generalised wave excitation force, and $\mathbf{K}_{rad}(t)$ is a retardation function.

Note that the wave drift forces and ocean current effects are not considered in the model.

5.4.1.4 Frequency domain

Equation (5.18) can be linearised, assuming small angular motions of the system and mapping the change in the tether length to the Cartesian coordinates of the body. Thus, the linearised frequency domain model of a generic point absorber can be written as [50], where only the surge, heave, and pitch modes are considered:

$$\left[-\omega^2 \left(\mathbf{M} + \mathbf{A}_m(\omega) \right) - i\omega \left(\mathbf{B}_{pto} + \mathbf{B}_{rad}(\omega) \right) + \mathbf{K}_{pto} \right] \mathbf{x}(\omega) = \mathbf{F}_e(\omega), \tag{5.19}$$

where

$$\mathbf{M} = \begin{pmatrix} m_b & 0 & 0 \\ 0 & m_b & 0 \\ 0 & 0 & I_y \end{pmatrix}, \quad \mathbf{A}_m(\omega) = \begin{pmatrix} A_{11} & 0 & A_{15} \\ 0 & A_{33} & 0 \\ A_{51} & 0 & A_{55} \end{pmatrix}, \tag{5.20}$$

$$\mathbf{B}_{rad}(\omega) = \begin{pmatrix} B_{11} & 0 & B_{15} \\ 0 & B_{33} & 0 \\ B_{51} & 0 & B_{55} \end{pmatrix}, \tag{5.21}$$

$$\mathbf{B}_{pto} = \begin{pmatrix} \frac{3}{2} B_{pto} \sin^2 \alpha_t & 0 & 0 \\ 0 & 3 B_{pto} \cos^2 \alpha_t & 0 \\ 0 & 0 & 0 \end{pmatrix}, \tag{5.22}$$

$$\mathbf{K}_{pto} = \begin{pmatrix} K_{pto,11} & 0 & K_{pto,15} \\ 0 & K_{pto,33} & 0 \\ K_{pto,51} & 0 & K_{pto,55} \end{pmatrix}, \tag{5.23}$$

$$K_{pto,11} = \frac{3 \sin^2 \alpha_t}{2} \left(K_{pto} + \frac{C_{pto}}{\ell_0} \right) - \frac{3 C_{pto}}{\ell_0},$$

$$K_{pto,15} = K_{pto,51} = \frac{3 C_{pto} h_c}{2 \ell_0},$$

$$K_{pto,33} = 3 \cos^2 \alpha_t \left(K_{pto} + \frac{C_{pto}}{\ell_0} \right) - \frac{3 C_{pto}}{\ell_0},$$

$$K_{pto,55} = -\frac{3 C_{pto} h_c (h_c + 2 \ell_0 \cos \alpha_t)(\cos^2 \alpha_t + 1)}{8 \ell_0 \cos^2 \alpha_t},$$

where ℓ_0 denotes the initial tether length.

5.4.1.5 Power absorption

Energy can be generated by the PTO machinery only when the tether, attached to it, is under tension. As a result, the total instantaneous mechanical power absorbed by the WEC is calculated as:

$$P_\Sigma(t) = \sum_{i=1}^{3} P_i(t) = \sum_{i=1}^{3} F_{t,i}(t) \Delta \dot{\ell}_i(t) \quad \text{if} \quad F_{t,i} < 0. \tag{5.24}$$

5.4.2 Modal analysis

Modal analysis is a tool commonly employed for determining, improving and optimising dynamic characteristics of complex vibration systems. Therefore, it is a suitable approach for the conceptual design of multi-mode WECs that harvest energy from several oscillation modes of the body.

5.4.2.1 Matrix eigenvalue problem for an undamped multiple DoF system

The eigenvalue problem is a problem commonly encountered in engineering and is the basis of modal analysis. The general motion equation for the free vibration of an undamped multiple DoF system is given by:

$$\mathbf{M}\ddot{\mathbf{x}}(t) + \mathbf{K}\mathbf{x}(t) = 0, \tag{5.25}$$

where $\mathbf{M}$ and $\mathbf{K}$ are the mass and stiffness matrices respectively. Defining an eigenvector $\mathbf{v}$ denoting the mode shape of the system under harmonic sinusoidal motion, the eigenequation can be obtained:

$$(\mathbf{K} - \omega^2 \mathbf{M})\mathbf{v} = 0. \tag{5.26}$$

The non-trivial solution of Eq. (5.26) is given by:

$$\det(\mathbf{K} - \omega^2 \mathbf{M}) = 0, \tag{5.27}$$

where $\det(\cdot)$ is the determinant of the matrix, which can be expanded, forming an nth order polynomial for ω^2. The roots of this polynomial are the eigenvalues of $\mathbf{K}^{-1}\mathbf{M}$, whose square roots are the system natural frequencies. Substituting each eigenvalue into Eq. (5.26), a corresponding eigenvector $\mathbf{v}$ can be derived. Therefore, the system has n eigenvectors, denoting the mode shapes of the system, also referred to as normal modes.

An oscillating body WEC operating at optimal conditions is lightly damped and thus can be approximated as an undamped system for modal analysis. The matrix eigenvalue problem can be used to solve the natural frequencies and mode shapes of the multi-mode WEC analytically, which are fundamental to understanding of the modal behaviour of a multi-mode WEC for design optimisation and control analysis.

5.4.2.2 Natural frequencies and mode shapes

Substituting the mass and stiffness matrices $\mathbf{M}$, $\mathbf{A_m}(\omega)$, and $\mathbf{K}_{pto}$, as defined in Section 5.4.1.4, into Eqs. (5.26), and (5.27), the eigenvalue problem for the three-tether WEC can be solved for varying tether inclination angles, α_t, and varying PTO stiffnesses, K_{pto}. The resulting natural frequencies and mode shapes of the 3-DoF system are displayed in Figure 5.9, as functions of the tether inclination angle and the PTO stiffness. Figure 5.9a shows two convex surfaces, denoting the natural frequencies of Mode 1 and Mode 3, respectively. Mode 1 is surge dominant as can be seen from its mode shapes shown in Figure 5.9b. Mode 2 is pitch dominant as evident in Figure 5.9c. Mode 3 is heave dominant as evident in Figure 5.9d. A graphical representation of Modes 1, 2, and 3 is shown in Figure 5.16, marked as modes 1, 4, and 3, respectively for a 6-DoF system. The line of intersection of the natural frequency surfaces of Modes 1 and 3 in Figure 5.9a indicates that at a tether inclination angle of 45°, both modes can be tuned simultaneously to resonance (e.g. the optimal phase condition) across a wave frequency range between 0.3 and 1.5 rad/s by varying the PTO stiffness. Figure 5.9e shows the contribution of Mode 1 and Mode 3 to the elongations of the three tethers, mapped from the system mode shapes by the inverse kinematic Jacobian matrix defined by Eq. (5.9). When the tether inclination angle increases from 0 to 90°, the contribution gradually shifts from Mode 3 (heave) dominant to Mode 1 (surge) dominant. At a tether inclination angle of 45°, Mode 1 and Mode 3 contribute to approximately equal tether elongations, thus both modes are able to cause considerable PTO power generation.

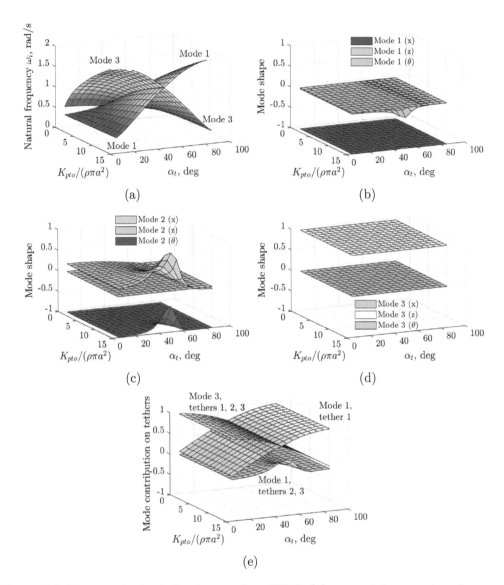

Figure 5.9: Eigenanalysis of the three-tether WEC: (a) natural frequencies of normal modes, (b) mode shape of Mode 1, (c) mode shape of Mode 2, (d) mode shape of Mode 3, (e) contribution of Modes 1 and 3 to tether elongation, vs. the tether inclination angle α_t and PTO stiffness K_{pto} (normalised by $\rho\pi a^2$). In subplots (b)–(e), the yellow colour indicates values close to 1, green colour indicates values close to 0, and purple colour indicates values close to -1.

5.4.3 Power absorption potential

The power performance of the three-tether WEC is demonstrated in comparison with the single-tether system, which is able to absorb power from the heaving mode only. The power output of both converters has been analysed in terms of the peak-to-average ratio, the contribution of individual PTO systems and different motion modes to the averaged absorbed power.

5.4.3.1 Power matrix

The averaged absorbed power and the statistical peak power for single-tether and three-tether WECs are shown in Figure 5.10, where control parameters have been optimised for each sea state. The power output of the WEC attached to three tethers is approximately two times higher than that of the single-tether device across all sea states. For example, at a wave climate with $H_s = 2$ m and $T_p = 9$ s, the power output of a converter with multiple mooring lines reaches 82 kW, while the same buoy with a single tether shows 41 kW of absorbed power. However, this factor of 2 is a feature of the particular buoy geometry considered in this study. For other shapes and aspect

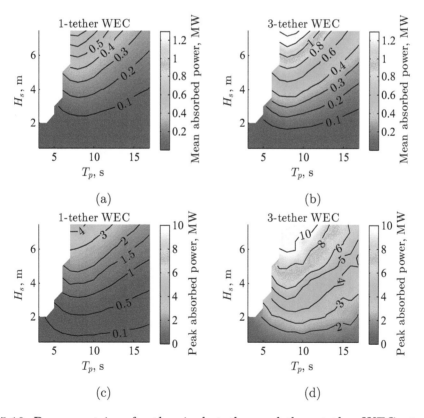

Figure 5.10: Power matrices for the single-tether and three-tether WECs: top plots correspond to the mean absorbed power, while plots on the bottom show the statistical peak values. The PTO parameters have been optimised for each sea state using grid search.

ratios, the increase in power production can be in the range of 1.6–2.5 times [703]. The level of instantaneous power for both converters is approximately 6–10 times higher than the mean power value, meaning that the designed capacity of the PTO should be an order of magnitude higher than the expected average energy output.

5.4.3.2 Power per power take-off system

As stated, the power take-off for a three-tether WEC can be implemented in several ways. For cases where there are individual PTO units for each mooring line, it can be observed that they make different contributions to the total power output. The tether arrangement shown in Figure 5.8 demonstrates only one extreme case where tether 1 is aligned with the direction of wave propagation. For this configuration, the PTO attached to tether 1 generates approximately 45% of the total mean power for most sea states, as shown in Figure 5.11a. The other two PTOs convert the rest of the energy with an equal share due to the symmetrical arrangement of the tethers with

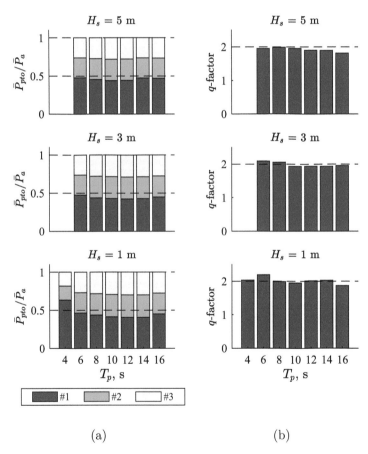

(a)   (b)

Figure 5.11: (a) Contribution of each power take-off system ($\overline{P}_{pto}$) to the total power absorption ($\overline{P}_a$) of the three-tether WEC for the selection of sea states and (b) q-factor as an efficiency improvement ratio between power absorption from the three-tether and the single-tether WECs.

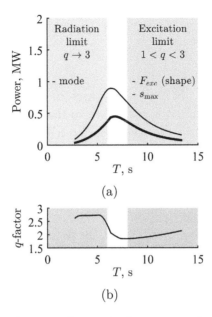

Figure 5.12: Comparison of power absorption between a single-tether and three-tether WECs in regular waves of 1 m amplitude assuming optimal control: (a) power levels and (b) q-factor as a ratio between power absorption from the three-tether and the single-tether WECs.

respect to the y-axis. Thus, depending on the wave direction statistics, the first PTO system may be designed to be larger than the other two. Another extreme situation is when one of the tethers operates perpendicular to the wave front with almost zero power output. However, according to the preliminary results in the frequency domain [702], the total power output of the WEC is largely unaffected by the wave direction relative to the arrangement of the mooring lines.

Comparison of the power absorption between both converters is demonstrated in Figure 5.11b, showing that the three-tether WEC can generate approximately 1.8–2.2 times more power than its single-tether counterpart. To explain the main reason why the improvement is only twofold and not three times as derived in [571], the Budal diagram (as cited in [229]) for single- and three-tether WECs is shown in Figure 5.12a, assuming optimal control at each wave period. The blue area on the plot corresponds to the radiation limit where buoy displacements are not constrained and power production is only governed by the mode of oscillation. As a result, the maximum increase in WEC efficiency can be up to three times if both heave and surge contribute to the power take-off. The red area in Figure 5.12a shows the range of wave periods where the buoy displacements reach constraint and power absorption limits, depending on the buoy shape and maximum allowed displacements. The q-factor in Figure 5.12b corresponds to the increase in power production for a three-tether case when compared with its single tether counterpart. Thus, its values vary between 1.85 and 2.7, depending on the wave period. As the irregular waves are composed from a number of waves with different frequencies, the overall effect of having three tethers (the quality factor) is averaged to the level of 2–2.2, as shown in Figure 5.11.

5.4.3.3 Power per mode

Since the main idea of the application of three mooring lines is to capture power from the buoy motion in surge, it is important to show how much energy is converted from each motion mode. The power contribution from one mode is calculated as the difference between excitation power, radiated power and viscous losses:

$$\overline{P}_{mode} = \overline{P}_e - \overline{P}_r - \overline{P}_v, \tag{5.28}$$

where $\overline{P}$ denotes the time average power.

The relative contribution from surge, heave, and pitch modes is shown in Figure 5.13 for the range of sea states. As expected for the single-tether WEC, the heave motion provides around 90–95% of the total power, while the remaining 5%

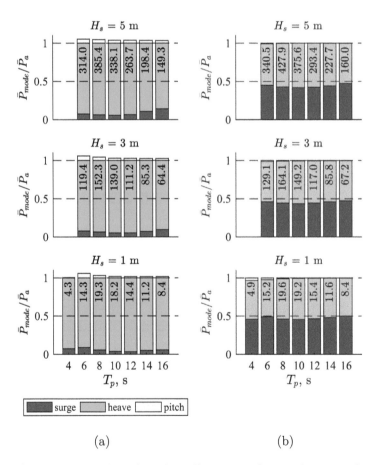

Figure 5.13: Contribution of modes of oscillation to the total power absorption of the (a) single-tether and (b) three-tether WECs. Numbers on bars indicate the absolute value of absorbed power by heave only.

are produced by surge. Interestingly, the buoy oscillations in pitch have a negative impact on the power production, which is indicated by the bars placed above 1 on the plot. This loss is presumably due to extremely poor coupling between the pitch motion of the buoy and the PTO, so useful work is minimal; whereas losses due to radiation and drag remain high. In contrast, in the three-tether configuration, pitch has a negligible effect on the WEC's performance in terms of the absorbed power. In general, slightly more than half of the energy (50–55%) is generated from heave while the remaining 45–50% is produced from surge. The contribution from surge decreases by several percents with an increasing wave height. Interestingly, the three-tether WEC demonstrates slightly higher (by 5%) efficiency in heave mode than the single-tether device, as indicated by the numbers on the bars.

5.4.4 Design optimisation

The preliminary analysis in Section 5.4.3 and the economic assessment presented in [701] demonstrated that a multi-mode WEC can produce up to two times more power and generate cheaper electricity than the same buoy with a single-tether configuration. The concept of the three-tether WEC has been adapted by Carnegie Clean Energy, Australia, to build their CETO-6 multi-moored system. This WEC will be installed at the so-called Albany test site in Western Australia, which has a wave climate specified in Figure 5.14. This section demonstrates the procedure for the design optimisation of a multi-mode wave energy converter, using the CETO-6 multi-moored system as an example. Two different objective functions are considered: maximisation of mechanical power production, and minimisation of the levelised cost of energy. In order to reduce the number of sea states used to assess the WEC performance, a sub-set of 34 sea states with a total probability of 99% are chosen to represent the deployment climate (outlined by a black line).

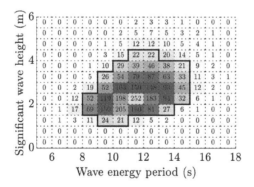

Figure 5.14: The wave climate at the Albany deployment site located in Western Australia (117.7547°E, 35.1081°S, 33.9 kW/m mean annual wave power resource) [348].

5.4.4.1 Approximation of levelised costs of electricity value

Due to the lack of commercial wave power plants, the calculations of the actual LCOE value are based on many uncertainties associated with power production and related costs. Using Eq. (1.11) in Chapter 1 [163], the LCOE can be approximated as:

$$
\text{LCoE}\left(\frac{\text{€}}{\text{kWh}}\right) = \text{RDC} \times \left(\frac{\text{Energy (MWh)}}{\text{Mass (kg)}}\right)^{-0.5}, \tag{5.29}
$$

where the mass corresponds to a significant mass of the WEC including the mass of the buoy and the anchoring system, and RDC is a site-dependent coefficient that will be omitted in calculations (RDC = 1) because optimisation is done for the same deployment site.

In this study, the calculation of the significant mass of the WEC is based on the following assumptions:

- the buoy mass is $m_b = 0.5\rho_w \pi a^2 h_c$;

- the required mass of the anchoring system (three piles) depends on the tether tension associated with buoyancy and the wave force. Therefore, it can be approximated by:

$$
m_{as} = \frac{m_{as}^{\text{ref}}}{(F_t^{peak})^{\text{ref}}} F_t^{peak}, \tag{5.30}
$$

where a three-tether WEC with a $a = 5.5$ m radius and $h_c = 5.5$ m height is taken as a reference case from [698] (the mass of three piles is estimated as $m_{as}^{\text{ref}} = 225 \times 10^3$ kg, and the statistical peak force is $(F_t^{peak})^{\text{ref}} = 1.94$ MN). Therefore, in order to estimate m_{as} for any given cylinder dimensions, it is required to calculate the tether peak force ($99\% = 2.57\sigma_{F_t}$) using the spectral domain model.

As a result, the LCOE model used in this study is:

$$
\text{LCOE} = \left(\frac{8760 P_{AAP}}{m_b + m_{as}}\right)^{-0.5}. \tag{5.31}
$$

5.4.4.2 Optimisation routine

The parameters of the WEC that are considered for optimisation are: the buoy radius a, the buoy aspect ratio defined as the ratio of the buoy height to its radius (h_c/a), the tether inclination angle α_t, the tether attachment angle α_{ap}, the vector of PTO stiffness coefficients for each of the $N = 34$ sea state considered $\mathbf{k}_{pto} = [K_{pto}^{(1)}, K_{pto}^{(2)}, \ldots, K_{pto}^{(N)}]^{\text{T}}$, the vector of PTO damping coefficients $\mathbf{b}_{pto} = [B_{pto}^{(1)}, B_{pto}^{(2)}, \ldots, B_{pto}^{(N)}]^{\text{T}}$. In total, there are 72 parameters that should be optimised:

$$
\mathbf{z} = [a, (h_c/a), \alpha_t, \alpha_{ap}, \mathbf{k}_{pto} \in \mathbb{R}^{N \times 1}, \mathbf{b}_{pto} \in \mathbb{R}^{N \times 1}]. \tag{5.32}
$$

The two objective functions considered in this study are to:

(i) maximise the average annual power output calculated using Eq. (1.4):

$$f_{O1} = \arg\max_{\mathbf{z}} P_{AAP}(\mathbf{z}), \text{subject to: } \mathbf{z} \in [\mathbf{z}_{\min}, \mathbf{z}_{\max}] \tag{5.33}$$

(ii) minimise the LCOE value specified in Eq. (5.31):

$$f_{O2} = \arg\min_{\mathbf{z}} \text{LCOE}(\mathbf{z}), \text{subject to: } \mathbf{z} \in [\mathbf{z}_{\min}, \mathbf{z}_{\max}] \tag{5.34}$$

The range of the design and PTO parameters used in the optimisation is specified in Table 5.2.

Table 5.2: Constraints on the design parameters.

Parameter	Unit	Min	Max
Buoy radius, a	m	5	20
Buoy aspect ratio, (h_c/a)		0.4	1.5
Tether inclination angle, α_t	deg	10	80
Tether attachment angle, α_{ap}	deg	10	80
PTO stiffness, K_{pto}	N/m	10^3	10^8
PTO damping, B_{pto}	N/(m/s)	10^3	10^8

5.4.4.3 Optimisation results

The optimised design parameters of the three-tether WEC obtained using two different objective functions are demonstrated in Table 5.3. The results demonstrate that to maximise the power production, the three-tether WEC should be designed with a relatively large radius (20 m) and height (30 m). With such geometry, the WEC effectively utilises all of the hydrodynamic modes: surge, heave, and pitch. However, this design is not cost effective, and in order to minimise the energy cost, the buoy radius should be of 11–14 m, with a height of 4.8–6 m.

Table 5.3: Optimised WEC design parameters.

Parameter	f_{O1}	f_{O2}
a [m]	19	13
h_c/a	1.5	0.4
α_t [deg]	26	18
α_{ap} [deg]	53	10
LCOE$\times 10^3$	–	18
Power [MW]	1.8	0.68

Also, the sensitivity of the power output and LCOE to the buoy radius and height is demonstrated in Figure 5.15. As expected, the power increases with increasing radius and height. However, a change in radius has a much stronger effect on the power output than a change in height. Also, the energy cost increases with buoy height, while the minimum LCOE can be achieved with a buoy of 11–14 m radius regardless of its height. Other studies dedicated to the geometry optimisation of the three-tether WEC for different deployment sites can be found in [707, 568, 569].

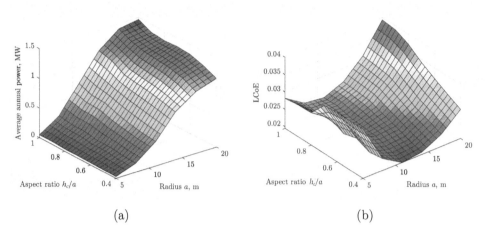

Figure 5.15: (a) Average annual power and (b) LCOE as a function of the buoy radius and aspect ratio.

5.4.5 Control system design

The fundamental theory for WEC control design, outlined in Section 3.2.1, has been developed for WECs that absorb power from one hydrodynamic mode. However, the hydrodynamic control of multi-mode WECs has two distinctive features from this theory. Firstly, the majority of multi-mode WECs do not have independent control of all hydrodynamic modes due to the use of inclined tethers, or mooring lines. Such power take-off configurations introduce additional coupling between rigid body modes, while making some degrees of freedom more controllable than others. Secondly, the hydrodynamic coupling between surge and pitch modes may introduce a destructive interference in the system, negatively affecting the total power production of a WEC [782, 781].

5.4.5.1 Controllability

The converter shown in Figure 5.7 belongs to the class of underactuated mechanisms as it has six DoFs but only three independent control inputs (i.e., is a non-square plant). In order to identify which DoFs can be controlled and quantify interactions in this MIMO system, it is necessary to linearise the nonlinear plant model from Eq. (5.18) and represent it in terms of the transfer function $\mathbf{G}(s)$ (for more details refer to [704]). Then, the transfer function of the WEC $\mathbf{G}(j\omega) \in R^{6\times3}$ should be decomposed using singular value decomposition (SVD):

$$\mathbf{G}(j\omega) = \mathbf{U}\boldsymbol{\Sigma}\mathbf{V}^{H}, \tag{5.35}$$

where $\mathbf{U} \in R^{6\times6}$ and $\mathbf{V} \in R^{3\times3}$ are the singular vectors that form orthonormal bases for the output and input spaces, correspondingly, and $\boldsymbol{\Sigma} \in R^{6\times3}$ is a matrix with three non-negative singular values $\bar{\sigma} \triangleq \sigma_1 \geq \sigma_2 \geq \sigma_3 \triangleq \underline{\sigma}$.

Singular vectors correspond to the six rigid body modes of vibration (not to be confused with the hydrodynamic modes) each of which has its own natural frequency and mode shape [260]. The graphical representation of the mode shapes that correspond to $U_i (i = 1 \ldots 6)$ are shown in Figure 5.16. Since the system has three control inputs, only the first three modes out of the six, namely U_1, U_2, and U_3 can be actively controlled by the power take-off machinery. It can be seen that both output modes U_1 and U_4 involve buoy motion in surge and pitch, but: (i) these DoFs are included in U_1 and U_4 with different phases and amplitudes; (ii) U_1 and U_4 have different resonant frequencies, so the modes primarily responsible for buoy motion depend on the period of the incoming wave; and (iii) U_1 is controllable while U_4 is not.

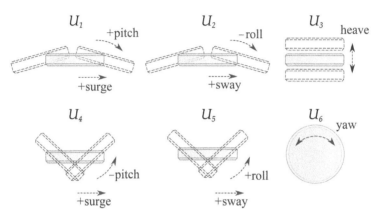

Figure 5.16: Graphical representation of the rigid body modes of the three-tether WEC.

In control theory, singular values show the output directions in which the system inputs are most effective [725]. The frequency-dependent singular values of the WEC that correspond to the controllable modes $U_i (i = 1 \ldots 3)$ are demonstrated in Figure 5.17a. It is clear from the plot that in the frequency range 0.5–0.75 rad/s (wave period 8–12 s) the coupled surge/pitch U_1 and sway/roll U_2 modes have singular values an order of magnitude higher than those for heaving mode U_3. This implies that less effort is required to control the buoy in surge and sway than in heave at this range of frequencies. Moreover, if the ratio between the maximum and minimum singular values (condition number) is larger than 10, inverse-based controllers may be sensitive to 'unstructured' input uncertainty, which is undesirable for practical applications [725].

Another measure that is widely used to identify the control properties of the plant is the relative gain array $\mathrm{RGA}(G) = \Lambda(G) \triangleq G \circ (G^{-1})^{\mathsf{T}}$, where $\circ$ denotes the Hadamard product (element-by-element multiplication). For non-square plants, small elements ($\ll 1$) in a row of RGA correspond to the output which cannot be controlled. The frequency dependent sum of each row of RGA for the three-tether WEC shown in Figure 5.17b indicates that the PTO system has a strong control authority in surge, sway and heave, but poor controllability over pitch and roll. However, it should be noted that this analysis has been performed for the nominal position of the buoy $\mathbf{x} = \mathbf{0}_{6 \times 1}$ and the control property of this WEC is subject to the buoy orientation.

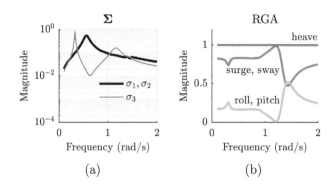

Figure 5.17: Controllability measures of the three-tether WEC: (a) singular values, and (b) Relative Gain Array.

5.4.5.2 Surge-pitch compromise

According to linear potential theory, the optimal velocity of a WEC that is able to absorb power from several degrees-of-freedom should satisfy the following equation [227]:

$$\mathbf{B_{rad}}(\omega)\hat{\mathbf{u}}_{\mathrm{opt}}(\omega) = \frac{1}{2}\hat{\mathbf{F}}_e(\omega), \qquad (5.36)$$

where the 'hat' symbol, ' ^ ', denotes the complex amplitude, $\mathbf{B_{rad}}(\omega) \in R^{6\times 6}$ is the matrix of radiation damping coefficients, $\hat{\mathbf{u}}_{\mathrm{opt}}(\omega) \in R^{6\times 1}$ is the vector of optimal velocity, and $\hat{\mathbf{F}}_e(\omega) \in R^{6\times 1}$ is the excitation force vector. If $\mathbf{B_{rad}}(\omega)$ is non-singular, then Eq. (5.36) has a unique solution $\hat{\mathbf{u}}_{\mathrm{opt}}(\omega) = \frac{1}{2}\mathbf{B_{rad}}^{-1}(\omega)\hat{\mathbf{F}}_e(\omega)$. However, $\mathbf{B_{rad}}^{-1}(\omega)$ does not exist for an axisymmetric body if surge and pitch are involved in power absorption, and Eq. (5.36) cannot be solved simultaneously for $\hat{u}_{1,\mathrm{opt}}$ and $\hat{u}_{5,\mathrm{opt}}$. Therefore, if one of the velocities is known, the other one can be found from the following equation:

$$\begin{bmatrix} B_{11} & B_{15} \\ B_{51} & B_{55} \end{bmatrix} \begin{bmatrix} \hat{u}_1 \\ \hat{u}_5 \end{bmatrix} = \frac{1}{2} \begin{bmatrix} \hat{F}_{e,1} \\ \hat{F}_{e,5} \end{bmatrix}. \qquad (5.37)$$

For instance, for the known value of the pitch velocity $\hat{u}_5$, the optimal buoy velocity in surge can be calculated as [227]:

$$\hat{u}_{1,\mathrm{opt}} = \frac{\hat{F}_{e,5} - 2B_{55}\hat{u}_5}{2B_{51}} = \frac{\hat{F}_{e,1} - 2B_{15}\hat{u}_5}{2B_{11}}, \qquad (5.38)$$

and vice versa. So according to Eq. (5.38), the requirement that the buoy velocity should be in phase with the excitation force does not apply if both surge and pitch modes are involved in power generation. However, if $\hat{F}_{e,1} \gg 2B_{15}\hat{u}_5$, the coupling between surge and pitch can be neglected in the design of the controller. In addition, due to the presence of the viscous damping force, the formulation of the optimal buoy velocity is modified as [265]:

$$\hat{u}_{1,\mathrm{opt}} = \frac{\hat{F}_{e,1}}{2(B_{11} + B_{v,1})}, \quad \hat{u}_{3,\mathrm{opt}} = \frac{\hat{F}_{e,3}}{2(B_{33} + B_{v,3})}, \qquad (5.39)$$

where $B_{v,1}$ and $B_{v,3}$ are linearised damping coefficients in surge and heave modes, respectively.

5.4.5.3 Review of advanced control strategies applied to multi-mode wave energy converters

The majority of control strategies developed for heaving WECs have been extended for the multi-mode prototypes. One of the first MIMO controllers for wave energy applications was developed in [428, 696], based on linear quadratic Gaussian optimal control theory. The controller has been tested on a three-tether cylindrical buoy, where each tether is connected to an individual electric generator. This study revealed that in such multi-mode WECs with inclined tethers, some generators deliver power to the electrical system while other generators act as motors actuating the WEC. Whether the generator will deliver or consume energy directly depends on the tether arrangement (e.g., inclination angles, number of tethers) and the applied control strategy, as shown in [699].

Several studies [4, 883, 5, 882, 416] developed MIMO controllers for WECs assuming full control authority over each hydrodynamic mode. Such studies enable analysis of the significance of individual degrees-of-freedom on overall power generation, but do not take into account any additional geometric coupling that exists in real multi-mode prototypes. Ref. [4] developed an approximate complex-conjugate control for a three-DoF WEC and demonstrated how to shift energy production from one dipole mode (surge or pitch) to another by changing their natural frequencies. Also, the results showed that power generation from both dipole modes can be detrimental to the power production of the entire WEC (as already explained in Section 5.2.2). Two of the most widely used hydrodynamic controllers are discussed in more details below.

Velocity tracking control. Fusco and Ringwood [265] have developed a "simple but effective" velocity tracking controller, sometimes called an approximate complex conjugate control [326], for a heaving WEC. The main idea of this controller is to measure an instantaneous excitation force acting on the WEC, estimate its amplitude and dominant frequency, and set the reference buoy velocity following the optimality conditions specified in Eq. (3.58). This controller has been extended to the three-tether system in [704, 699]. The control structure has two major loops (refer to Figure 5.18): (i) a high level loop sets the reference (desired) velocity of the buoy and (ii) a low level loop provides the required machinery (power take-off) force to achieve this velocity.

The differences between SISO and MIMO velocity tracking controllers lie within both control loops. In [699], the main focus of the controller was given to the surge and heave modes acknowledging the fact that the three-tether configuration provides poor control authority over pitch. As a result, the reference buoy velocities were set according to Eq. (5.39) while completely ignoring the surge-pitch hydrodynamic coupling. The velocity tracking control loop in [265] was implemented using an internal model controller (IMC). It is possible to use a decoupled version of the IMC strategy for a multi-mode WEC but only if the natural frequency of the pitch mode is

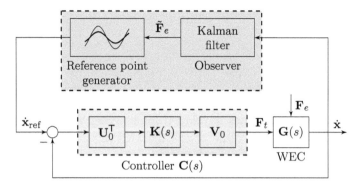

Figure 5.18: The block-diagram of the velocity tracking control designed for the three-tether WEC. The "tilde" symbol, '~', denotes an estimated value of the variable.

higher than that of the other modes, as shown in [704]. However, for some buoy geometries, the inverse-based controllers cannot be implemented and it is recommended to use other control structures for the inner control loop. For example, [699] used an SVD-based controller to decouple the control problem of a multi-variable system accompanied by a simple PI controller for the velocity tracking purposes.

The results demonstrate that a velocity tracking controller can significantly improve the power output of a multi-mode WEC (up to 40%) over a spring-damper controller. However, the elimination of the pitch mode from the control problem may lead to a significant power loss due to uncontrolled oscillations in pitch, as shown in Figure 5.19a. Moreover, the findings documented in [696] for the LQG controller have been confirmed for this type of the controller as well stating that some generators should operate as actuators while power to the grid is delivered by other power take-off units (refer to Figure 5.19b)

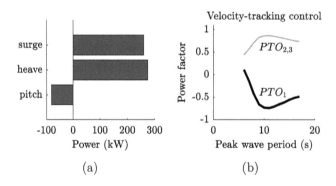

Figure 5.19: The contribution of (a) each hydrodynamic mode and (b) each PTO unit to the total power output of the three-tether WEC using the velocity tracking control strategy.

Alternative ways to implement the MIMO velocity tracking controller have been demonstrated in [353] for the WaveSub WEC. The reference velocities have been set to all six degrees of freedom. The singularity problem explained in Section 5.4.5.2

has been avoided by introducing a matrix of linearised viscous damping coefficients. The tracking control loop has been achieved using a linear quadratic regulator.

Model predictive control. Another popular control strategy that found its application in the wave energy industry is a mode predictive control (MPC). It has been studied extensively for a single-DoF WECs [327], and only recently has been extended to the multi-mode systems. The controller's formulation and its modification from SISO to MIMO systems are straightforward and do not require any special considerations. The only challenge with the MIMO version of the MPC is that the number of states increases proportionally to the number of degrees-of-freedom, therefore requiring higher computational resources for online optimisation. So MPC applied to multi-mode WECs might not be feasible for practical applications.

MPC has been applied to the multi-mode WaveBot which is able to absorb power from all the individually controlled hydrodynamic modes in [134]. Later, Ref. [882] developed MPC for the same WEC geometry while considering the nonlinear dynamics and parametric resonance due to coupling between pitch and heave modes. [355] tested the MPC on the WaveSub device, and [705] used the CETO WEC to identify how the MPC tuning affects the design of the power take-off unit. All these studies confirm that the MPC outperforms other control strategies in terms of power production. However, this model-based controller is very sensitive to model uncertainties and excitation force prediction errors [673, 355]. Moreover, the modelling errors accumulate with increasing numbers of degrees-of-freedom.

5.4.6 Parametric instability

As demonstrated in Section 5.3, the prevailing majority of multi-mode WECs utilises inclined tethers, or mooring lines, to absorb power from multiple hydrodynamic modes. As a result, all modes become geometrically coupled and the motion in one degree-of-freedom may excite oscillations in another. Thus, any weakly damped modes that do not have strong coupling with the power take-off (e.g. pitch or yaw) could be prone to dynamic instability [599]. Thus, analytical and experimental investigation [600] of a multi-moored CETO system demonstrated that the large motion amplitudes in heave parametrically excite parasitic yaw motion due to the nonlinear tether geometry. This parametric instability may occur when the natural frequency of the weakly coupled mode (yaw) is half or equal to the natural frequency of the dominant mode (heave). Therefore, all modes of motion need to be considered in the design process of multi-mode prototypes, not just those that are power-producing. During the design process, it is necessary to ensure that the natural frequencies of the susceptible modes are outside the range of the parametric forcing, or wave excitation, frequencies.

5.5 SUMMARY

This chapter focuses on multi-mode wave energy converters, a class of oscillating WECs that absorb power from multiple hydrodynamic modes. It is generally assumed that oscillating bodies controlled in surge, heave, and pitch modes are able to absorb three times more power than just heaving devices. However, the analysis in this chapter demonstrates that this assertion is true for both submerged and floating converters but only for relatively short wave periods. In ocean waves with longer wavelengths, the contribution of the surge mode to the power absorption of floating systems is marginal, while its effect is still significant for fully submerged WECs.

A review of the existing multi-mode prototypes reveals that they all employ inclined mooring lines/legs to control all of the hydrodynamic modes simultaneously, leading to higher power production. However, the type and shape of the oscillating body, their operating position within the water column, and targeted installation position from the shore differ between prototypes. Thus, some multi-mode WECs operate on the water surface, while others operate under the water surface being fully submerged; some prototypes are designed as point-absorbers, while others employ a terminator-type oscillating body; some target near-shore deployment, being tied to the seafloor, while others are designed for deep-water installation utilising an additional submerged support structure.

Numerical analysis of multi-mode WECs is demonstrated using the example of a fully submerged three-tether WEC. The performance and behaviour of this WEC are analysed using frequency and time-domain models based on linear wave theory with included second-order effects due to the viscous drag force. Each power take-off unit is modelled as a linear spring-damper system with tunable control parameters. To demonstrate the potential benefits of multi-mode WECs, a comparison is made between the single-tether and three-tether technologies showing that the power production of the latter is up to 2.2 times higher. The contribution of surge and heave motion modes to the power output is approximately equal, while pitch has a negligible effect on power for a three-tether WEC.

Investigating the effect of the design parameters on the performance of the three-tether WEC, it is demonstrated that a cylindrical buoy with a relatively small (¡0.4) height to radius ratio has the best economic measures, as compared with other shapes and aspect ratios. A discussion of the design of a control system for multi-mode WECs is provided, demonstrating how to extend the control approach developed for a single-degree-of-freedom system to a three-tether converter. Moreover, it is demonstrated that special caution should be taken when designing individual PTO units for each tether. Advanced control strategies (velocity-tracking, MPC, etc.) require one out of three PTO units to operate as an actuator while not generating any power and using energy from the local storage or electrical network.

CHAPTER 6

Attenuator wave energy converters

Siming Zheng

University of Plymouth, siming.zheng@plymouth.ac.uk

6.1 INTRODUCTION

This chapter is dedicated to attenuator devices. Attenuator WECs lie parallel to the wave direction (i.e., have their principal axis perpendicular to the wavefront) and effectively "ride" the waves. Attenuator devices have the advantage of high-power capture capability, which is conducive to large-scale power generation. The offshore device closest to commercial operation, Pelamis [853], is an attenuator WEC. As one of the most promising WECs, attenuator type devices have been extensively investigated and implemented.

As mentioned in the previous chapters (e.g., Chapter 2), the hydrodynamics of WECs are one of the most fundamental problems affecting the performance of the devices. Compared with the other WECs, the hydrodynamics of attenuator devices are more complicated. Attenuator devices are mostly composed of several bodies, therefore the hydrodynamic interactions between multi-bodies should be considered. Moreover, those bodies are connected by joints, i.e., there are constraints between the adjacent bodies, rather than free-floating, making their dynamic motion more complex and difficult for analysis.

This chapter gives an introduction to attenuator WECs. The remainder of this chapter is structured as follows: the working principles of attenuator WECs are presented in Section 6.2. An overview of existing attenuator WECs is given in Section 6.3. The modelling methods and the case studies of a two-hinged-float device are given in Section 6.4 to demonstrate the hydrodynamic characteristics of attenuator WECs. Some selected conclusions are drawn in Section 6.5.

DOI: 10.1201/9781003198956-6

6.2 WORKING PRINCIPLE

Attenuator WECs lie parallel to the wave direction (i.e., have their principal axis perpendicular to the wavefront) and effectively "ride" the waves [185]. From the perspective of the power capture type, attenuator WECs can be classified into two types: hinged-float-based devices and flexible-tube-based devices (see Figure 6.1).

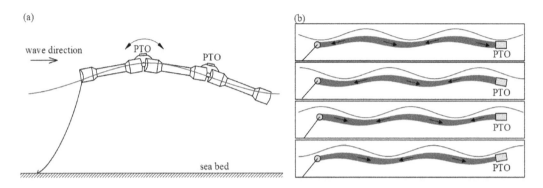

Figure 6.1: Working principles of attenuator devices: (a) hinged-float-based and (b) flexible-tube-based.

6.2.1 Hinged-float-based devices

Most of the attenuator WECs that have been proposed so far are hinged-float-based devices (also called the "raft"-type WECs by many writers, e.g., [875, 181]), a subclass of oscillating type WECs, which are generally composed of a series of semi-submerged articulated floaters (Figure 6.1a). The hinged floaters use relative rotations around connection joints to drive a power take-off (PTO) system, such that the ocean wave energy can be converted into useful energy. The PTO system of a hinged-float-based device generally consists of hydraulic cylinders and some other components. As ocean waves pass through the device, the movement around each hinge is excited, which in turn drives the hydraulic cylinders. The pressurised hydraulic fluid can be further used to drive a motor to generate electricity or pressurise water for sea-water desalinisation.

6.2.2 Flexible-tube-based devices

The flexible-tube-based attenuator WEC, which is also called a "bulge wave" device, is mainly composed of a fully submerged flexible tube. The tube is closed at both ends and filled completely with water (Figure 6.1b). A "bulge wave" can be formed inside the tube under wave action. The bulge wave then drives a standard low-head turbine located at the far end of the device to generate electricity. An alternative PTO system is the distributed PTO system deployed all along the tube. The distributed PTO system is expected to be able to capture more wave power and provide benefits to survivability.

6.3 EXISTING PROTOTYPES

An overview of existing attenuator WECs is given in this section.

6.3.1 Cockerell raft

The Cockerell raft (Figure 6.2a), proposed by Cockerell in 1974, could be the earliest attenuator device [840]. It is comprised of a series of floating rafts connected via hinges. The wave-induced relative rotation of interconnected rafts drives the hydraulic motor installed at the joint to generate electricity. To evaluate the performance of the Cockerell raft, Haren and Mei [335] developed a two-dimensional (assuming that the raft width is larger than the wavelength) hydrodynamic model based on shallow water linear wave theory. They revealed that, for a device with fixed total length, the three-raft device performed almost as well as four or five rafts, and most of the energy was captured at the leading hinge. Thus, a two or three-raft device was found adequate and further addition of rafts did not help materially.

Figure 6.2: Several selected hinged-float-based attenuator devices: (a) Cockerell raft [517, 870]; (b) MWP [518, 809]; (c) Pelamis [853]; (d) DEXA [502]; (e) Seapower Platform [256] (image courtesy of Sea Power Ltd.); (f) M4 (image courtesy of Prof. Peter Stansby); (g) Mocean Energy device [547] (image courtesy of Mocean Energy Ltd.).

6.3.2 McCabe Wave Pump

The McCabe Wave Pump (MWP) mainly comprises three-hinged pontoons (Figure 6.2b). The fore and aft pontoons are symmetrically connected to the central small one, which is attached to a submerged damper plate. Hydraulic pumps are attached at each hinge. As ocean waves pass through MWP, the movement around each hinge is excited, which in turn drives the hydraulic pumps. The pressurised hydraulic fluid can be further used to pressurise water for seawater desalinisation or to drive a motor to generate electricity. An experimental study demonstrated that the pitching motions of the pontoons were enhanced by the submerged damper plate at all of the wave periods studied, which would lead to higher volume rates of pumped water in the operational system [418]. The power output of MWP was shown to increase significantly by modifying the length of the device to be compatible with the wavelength. Very recently, Wang and Ringwood [816] optimised the geometry of MWP to maximise the energy extraction in given sea states and in site-specific wave climates. It was reported that there was no obvious performance benefit in three pontoons, over a two-pontoon system, indicating the potential for significant capital cost savings.

6.3.3 Pelamis

The most advanced of attenuator devices, across all kinds of WECs, is probably Pelamis (Figure 6.2c), which has already reached TRL7 (i.e., Technology Readiness Level 7, which corresponds to "full-scale prototype tested at sea"). The Pelamis comprises several tube sections connected by universal joints, allowing the adjacent two tube sections to rotate around one another in both pitch and yaw modes. The movements around the two axes at each joint are resisted by four hydraulic cylinders, pumping high-pressure oil through hydraulic motors to drive electrical generators. To provide energy storage and to maintain constant flow to the hydraulic motor, accumulators are equipped in the circuit. In 2004, the first full-scale prototype of Pelamis, the P1 (750 kW), was demonstrated at EMEC's wave test site at Billia Croo and connected to the UK grid. The device was 120 m long, 3.5 m in diameter, 700 t in displacement, and consisted of four tube sections. In 2009, three Pelamis P1 machines (2.25 MW) were tested in Portugal. After the testing at EMEC, a second-generation device, the P2, was proposed. The P2 was 180 m long, 4.0 m in diameter, 1350 t in total displacement, and was composed of five tube sections. Apart from the changes in dimension, the PTO system was improved, and the joint connection was simplified to enhance wave power absorption and reduce risks at the same time. From 2010 to 2014, two Pelamis P2 machines were tested off Orkney, accumulating over 15,000 h of operation [377]. Dalton et al., [155] investigated the performance and economic viability of the Pelamis WEC by using 2007 wave energy data from various global locations: Ireland, Portugal, USA, and Canada. The results, as given in Figure 6.3a, show that the annual wave energy output (AEO) and capacity of the Pelamis were highest from the Irish location, followed by the USAWC, Canada EC, Portugal, Canada WC, and finally USA WC. Figure 6.3b presents the LCOE for Pelamis device developments [201], in which the LCOE has been estimated for these

(a)

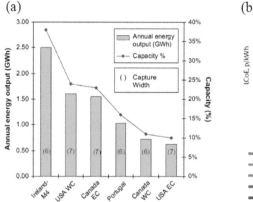

(b)

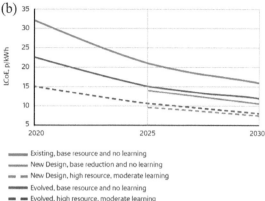

Figure 6.3: Performance and economic viability of the Pelamis WEC: (a) the annual wave energy output (AEO) and capacity [155] and (b) LCOE [201] (image courtesy of Energy Technologies Institute).

scenario conditions: "base" and "high" energy sites; no, or moderate learning; and new design deployment in the mid-2020s. Unfortunately, the Pelamis wave power company went into administration in November 2014 after failing to find further funding to develop its technology. Wave Energy Scotland now owns their assets and IP. A closer analysis revealed that a range of root causes underlay the failed development of Pelamis [206]. Technological development was characterised by too high ambitions and a race through technology readiness levels (TRLs), rather than actual technological performance. The development of the P1 machine went through a procedure of scaled development (e.g., testing of scale models followed by full-scale testing of hinges and other components before, finally, testing of a full-scale device), whereas the P2 did not repeat this process and went straight to full scale instead. Due to pressure on the hinges, the reliability of the device was also an issue. This issue gets more challenging with the device consisting of more floats and hinges. The lack of trust between the funders and the developer might have been more important than the technological challenges. The failure of Pelamis serves as an example, where a mix of technological and non-technological barriers put a strong brake on the project's advancement. Although there can still be concerns about the technological performance and LCOE potential, this type of failure does not prove that the concept has failed.

6.3.4 The DEXA

The DEXA (Figure 6.2d) developed in Denmark is another attenuator device. It consists of two rigid pontoons with a hinge in between, which allows each pontoon to pivot in relation to the other, driving a PTO system placed close to the centre of the device with a rated power of 250 kW. To evaluate the influence of DEXA on wave attenuation, Zanuttigh et al., [862] introduced the concept of a floating breakwater wave transmission coefficient and examined the protective effect of DEXA devices on the shoreline numerically and experimentally. Within a specified range of wave conditions, both the power output and wave transmission coefficient were found to increase, whereas the wave power absorption efficiency decreased with the increase in wavelength.

6.3.5 Seapower Platform

The Seapower Platform (Figure 6.2e), fundamentally a hinged attenuator device, is composed of hollow pontoons making up two separate bodies. The simple hull structures are believed to result in low cost and long lifetime, leading to the cost-effectiveness of the device. A PTO system can be mounted either on the deck of the main body or onto the main hinge of the unit. Two separate PTO streams are being developed [697]: (a) a Hydraulic Seawater to Freshwater PTO System and (b) a Mechanical Direct Drive PTO module – a 1 MW rated PTO with a 25% Capacity Factor, depending on the requirements for desalination and electricity generation. After carefully sizing the pontoon's properties and their spacing, the optimised Seapower Platform can maximise the average wave power capture for a site.

6.3.6 M4

In recent years, M4 has become one of the most developed attenuator devices. The M4 device extends the concept of the Seapower Platform further by using cylindrical (vertical axis) floats to avoid losses due to external sloshing, and the floats have hemispherical or rounded bases to give minimal drag losses [106]. The three cylindrical vertically axisymmetric floats have different sizes, with the smallest float for the bow and the largest for the stern. It is expected that different sizes of float give a range of natural periods, which in turn enables power capture across a range of wave frequencies for a sea site. A series of numerical and experimental studies were carried out to study the performance of M4 in both regular and irregular waves [743, 745]. It was revealed that the energy loss caused by drag force could be reduced by adopting rounded base floats. The maximum wave energy capture width with rounded base floats was nearly 60% larger than that with flat based floats. Wave power absorption of the M4 could be improved by increasing the bow to mid float spacing to be more than 50% greater than the mid to stern float spacing. To further increase wave power absorption and potentially reduce electricity generation costs, Stansby et al. [744] extended the three-float M4 concept into the four and even more-float M4 concept, with the number of mid and stern floats increased, while maintaining a single bow float for mooring connection (Figure 6.2f). The six-float configuration with one bow float, three mid floats and two stern floats was found to be marginally the most cost-effective version. Some specific studies on the performance of the six-float M4 device in terms of wave power absorption, dynamic response, energy yield, mooring forces, etc. can be found in [742, 106].

There are also some other hinged-float-based attenuator devices, e.g., the Crest Wing WEC [408] and the Mocean Energy WEC (Figure 6.2g). The unique geometry of the WECs developed by Mocean Energy was indicated to significantly improve

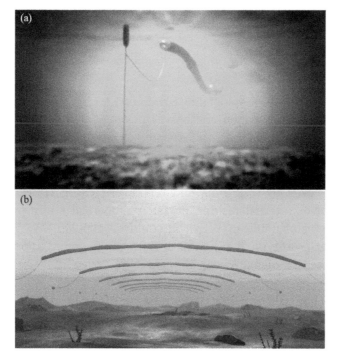

Figure 6.4: Two selected flexible-tube-based attenuator devices: (a) Anaconda [131] (image courtesy of Checkmate Seaenergy Ltd) and (b) WEC S3 [693] (image courtesy of SBM Offshore).

performance over traditional articulated WECs, and increase the survivability by diving through the largest waves [377].

6.3.7 Anaconda

The most well-known flexible-tube-based attenuator device is the Anaconda device (Figure 6.4a), which essentially consists of a rubber-made flexible tube filled with water and is placed under the water surface in the sea. The rubber tube can range from 50 m in length with a 3 m diameter to 150 m length with a diameter of 5 m, depending on the wave climate and the power required [788, 116]. Both ends of the rubber tube are sealed and it is anchored with its head to the waves. As waves propagate through the device, a running "bulge wave" is generated by squeezing the water-filled rubber tube, causing pressure variations along its length. Anaconda is designed to satisfy a resonance condition, i.e., its bulge wave speed is close to the speed of the external water waves above, hence the bulges grow as they travel along the tube, gathering wave energy. Ultimately, the series of bulge waves inside the tube feeds a power-generating turbine at Anaconda's far end to generate electricity. The average output of a full-scale Anaconda 7 m diameter and 150 m long was predicted at 1 MW in typical Atlantic conditions [233]. Fabriconda, an adaptation of the Anaconda WEC, is also a flexible-tube-based device, but consists of a tube, which is made from a series of smaller fabric tubes, called cells, joined together longitudinally to form a larger central tube [331].

6.3.8 WEC S3

The WEC S3 (Figure 6.4b) developed by SBM Offshore is a flexible-tube-based device similar to the Anaconda, which consists of a submerged flexible tube filled with seawater, and the wave action causes a bulge wave in the same way. Instead of a mechanical PTO, the WEC S3 contains a series of rings of electroactive polymers (EAPs) distributed along the tube, working as a distributed PTO system. Energy generation is achieved by stretching and relaxing of the EAP system without any need for external or internal moving parts. A numerical study predicted that the annual mean power of a 100 m WEC S3 could be 100-200 kW [54]. A 60 m long, 1.2 m in diameter prototype of WEC S3 device, submerged at approximately 4 m, is to be tested in the sea, offshore near Monaco [693].

6.4 CASE STUDY – TWO HINGED FLOATS

Many attenuator devices, as reviewed in the previous section, including DEXA, Seapower Platform, three-float M4, and Mocean Energy WEC, are mainly composed of two hinged floats. As reported by [335] and [816], an attenuator device, which consists of two hinged floats, is adequate for wave power absorption and further addition of floats does not help materially. Instead a greater number of hinges decreases survivability. The remainder of this chapter is focused on the modelling of an attenuator device consisting of two hinged floats. However, some of the derivations presented in this chapter can be extended for the case with multiple hinged floats.

6.4.1 Modelling methods of two hinged floats

6.4.1.1 Frequency domain

At least two methods can be employed to predict the dynamics of two hinged floats: one is treating the two hinged floats as two individual free floating structures with a displacement constraint matrix and a vector of constraint forces at the joint considered in the motion matrix equation (e.g., see [750, 875]), which is called the *Lagrange multiplier technique*; the other is by treating the two hinged floats as a whole structure and introducing a generalised mode to represent the relative rotation between the two floats (e.g., see [573]), which is called the *generalised mode technique*.

In the numerical simulation, all amplitudes are assumed to be small enough that linear theory applies, and the fluid is assumed to be inviscid, incompressible, and irrotational. It is further assumed that all motion is time-harmonic, with angular frequency ω.

An attenuator device consisting of two-hinged barges (see Figure 6.5) is taken as an example to demonstrate how both of these techniques work. The dimensions of the device are as follows: $a = 10$ m; $b = 10$ m; $L = 40$ m; $l_g = 10$ m; the device floats in the water at infinite water depth and is subjected to regular waves propagating

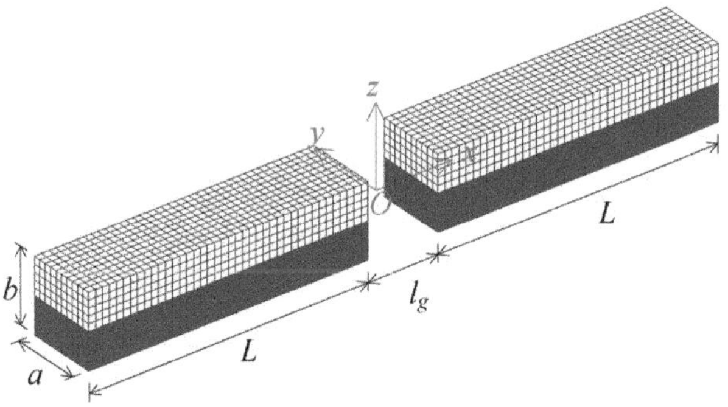

Figure 6.5: Schematic diagram of two hinged floats.

along its length. There is a hinge at the middle of the gap between the two floats connecting them together. The effect induced by a linear PTO system at the hinge is represented by a rotary damping coefficient c_{pto}, a rotary stiffness k_{pto}, and a rotary inertia I_{pto}. For the sake of simplification, the rotary stiffness and inertia effect can be represented by a new parameter $z_{pto} = k_{pto} - \omega^2 I_{pto}$. It is assumed that the PTO associated mass is negligible compared with the hinged floats.

Generalised mode technique

The two hinged floats can be treated as a whole structure. Figure 6.6 presents a schematic diagram of the definition of the modes for the generalised mode technique. In this case the six rigid-body modes are defined as if the hinge were rigid, and the new "tent"-shaped mode ($j = 7$), which is employed to denote the relative rotation between the two hinged floats, is specified by the vectors:

$$
\begin{cases}
u_7 = -\frac{z - z_0}{x_1 - x_0}, & x \in [x_0, x_1] \\
w_7 = \frac{x - x_0}{x_1 - x_0}, & x \in [x_0, x_1]
\end{cases}
\tag{6.1}
$$

and

$$
\begin{cases}
u_7 = \frac{z - z_0}{x_2 - x_1}, & x \in [x_1, x_2] \\
w_7 = \frac{x_2 - x}{x_2 - x_1}, & x \in [x_1, x_2]
\end{cases}
\tag{6.2}
$$

where (u_7, v_7, w_7) represents the displacement vector of the generalised mode, in which v_7 vanishes for the present case; x_0 and x_2 denote the positions of the very left and very right of the hinged floats in the x-axis, respectively; and x_1 denotes the position of the hinge in the x-axis.

When incident waves propagate along the Ox–axis, only the motion in surge, heave, pitch, and tent modes are excited. Oyz is a plane of symmetry for both the geometries and the physical properties of the device. The numerical simulation is focused on the responses in the heave and tent modes. These two modes are coupled with one another. The surge and pitch motions are not considered because they are decoupled with the heave and tent modes.

The motion response matrix equation of the hinged floats may be expressed as:

$$
[-\omega^2 (\mathbf{M} + \mathbf{A}_m) - i\omega (\mathbf{B}_{rad} + \mathbf{B}_{pto}) + \mathbf{K}_s + \mathbf{Z}_{pto}] \boldsymbol{X}(\omega) = \boldsymbol{F}_e(\omega),
\tag{6.3}
$$

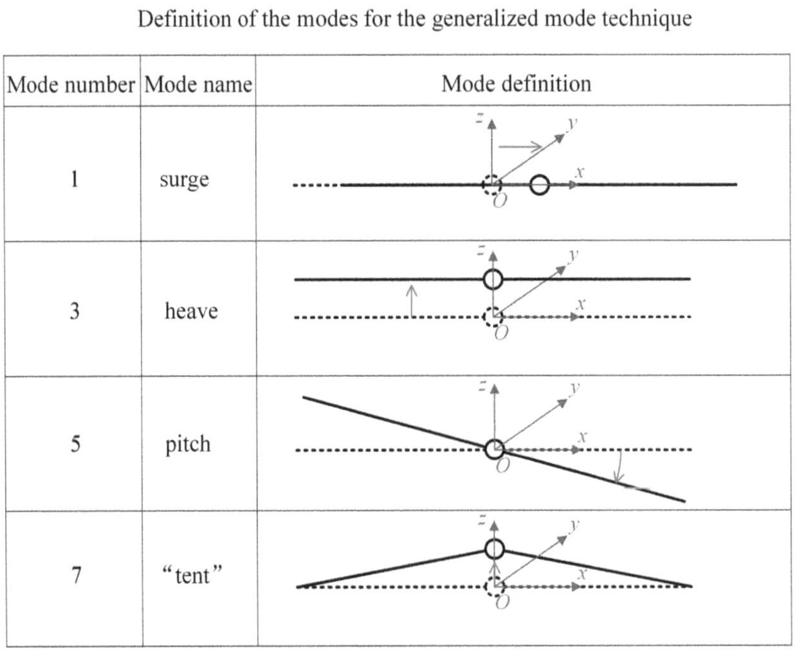

Definition of the modes for the generalized mode technique

Mode number	Mode name	Mode definition
1	surge	
3	heave	
5	pitch	
7	"tent"	

Figure 6.6: Schematic diagram of the definition of the modes for the generalised mode technique.

in which $\mathbf{M}$ and $\mathbf{K}_s$ are the device mass and hydrostatic stiffness matrices, respectively, of (2×2); $\mathbf{A}_m$ and $\mathbf{B}_{rad}$ represent the added-mass and wave radiation damping matrices, respectively, of (2×2); $\boldsymbol{X}(\omega) = [X_3, X_7]^T$ and $\boldsymbol{F}_e(\omega) = [f_3, f_7]^T$ represent the vector of the complex amplitude of the displacements of the device and the vector of the wave excitation forces acting on the device in heave mode and tent mode, respectively. $\mathbf{B}_{pto}$ and $\mathbf{Z}_{pto}$ represent the PTO associated damping matrix and the matrix of the PTO stiffness/inertia, respectively. $\mathbf{A}_m$, $\mathbf{B}_{rad}$, and $\boldsymbol{F}_e$ can be obtained by using a BEM-based code, e.g., WAMIT. The specific expressions of the matrices at the left-hand side of Eq. (6.3) are given as follows:

$$\mathbf{M} = \begin{bmatrix} m_{3,3}^{(0)} & m_{3,7}^{(0)} \\ m_{7,3}^{(0)} & m_{7,7}^{(0)} \end{bmatrix}, \qquad \mathbf{K}_s = \begin{bmatrix} k_{3,3} & k_{3,7} \\ k_{7,3} & k_{7,7} \end{bmatrix}, \tag{6.4}$$

$$\mathbf{A}_m = \begin{bmatrix} m_{3,3} & m_{3,7} \\ m_{7,3} & m_{7,7} \end{bmatrix}, \qquad \mathbf{B}_{rad} = \begin{bmatrix} c_{3,3} & c_{3,7} \\ c_{7,3} & c_{7,7} \end{bmatrix}, \tag{6.5}$$

$$\mathbf{B}_{pto} = \begin{bmatrix} 0 & 0 \\ 0 & 4c_{pto}/l_0^2 \end{bmatrix}, \qquad \mathbf{Z}_{pto} = \begin{bmatrix} 0 & 0 \\ 0 & 4z_{pto}/l_0^2 \end{bmatrix}, \tag{6.6}$$

where $l_0 = L + l_g/2$.

Modes 3 and 7 are not only coupled in terms of hydrodynamic coefficient matrices $\mathbf{A}_m$ and $\mathbf{B}_{rad}$, but also with regard to the matrices of the device mass and hydrostatic stiffness, i.e., $\mathbf{M}$ and $\mathbf{K}_s$. $k_{3,3} = 2\rho g L a$, $k_{3,7} = k_{7,3} = \rho g a L^2/l_0$, $k_{7,7} = 2\rho g a L(L^2/3 - b^2/8)/l_0^2$. $m_{3,3}^{(0)} = 2m_0$, $m_{7,3}^{(0)} = m_{3,7}^{(0)} = m_0 L/l_0$, $m_{7,7}^{(0)} = 2(I_0 + m_0 L^2/4)/l_0^2$, where $I_0 = m_0(L^2 + b^2)/12$ represents the rotary inertia of the barge around its mass centre.

The power absorbed in the PTO damper (in time average) is:

$$\overline{P}_{pto} = \frac{1}{2} c_{pto} \omega^2 \left| \frac{2X_7}{l_0} \right|^2 = \frac{2c_{pto}\omega^2 |X_7|^2}{l_0^2}, \tag{6.7}$$

in which $2X_7/l_0$ denotes the relative rotation between the hinged floats.

The generalised mode technique can also be applied to the case of flexible-tube-based attenuator devices, for which the shape of the generalised modes would be more complicated (e.g., see [54]).

Lagrange multiplier technique

The Lagrange multiplier technique is convenient for the analysis of multiple rigid bodies joined by rigid or flexible connections. Once the conventional diffraction and radiation analyses of free-floating rigid bodies have been performed, the responses of the bodies under a variety of constraints, e.g., rigid or flexible connections, may be very simply obtained [750, 875]. For an attenuator device consisting of two-hinged floats, the motion matrix equation in the frequency domain can be written as:

$$\begin{bmatrix} -\omega^2(\mathbf{M} + \mathbf{A}_m) - i\omega(\mathbf{B}_{rad} + \mathbf{B}_{pto}) + \mathbf{K}_s + \mathbf{Z}_{pto} & \mathbf{A}_J^T \\ \mathbf{A}_J & \mathbf{0} \end{bmatrix} \left\{ \begin{array}{c} \mathbf{X}(\omega) \\ \mathbf{F}_J \end{array} \right\} = \left\{ \begin{array}{c} \mathbf{F}_e(\omega) \\ \mathbf{0} \end{array} \right\}, \tag{6.8}$$

where $\mathbf{M}$ is the rigid body mass matrix for the two bodies of (6×6); $\mathbf{X}(\omega) = [X_1^{(1)}, X_3^{(1)}, X_5^{(1)}, X_1^{(2)}, X_3^{(2)}, X_5^{(2)}]^T$ and $\mathbf{F}_e(\omega) = [f_1^{(1)}, f_3^{(1)}, f_5^{(1)}, f_1^{(2)}, f_3^{(2)}, f_5^{(2)}]^T$ denote the vectors of the displacement of the floats and the wave excitation forces acting on the floats, respectively, in which $X_j^{(i)}$ and $f_j^{(i)}$ are associated with the ith float in the jth mode ($i = 1, 2$ corresponds to the fore and aft floats; $j = 1, 3, 5$ corresponds to the surge, heave, and pitch modes, respectively). $\mathbf{A}_m$ and $\mathbf{B}_{rad}$ are the added mass and wave radiation damping matrices, which can be obtained by solving six radiation problems (i.e., the surge, heave, and pitch motions of each float). $\mathbf{B}_{pto}$ is a matrix of (6×6) representing the effect of the PTO damping, and $\mathbf{Z}_{pto}$ is a matrix of (6×6) denoting the effect of PTO stiffness/inertia. $\mathbf{A}_J$ is the displacement constraint matrix of $(N \times 6)$, in which N represents the number of constraints. For the present "hinge" constraint, $N = 2$; $\mathbf{F}_J$ is the joint force vector of $(N \times 1)$. The expressions of $\mathbf{A}_J$ and $\mathbf{B}_{pto}$ may read:

$$\mathbf{A}_J = \begin{bmatrix} 1 & 0 & 0 & -1 & 0 & 0 \\ 0 & 1 & -l_0 & 0 & -1 & -l_0 \end{bmatrix}, \tag{6.9}$$

and

$$\mathbf{B}_{pto} = \begin{bmatrix} 0 & 0 & 0 & 0 & 0 & 0 \\ 0 & 0 & 0 & 0 & 0 & 0 \\ 0 & 0 & c_{pto} & 0 & 0 & -c_{pto} \\ 0 & 0 & 0 & 0 & 0 & 0 \\ 0 & 0 & 0 & 0 & 0 & 0 \\ 0 & 0 & -c_{pto} & 0 & 0 & c_{pto} \end{bmatrix}, \tag{6.10}$$

respectively. If c_{pto} is replaced by z_{pto}, $\mathbf{B}_{pto}$ becomes $\mathbf{Z}_{pto}$.

The (time-average) power absorbed by the two hinged floats can be written as:

$$\overline{P}_{pto} = \frac{1}{2} c_{pto} \omega^2 |X_5^{(1)} - X_5^{(2)}|^2 \tag{6.11}$$

Results comparison

The average wave energy capture width ratio (or capture factor) η_{eff} in a three dimensional problem can be expressed as:

$$\eta_{\text{eff}} = \frac{k\overline{P}_{pto}}{J}, \tag{6.12}$$

where J denotes the incoming wave power per unit width of the wave front [875].

Figure 6.7 presents the comparison of the results of the vertical motion at the hinge, the relative rotation of the barges, and the wave power capture width ratio by using the generalised mode method and the Lagrange multiplier method. As expected, an excellent agreement between the results by using different methods is obtained.

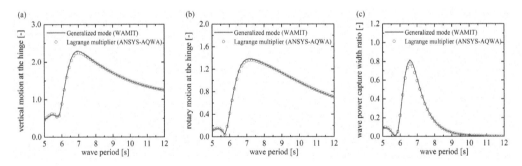

Figure 6.7: Comparison of the results by using the generalised mode method and the Lagrange multiplier method for the two-hinged barge case with $c_{pto} = 10^7$ Nm/(rad/s), $z_{pto} = 0$: (a) dimensionless vertical motion at the hinge, D_z/A, where D_z represents the vertical displacement amplitude of the hinge; (b) dimensionless relative rotation of the barges, $\theta_y/(2kA)$, where θ_y is the amplitude of the relative rotary displacement of the barges; and (c) wave power capture width ratio, η_{eff}.

6.4.1.2 Time domain

To study the effect of a nonlinear PTO system, such as a Coulomb damping or a control strategy, a time domain analysis is required. The matrix equation of an attenuator device consisting of two-hinged floats in the time domain may be written as [875]:

$$(\mathbf{M}+\mathbf{A}_\infty)\ddot{\boldsymbol{x}}(t) + \int_{-\infty}^{t} \mathbf{K}_{rad}(t-\tau)\dot{\boldsymbol{x}}(t)\mathrm{d}\tau + \mathbf{K}_s \boldsymbol{x}(t) + \boldsymbol{F}_{joint} + \boldsymbol{F}_{pto}(t) = \boldsymbol{F}_e(t), \quad (6.13)$$

where $\mathbf{A}_\infty$ is the hydrodynamic added mass matrix for $\omega \to \infty$; x, $\dot{\boldsymbol{x}}$ and $\ddot{\boldsymbol{x}}$ are the generalised displacement, velocity and acceleration vectors in the time domain, respectively; $\mathbf{K}_{rad}$ is the retardation function matrix, which can be obtained from the convolution integrals of the frequency-dependent damping matrix $\mathbf{B}_{rad}$, presented by [146]; $\boldsymbol{F}_{joint}$ is the resistant force/moment vector due to the joint connection and control strategy in the time domain; $\boldsymbol{F}_{pto}$ is the resistant force/moment vector due to the PTO system in real time.

The displacement hinged constraint equations in time domain can be expressed as:

$$\mathbf{A}_J x(t) = \mathbf{0}, \quad (6.14)$$

and

$$\mathbf{A}_J^T \boldsymbol{f}_J(t) = \boldsymbol{F}_{joint}(t), \quad (6.15)$$

where $\mathbf{A}_J$ is the same as that employed in the frequency domain; and $\boldsymbol{f}_J$ is the joint force vector in the time domain.

When the two floats are forced to rotate as a whole, i.e., the hinge becomes a rigid connection between the floats, the number of constraints increases from 2 to 3 with a pitch moment M_{check} added to $\boldsymbol{f}_J$ to restrict the relative pitch motion between the two floats. This additional constraint is necessary to calculate for some PTO/control moment, such as the Coulomb damping associated PTO moment and the additional moment required for the latching control, which will be discussed later. The rigid constraint matrix can be written as:

$$\mathbf{A}_J = \begin{bmatrix} 1 & 0 & 0 & -1 & 0 & 0 \\ 0 & 1 & -l_0 & 0 & -1 & -l_0 \\ 0 & 0 & 1 & 0 & 0 & -1 \end{bmatrix}. \quad (6.16)$$

The wave power captured by the device in real time, $P_{pto}(t)$, may be expressed as:

$$P_{pto}(t) = M_{pto}(t)(\dot{x}_5^{(2)}(t) - \dot{x}_5^{(1)}(t)), \quad (6.17)$$

where $M_{pto}(t)$ represents the instantaneous PTO resist moment at the hinge due to PTO damping; and $\dot{x}_5^{(1)}(t)$ and $\dot{x}_5^{(2)}(t)$ denote the pitch velocities of the fore and aft floats, respectively.

The time-average power absorbed by the device, $\overline{P}_{pto}$, can be written as:

$$\overline{P}_{pto} = \frac{\int_{t_0}^{t_0+T} P_{pto}(t)\mathrm{d}t}{T}, \quad (6.18)$$

where t_0 is a moment when the device has already entered the steady state of motion. The corresponding time-average wave power capture factor in the time domain can be calculated with Eq. (6.12).

Coulomb damping

The effect of a hydraulic PTO system can be modelled as Coulomb damping. A challenge for the numerical implementation of the Coulomb damping force is that the derivative is infinite at the vanishing relative rotary velocity, leading to non-physical oscillations of the relative velocity around 0 and erroneous values for wave power absorption [52]. To solve this issue, a criterion condition $|\dot{x}_5^{(2)} - \dot{x}_5^{(1)}| < 0.001$ can be introduced, hence the corresponding PTO moment at the hinge due to the Coulomb damping is expressed as:

$$
M_{pto} = \begin{cases} c_{coulomb} \cdot \text{sign}[\dot{x}_5^{(2)}(t) - \dot{x}_5^{(1)}(t)], & |\dot{x}_5^{(2)}(t) - \dot{x}_5^{(1)}(t)| \geq 0.001 \\ \min\{c_{coulomb}, |M_{check}(t)|\} \cdot \text{sign}[M_{check}(t)], & |\dot{x}_5^{(2)}(t) - \dot{x}_5^{(1)}(t)| < 0.001 \end{cases}
$$

$$(6.19)$$

in which $c_{coulomb}$ denotes the Coulomb damping; M_{check} represents the moment required at the hinge to make the floats rotate as a whole without any effect from the PTO system. When $M_{pto} = M_{check}$, the hinge between the two floats is seen as a rigid connection.

It is believed that the relative rotary velocity $|\dot{x}_5^{(2)}(t) - \dot{x}_5^{(1)}(t)| < 0.001$ is too small for a PTO system to generate electricity, hence the wave power captured by the Coulomb damping may be calculated by:

$$
P_{pto} = \begin{cases} c_{coulomb} \cdot [\dot{x}_5^{(2)}(t) - \dot{x}_5^{(1)}(t)], & |\dot{x}_5^{(2)}(t) - \dot{x}_5^{(1)}(t)| \geq 0.001 \\ 0, & |\dot{x}_5^{(2)}(t) - \dot{x}_5^{(1)}(t)| < 0.001 \end{cases}.
$$

$$(6.20)$$

Latching control

Latching control is a control strategy suitable for a device whose natural period is smaller than the exciting wave period (and hence the attenuator device may have a smaller rotary inertia). The objective behind latching control is to latch the relative rotation of the two floats at the extremes of their relative movement (when the relative rotary velocity is zero), and release them when the wave excitation forces/moments are in a good phase to maximise energy extraction. This may be equivalent to extending the natural period of the device and making it closer to the wave excitation period. When latching control is applied, an additional moment $M_{control} = M_{check}$ must be introduced into the dynamics of the device to cancel the relative acceleration of the controlled motion.

Latching time, $t_{latched}$, which is defined as the latching time duration after the relative rotary velocity of the two floats reaches zero, could be the most important factor for latching control affecting the performance of the device. When $t_{latched}$ is too small, the effect of latching control would be rather limited; whilst if $t_{latched}$ is too large, the natural period of the device would be extended so much that it could be even larger than the wave period, going against the optimisation of wave power extraction. There is an optimised value of $t_{latched}$ for maximising the wave power absorption of the device.

6.4.2 Theoretical model for evaluating the maximum wave power absorption

The maximum wave power absorption by an attenuator consisting of two hinged floats subjected to regular waves can be evaluated by using a trial-and-error method, which is rather time consuming for obtaining accurate results and, hence, goes against the rapid multi parameter optimisation analysis of the device required at the R&D stage. In this section, an analytical model is proposed to evaluate the maximum power absorption and the corresponding optimised PTO system for a two-hinged floats [874, 872].

Following Eq. (6.3), the vector of the displacement of the two floats, together with the vector of the force at the hinge, can be written as:

$$\left\{ \begin{array}{c} \boldsymbol{X} \\ \boldsymbol{F_J} \end{array} \right\} = \mathbf{S}^{-1} \left\{ \begin{array}{c} \boldsymbol{F_e} \\ \mathbf{0} \end{array} \right\}, \tag{6.21}$$

in which $\mathbf{S}$ represents the coefficient matrix of Eq. (6.3). Hence, the relative pitch motion between the two hinged floats and the time-average wave power absorption can be expressed as:

$$\Delta X = \boldsymbol{H_0^T} \mathbf{S}^{-1} \left\{ \begin{array}{c} \boldsymbol{F_e} \\ \mathbf{0} \end{array} \right\}, \tag{6.22}$$

and

$$\overline{P}_{pto} = \frac{1}{2} \omega^2 c_{pto} \left[\mathbf{S}^{-1} \left\{ \begin{array}{c} \boldsymbol{F_e} \\ \mathbf{0} \end{array} \right\} \right]^{*} \boldsymbol{H_0} \boldsymbol{H_0^T} \mathbf{S}^{-1} \left\{ \begin{array}{c} \boldsymbol{F_e} \\ \mathbf{0} \end{array} \right\}, \tag{6.23}$$

respectively, where $\boldsymbol{H_0} = [\begin{array}{ccccccccc} 0 & 0 & 1 & 0 & 0 & -1 & 0 & 0 \end{array}]^T$.

By means of a unitary transformation, the symmetric real matrix $\boldsymbol{H_0}\boldsymbol{H_0^T}$ can be further rewritten as:

$$\boldsymbol{H_0}\boldsymbol{H_0^T} = \mathbf{Q}\boldsymbol{\Pi}^{*}\boldsymbol{\Pi}\mathbf{Q}^{*}, \tag{6.24}$$

where

$$\mathbf{Q} = \begin{bmatrix} 1 & 0 & 0 & 0 & 0 & 0 & 0 & 0 \\ 0 & 1 & 0 & 0 & 0 & 0 & 0 & 0 \\ 0 & 0 & 0 & 0 & -2^{-0.5} & 0 & 0 & -2^{-0.5} \\ 0 & 0 & 0 & 1 & 0 & 0 & 0 & 0 \\ 0 & 0 & 1 & 0 & 0 & 0 & 0 & 0 \\ 0 & 0 & 0 & 0 & -2^{-0.5} & 0 & 0 & 2^{-0.5} \\ 0 & 0 & 0 & 0 & 0 & 1 & 0 & 0 \\ 0 & 0 & 0 & 0 & 0 & 0 & 1 & 0 \end{bmatrix} \tag{6.25}$$

and $\boldsymbol{\Pi} = [\begin{array}{cccccccc} 0 & 0 & 0 & 0 & 0 & 0 & 0 & \sqrt{2} \end{array}]$, for which the time-average power absorbed by the device can be written as:

$$\overline{P}_{pto} = \frac{1}{2} \omega^2 c_{pto} \left| \boldsymbol{\Pi}(\mathbf{S}\mathbf{Q})^{-1} \left\{ \begin{array}{c} \boldsymbol{F_e} \\ \mathbf{0} \end{array} \right\} \right|^2. \tag{6.26}$$

It is worth noting that the element at the bottom right corner of the complex matrix $\mathbf{SQ}$ vanishes. With the employment of the formulas for the block matrix inversion, the inverse of $\mathbf{SQ}$ may be written as:

$$(\mathbf{SQ})^{-1} = \begin{bmatrix} \mathbf{A}_0 & \mathbf{B} \\ \mathbf{C}_0 & 0 \end{bmatrix}^{-1} = \begin{bmatrix} \mathbf{A}_0^{-1} - \mathbf{A}_0^{-1}\mathbf{B}(\mathbf{C}_0\mathbf{A}_0^{-1}\mathbf{B})^{-1}\mathbf{C}_0\mathbf{A}_0^{-1} & \mathbf{A}_0^{-1}\mathbf{B}(\mathbf{C}_0\mathbf{A}_0^{-1}\mathbf{B})^{-1} \\ (\mathbf{C}_0\mathbf{A}_0^{-1}\mathbf{B})^{-1}\mathbf{C}_0\mathbf{A}_0^{-1} & -(\mathbf{C}_0\mathbf{A}_0^{-1}\mathbf{B})^{-1} \end{bmatrix},$$
$$(6.27)$$

where

$$\mathbf{B} = \mathbf{B}_0 + \sqrt{2}(z_{pto} - i\omega c_{pto})[\ 0\ \ 0\ \ -1\ \ 0\ \ 0\ \ 1\ \ 0\ ]^{\mathrm{T}}, \qquad (6.28)$$

$\mathbf{A}_0$, $\mathbf{B}_0$, and $\mathbf{C}_0$ are associated with the PTO system independent partitioned matrices.

Only the last element of $\mathbf{\Pi}$ is non-vanishing, hence the last row of the matrix $\mathbf{SQ})^{-1}$ is of interest, and can be expressed as:

$$[\ (\mathbf{C}_0\mathbf{A}_0^{-1}\mathbf{B})^{-1}\mathbf{C}_0\mathbf{A}_0^{-1}\ \ -(\mathbf{C}_0\mathbf{A}_0^{-1}\mathbf{B})^{-1}\ ] = (\mathbf{C}_0\mathbf{A}_0^{-1}\mathbf{B})^{-1}[\ \mathbf{C}_0\mathbf{A}_0^{-1}\ \ -1\ ], \quad (6.29)$$

where

$$\begin{aligned}
\mathbf{C}_0\mathbf{A}_0^{-1}\mathbf{B} &= \mathbf{C}_0\mathbf{A}_0^{-1}(\mathbf{B}_0 + \sqrt{2}(z_{pto} - i\omega c_{pto})[\ 0\ \ 0\ \ -1\ \ 0\ \ 0\ \ 1\ \ 0\ ]^{\mathrm{T}}) \\
&= a_0(z_{pto} - i\omega c_{pto}) + b_0,
\end{aligned} \qquad (6.30)$$

in which a_0 and b_0 are independent of the PTO system.

Using Eqs. (6.26)–(6.30), the time-average power absorbed by the device can be ultimately expressed as:

$$\overline{P}_{pto} = \frac{\left| \mathbf{C}_0\mathbf{A}_0^{-1}\left\{ \begin{matrix} \mathbf{F}_e \\ 0 \end{matrix} \right\} \right|^2 \omega^2 c_{pto}}{|a_0(z_{pto} - i\omega c_{pto}) + b_0|^2}, \qquad (6.31)$$

in which the PTO associated parameters, i.e., c_{pto} and z_{pto}, are separated from the complicated hydrodynamic coefficients and structure mass/stiffness properties. Expressions of the maximum wave power absorption and the corresponding optimised PTO parameters easily can be derived, based on Eq. (6.31).

6.4.2.1   Maximum wave power absorption without any constraints

For any certain value of z_{pto}, when $\partial \overline{P}_{pto}/\partial c_{pto} = 0$ is satisfied, which occurs if:

$$c_{pto} = \frac{|a_0 z_{pto} + b_0|}{|a_0|\omega} \equiv c_{opt}, \qquad (6.32)$$

the maximum of absorbed power:

$$P_{max} = \frac{1}{2}\left| \mathbf{C}_0\mathbf{A}_0^{-1}\left\{ \begin{matrix} \mathbf{F}_e \\ 0 \end{matrix} \right\} \right|^2 \frac{\omega}{|a_0||a_0 z_{pto} + b_0| + \mathrm{Im}(a_0 b_0^*)} \qquad (6.33)$$

can be achieved. The amplitude of the corresponding relative rotation displacements between the two floats for this optimisation is:

$$|\Delta X_{opt}| = \frac{\sqrt{2}\left|\mathbf{C}_0\mathbf{A}_0^{-1}\left\{\begin{array}{c} \mathbf{F}_e \\ 0 \end{array}\right\}\right|}{\left|(a_0 z_{pto} + b_0) - \frac{ia_0}{|a_0|}|a_0 z_{pto} + b_0|\right|}. \tag{6.34}$$

The wave power absorption can be further enhanced should z_{pto} can be arbitrarily selected. Ideally, if $\partial \overline{P}_{pto}/\partial c_{pto} = 0$ and $\partial \overline{P}_{pto}/\partial z_{pto} = 0$ can be satisfied simultaneously, for which:

$$z_{pto} = -\frac{\mathrm{Re}(a_0 b_0^*)}{|a_0|^2} \equiv z_{OPT}, \qquad c_{pto} = \frac{|\mathrm{Im}(a_0 b_0^*)|}{|a_0|^2 \omega} \equiv c_{OPT}, \tag{6.35}$$

then the maximum absorbed power and the amplitude of the corresponding relative rotation displacement between the floats are given by:

$$P_{MAX} = \frac{1}{4}\left|\mathbf{C}_0\mathbf{A}_0^{-1}\left\{\begin{array}{c} \mathbf{F}_e \\ 0 \end{array}\right\}\right|^2 \frac{\omega}{\mathrm{Im}(a_0 b_0^*)}, \tag{6.36}$$

and

$$|\Delta X_{OPT}| = \frac{\left|\mathbf{C}_0\mathbf{A}_0^{-1}\left\{\begin{array}{c} \mathbf{F}_e \\ 0 \end{array}\right\}\right||a_0|}{\sqrt{2}\mathrm{Im}(a_0 b_0^*)}, \tag{6.37}$$

respectively.

6.4.2.2 Maximum wave power absorption with constraints

The power absorptions derived above are all obtained without any consideration of displacement limitations. In practice, there are physical limitations between two hinged floats to restrict the amplitude of their relative rotation and protect the floats from striking each other. Moreover, the theoretical maximum wave power absorption for a small device oscillating in long waves may be unobtainable [211]. In long waves, a relatively small device requires large response amplitudes for optimum power absorption. Too large response makes the assumptions of linear potential flow theory invalid.

To consider the effect of physical limitations between two hinged floats, a constraint can be imposed as:

$$|\Delta X|^2 = \Delta X^* \Delta X \le \delta^2, \tag{6.38}$$

in which δ is the upper limit amplitude of the relative rotation displacement of the hinged floats.

The maximum wave power absorption with optimised c_{pto} under the relative rotation constraint is denoted by $P_{max,c}$. If the relative rotation amplitude predicted by Eq. (6.34) satisfies Eq. (6.38), then $\Delta X = \Delta X_{opt}$ gives the maximum wave power absorption $P_{max,c} = P_{max}$. If $|\Delta X_{opt}| > \delta$, then the wave power absorption should be maximised, subject to $|\Delta X| = \delta$. This may be achieved by introducing a Lagrange multiplier, μ, and a scalar, Q, expressed as:

$$Q = \overline{P}_{pto} - \frac{1}{2}\mu(\Delta X^*\Delta X - \delta^2) = \frac{1}{2}\omega^2 c_{pto}\Delta X^*\Delta X - \frac{1}{2}\mu(\Delta X^*\Delta X - \delta^2)$$

$$= \frac{\left|\mathbf{C}_0\mathbf{A}_0^{-1}\left\{\begin{matrix}\mathbf{F}_e\\0\end{matrix}\right\}\right|^2 (\omega^2 c_{pto} - \mu)}{|a_0(z_{pto} - i\omega c_{pto}) + b_0|^2} + \frac{1}{2}\mu\delta^2. \tag{6.39}$$

The maximum power absorbed with optimised c_{pto} under the relative rotation constraint can be achieved when:

$$\frac{\partial Q}{\partial c_{pto}} = 0, \qquad \frac{\partial Q}{\partial \mu} = 0 \tag{6.40}$$

are satisfied, which occurs if:

$$c_{pto} = \frac{1}{|a_0|^2\omega}\left[\sqrt{2\left|\mathbf{C}_0\mathbf{A}_0^{-1}\left\{\begin{matrix}\mathbf{F}_e\\0\end{matrix}\right\}\right|^2 \frac{|a_0|^2}{\delta^2} - [|a_0|^2 z_{pto} + \mathrm{Re}(a_0 b_0^*)]^2} - \mathrm{Im}(a_0 b_0^*)\right] \equiv c_{opt,c}, \tag{6.41}$$

for which

$$P_{max,c} = \frac{1}{2}\omega^2 c_{opt,c}|\Delta X|^2$$

$$= \frac{1}{2}\frac{\delta^2\omega}{|a_0|^2}\left[\sqrt{2\left|\mathbf{C}_0\mathbf{A}_0^{-1}\left\{\begin{matrix}\mathbf{F}_e\\0\end{matrix}\right\}\right|^2 \frac{|a_0|^2}{\delta^2} - [|a_0|^2 z_{pto} + \mathrm{Re}(a_0 b_0^*)]^2} - \mathrm{Im}(a_0 b_0^*)\right]. \tag{6.42}$$

The maximum wave power absorption with c_{pto} and z_{pto} optimised simultaneously under the relative rotation constraint, denoted as $P_{MAX,c}$, can be predicted in a similar manner. If $|\Delta X_{OPT}| \leq \delta$, then $\Delta X = \Delta X_{OPT}$ gives $P_{MAX,c} = P_{MAX}$. If $|\Delta X_{OPT}| > \delta$, the wave power absorption should be maximised subject to $|\Delta X| = \delta$, for which the condition becomes:

$$\frac{\partial Q}{\partial c_{pto}} = 0, \qquad \frac{\partial Q}{\partial z_{pto}} = 0, \qquad \frac{\partial Q}{\partial \mu} = 0, \tag{6.43}$$

resulting in:

$$\begin{cases} z_{pto} = -\dfrac{\mathrm{Re}(a_0 b_0^*)}{|a_0|^2} \equiv z_{OPT,c} \\ c_{pto} = \dfrac{1}{|a_0|^2\omega}\left[\sqrt{2}\left|\mathbf{C}_0\mathbf{A}_0^{-1}\left\{\begin{matrix}\mathbf{F}_e\\0\end{matrix}\right\}\right|\dfrac{|a_0|}{\delta} - \mathrm{Im}(a_0 b_0^*)\right] \equiv c_{OPT,c} \end{cases}, \tag{6.44}$$

and

$$P_{MAX,c} = \frac{1}{2}\omega^2 c_{OPT,c}|\Delta x|^2$$

$$= \frac{1}{2}\frac{\delta\omega}{|a_0|}\left[\sqrt{2}\left|\mathbf{C}_0\mathbf{A}_0^{-1}\left\{\begin{matrix}\mathbf{F}_e\\0\end{matrix}\right\}\right| - \frac{\delta}{|a_0|}\mathrm{Im}(a_0 b_0^*)\right]. \tag{6.45}$$

It is interesting to note that the conditions to maximise the power in terms of $z_{OPT,c}$ are the same regardless of the constraints (see Eq. (6.44)); while the PTO damping must account for the upper limit δ.

In this section, an attenuator device, consisting of two hinged floats with an elliptical section floating in the sea with water depth $h = 20$ m, is considered (see Figure 6.8). The detailed geometric and physical properties are as follows: the length of each float is $L = 20$ m; the major and minor axes of the elliptical section of each float are $a = 5$ m and $b = 2.5$ m, respectively, unless otherwise specified; the spacing between the floats is $l_s = 1$ m; and the density of water and acceleration of gravity are $\rho = 1025$ kg/m^3 and $g = 9.81$ m/s^2, respectively. The density of each float is half as large as that of the water, hence the device is semi-submerged in the water. The mass is uniformly distributed all over the floats unless otherwise specified. A train of regular waves with an amplitude $A = 1.0$ m and period $T = 5.0$ s (unless otherwise specified) passing along the length of the hinged floats is considered.

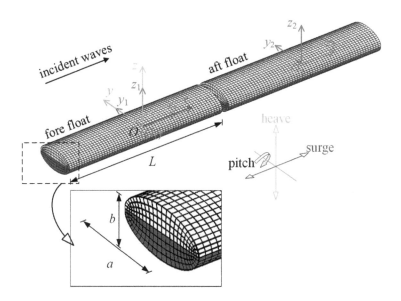

Figure 6.8: Schematic of the attenuator device, consisting of two hinged floats.

6.4.3 Frequency-domain analysis

For the device subjected to regular waves, the effects of a linear PTO, float rotary inertia, and axis ratio of the cross section on wave power absorption of the device can be studied in the frequency domain.

6.4.3.1 Float length and linear power take-off damping

To examine the effect of PTO damping and float length on wave power absorption, a wide range of the dimensionless linear PTO damping coefficient $c_{pto}\sqrt{gh}/(\rho g a L^4) = 0 \sim 0.05$ and float length $L/h = 0.25 \sim 2.0$ are examined.

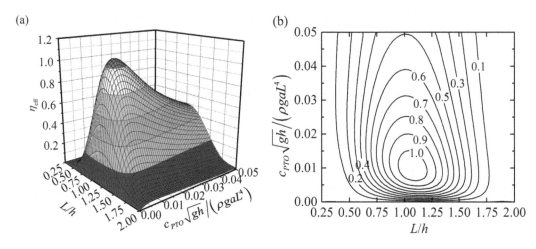

Figure 6.9: Variation of wave power capture factor η_{eff} with the dimensionless PTO damping coefficient and float length for $b/a = 0.5$ and $T = 5$ s: (a) 3D power capture factor and (b) power capture factor contour.

Figure 6.9 illustrates the variation in the wave power capture factor with the dimensionless PTO damping coefficient and dimensionless float length for $b/a = 0.5$ and $T = 5$ s. The wave power capture factor η_{eff} is found to increase first with both the dimensionless PTO damping coefficient or float length, and then monotonically decreases after reaching a maximum value, i.e., there is an optimal float length and an optimal PTO damping coefficient to maximise the wave power absorption of the device. The maximum wave power capture width factor reaches $\eta_{\text{eff}} = 1.07$ at $c_{pto} \sqrt{gh}/(\rho gaL^4) = 0.01$ and $L/h = 1.0$. Figure 6.10 presents the corresponding pitch excitation moments acting on each float as well as the pitch velocities of each float for this optimised circumstance. In spite of the optimal combination of float length and PTO damping coefficient, there still exist obvious phase lags between the pitch excitation moments and the pitch velocities, indicating that at some time the pitch excitation moments resist the float motions rather than drive them. There is still room for maximising the wave power absorption should more parameters be optimised, e.g., via float rotary inertia, the effect of which will be discussed next.

6.4.3.2 Float rotary inertia

The mass moment of inertia around the centre of the float mass, which is generally expressed in terms of the float radius of gyration r, is one of the vital parameters affecting the performance of the device.

Figure 6.11 shows the variation of η_{eff} with the dimensionless PTO damping coefficient for various dimensionless float radii of gyration r/L at $T = 4$ s, 5 s, and 6 s. For any specified wave condition, there is an optimal radius of gyration to significantly improve wave power absorption of the device all over the examined range of PTO damping coefficients. For a smaller wave period, the optimal radius of gyration for maximising wave power absorption is smaller, and vice versa. Too large or too small

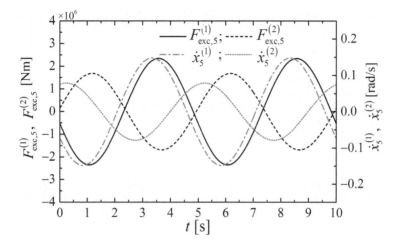

Figure 6.10: Variation of pitch velocities of the floats and pitch excitation moments acting on the floats over time for the optimised case $L/h = 1.0$, $T = 5$ s, and $c_{pto} \sqrt{gh}/(\rho g a L^4) = 0.01$.

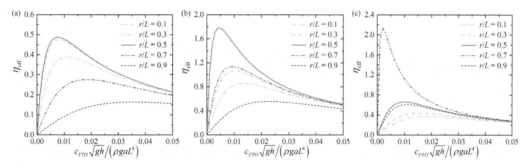

Figure 6.11: Variation of wave power capture factor η_{eff} with a dimensionless PTO damping coefficient for various radii of gyration $L/h = 1.0$, $b/a = 0.5$: (a) $T = 4$ s; (b) $T = 5$ s; and (c) $T = 6$ s.

a radius of gyration is not conducive to wave power absorption, resulting in a smaller η_{max}.

Figure 6.12 demonstrates a contour of the wave power capture factor η_{eff} versus the dimensionless PTO damping coefficient and radius of gyration for $T = 5$ s. It indicates that there is an optimal PTO damping coefficient, $c_{pto} \sqrt{gh}/(\rho g a L^4) = 0.0039$, and an optimal radius of gyration, $r/L = 0.55$, to maximise the power capture factor to 1.93, much larger than the maximum capture factor, 1.07, for $c_{pto} \sqrt{gh}/(\rho g a L^4) = 0.01$ and $r/L = 0.29$ (Figure 6.9) under the mass uniformly distributed assumption. Figure 6.13 shows the variation of the pitch velocities of the floats and the pitch excitation moments acting on the rafts with time for $c_{pto} \sqrt{gh}/(\rho g a L^4) = 0.0039$ and $r/L = 0.55$. Only slight phase lags between $F^{(1)}_{exc,5}$ and $\dot{x}^{(1)}_5$ as well as $F^{(2)}_{exc,5}$ and $\dot{x}^{(2)}_5$ are observed, which are rather different from the ones shown in Figure 6.10. It means that the pitch excitation moments nearly drive the float motions all over each wave period, providing benefits to wave power absorption.

(a)

(b)

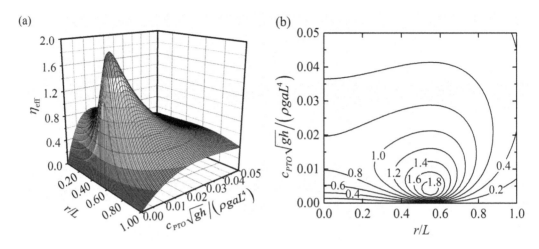

Figure 6.12: Variation of wave power capture factor η_{eff} with dimensionless PTO damping coefficient and radius of gyration for $L/h = 1.0$, $b/a = 0.5$, and $T = 5$ s: (a) 3D power capture factor and (b) power capture factor contour.

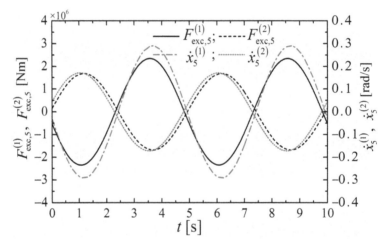

Figure 6.13: Variation of pitch velocities of the floats and pitch excitation moments acting on the floats with time for $L/h = 1.0$, $T = 5$ s, $c_{pto} \sqrt{gh}/(\rho gaL^4) = 0.0039$, and $r/L = 0.55$.

6.4.3.3 Axis ratio of cross section

Performance of the devices with $a/b = 0.4, 0.5, 0.6, 0.8$ and 1.0 is examined to investigate the effect of the axis ratio of cross section on wave power absorption. Figure 6.14 shows that a larger axis ratio results in a larger wave power capture factor for a small PTO damping coefficient. Nevertheless, for large PTO damping coefficients, increasing the axis ratio weakens wave power absorption. As indicated in Figure 6.14b, the larger the axis ratio b/a, the larger the maximum wave power capture factor η_{max} and the smaller the optional dimensionless PTO damping coefficient required.

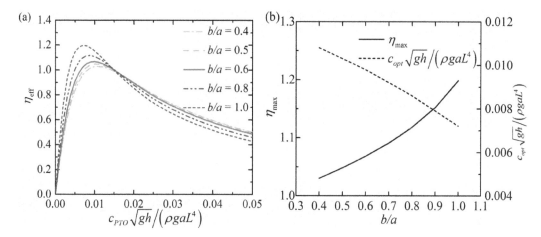

Figure 6.14: Effect of the axis ratio of cross-section b/a on the wave power capture factor for $T = 5$ s: (a) variation of η_{eff} with the dimensionless PTO damping coefficient for different b/a; (b) variation of the maximum wave power capture factor η_{max} and the corresponding optimal PTO damping coefficient with b/a.

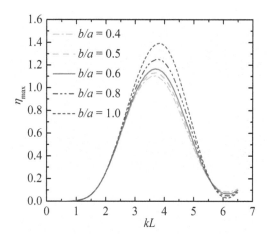

Figure 6.15: Frequency response of the maximum wave power capture factor η_{max} for $L/h = 1.0$, $b/a = 0.4$, 0.5, 0.6, 0.8, and 1.0.

Figure 6.15 shows the frequency response of the maximum wave power capture factor η_{max} for $L/h = 1.0$, $b/a = 0.4$, 0.5, 0.6, 0.8, and 1.0. For $kL = 2.5 \sim 5.0$, a larger b/a is found to give more advantage in the probability to increase power capture factor.

It should be pointed out that the results as plotted in Figures 6.14 and 6.15 are all based on the assumption that the mass is uniformly distributed over each float.

Figure 6.16 shows a comparison between the performance of the devices with cross section aspect ratios $b/a = 0.5$ and 1.0 for dimensionless optimal radii of gyration $r/L = 0.55$ and 0.4, and optimal dimensionless damping coefficients 0.0039 and 0.0031, respectively, which are associated with the optimisation for $T = 5$ s (corresponding

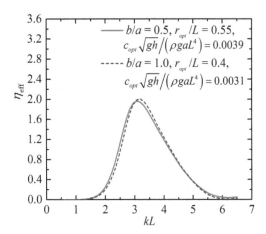

Figure 6.16: Frequency response of η_{eff} of the devices for $b/a = 0.5$ and 1.0 with PTO damping and the radius of gyration optimised at $T = 5$ s (corresponding to $kL = 3.2$).

to $kL = 3.2$). The difference in η_{eff} between these two curves for $b/a = 0.5$ and 1.0 at the same kL is found to be rather small and can be neglected.

6.4.4 Time-domain analysis

The effect of Coulomb damping and latching control on wave power absorption of the device is studied in the time domain. The device is subjected to regular waves with $T = 5$ s and $A = 0.5$ m.

6.4.4.1 Coulomb damping

In the case where a hydraulic PTO system is applied to the device, the PTO effect can be approximated as Coulomb damping. Figure 6.17 shows how the displacement and velocity of the floats in pitch, together with the PTO moment at the hinge and the pitch wave excitation moments varies with time for $c_{coulomb}/(\rho g A a L^2) = 0.06$ and $T = 5$ s. As expected, at some moments, there is no relative angular rotation displacement, and the two floats rotate at the same velocity like a unified whole moving in the waves. As shown in Figure 6.17b, there are obvious phase lags between the pitch excitation moments and the pitch velocities of each float, especially those of the aft float.

Variation of the time-average wave power capture factor, η_{eff}, with a dimensionless Coulomb damping coefficient for $r/L = 0.1$, 0.3, 0.5, 0.7, 0.9, and $T = 5.0$ s is shown in Figure 6.18. η_{eff} is found to increase with increasing $c_{coulomb}$, and then decrease after reaching a maximum value, i.e., there is a proper value of $c_{coulomb}$ to maximise the wave power capture factor. The time-average capture factor of the device with $r/L = 0.5$ and Coulomb damping $c_{coulomb}/(\rho g A a L^2) = 0.045$ is found to be much larger than those for the other cases examined. For $c_{coulomb}/(\rho g A a L^2) > 0.08$, the floats with a smaller radius of gyration tend to capture more power from the waves.

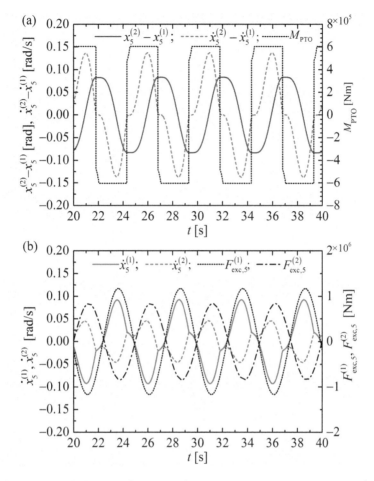

Figure 6.17: Response of the two floats in the time domain for $T = 5$ s, $L/h = 1.0$, and $c_{coulomb}/(\rho g A a L^2) = 0.06$: (a) pitch displacement of the floats and the PTO moment at the hinge and (b) pitch velocity of the floats and the pitch wave excitation moments acting on the floats.

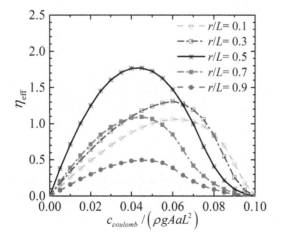

Figure 6.18: Variation of the time-average wave power capture factor with $c_{coulomb}/(\rho g A a L^2)$ for $T = 5$ s, $L/h = 1.0$, and different r/L.

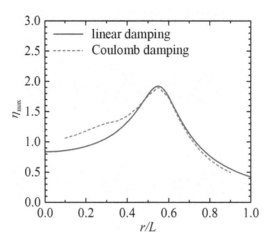

Figure 6.19: Variation of the maximum wave power capture factors with the float radius of gyration for devices using linear damping and Coulomb damping, $T = 5$ s.

To identify the difference in behaviour exhibited between linear and nonlinear (Coulomb) damping, a comparison of the peak wave power capture factors obtained by using linear damping and Coulomb damping is plotted in Figure 6.19. The peak maximum wave power capture factors obtained by using Coulomb damping, 1.91, is almost the same as that (1.93) obtained by using linear PTO damping. For $r/L <$ 0.4, the maximum wave power capture factor obtained by using optimised Coulomb damping is obviously larger than that found by using optimised linear damping; while for $r/L > 0.7$, the maximum wave power capture factor obtained by using linear PTO damping is slightly larger than that found by using Coulomb damping.

6.4.4.2 Latching control

In the study of the latching control, linear damping is adopted in the PTO system. Figure 6.20 shows the variation of the relative rotary displacement/velocity and the pitch wave excitation moments for $c_{pto} \sqrt{gh}/(\rho g a L^4) = 0.01$, $T = 5$ s, and $t_{latched} =$ 0.75 s. There is an obvious relative stall when the relative rotary displacement reaches the maximum. With the employment of the latching control with $t_{latched} = 0.75$ s, the pitch velocities of the floats are highly in phase with the pitch excitation moments, compared with those shown in Figure 6.10.

Figure 6.21 shows the variation of the average power capture factor with $t_{latched}$ for various dimensionless PTO damping coefficients and $T = 5$ s. As $t_{latched}$ increases from 0 to 2.5 s, the wave power capture factor first increases and then decreases after reaching a maximum value. For $t_{latched} < 1.2$, the larger the PTO damping coefficient is, the more slowly the wave power capture factor varies with $t_{latched}$. For $c_{pto} \sqrt{gh}/(\rho g a L^4) = 0.0025$, the wave power capture factor reaches 1.59, which occurs at $t_{latched} = 0.88$ s, almost three times as large (0.54) as that without any latching control. Nevertheless, as expected, when the latch time is too long, e.g., $t_{latched} > 1.3$ s,

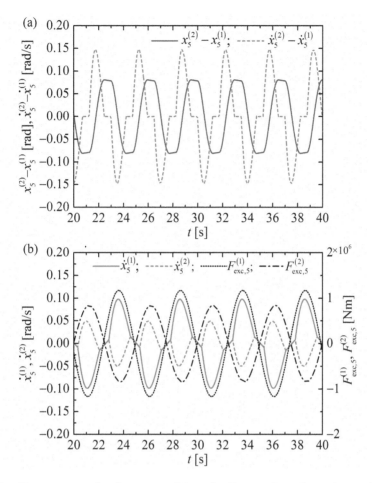

Figure 6.20: Response of the two hinged floats in the time domain for $c_{pto}\sqrt{gh}/(\rho gaL^4) = 0.01$, $T = 5$ s, and $t_{latched} = 0.75$ s: (a) pitch displacements of the floats and the PTO moment at the hinge and (b) pitch velocities of the floats and the pitch wave excitation moments.

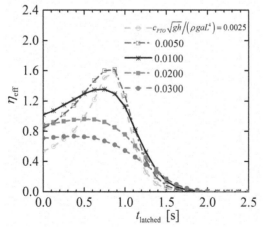

Figure 6.21: Variation of the time-average wave power capture factor with $t_{latched}$ for various c_{pto} with $T = 5$ s.

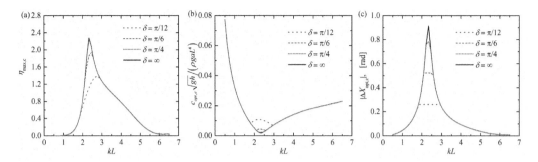

Figure 6.22: Frequency response of the maximum wave power capture factor, the corresponding optimised PTO damping, and the amplitude of the relative rotation between the hinged floats with $z_{pto} = 0$: (a) $\eta_{max,c}$; (b) $c_{opt,c}\sqrt{gh}/(\rho g a L^4)$; and (c) $|X_{opt,c}|$.

it is observed that the wave power capture factors are all smaller than those without any latching control.

6.4.5 Maximum wave power absorption

In this subsection, the maximum wave power absorption of the same device is investigated with the model, as introduced in Section 6.4. The device is subjected to regular waves of amplitude $A = 1.0$ m.

6.4.5.1 Maximum power absorption with optimised power take-off damping

Figure 6.22 presents the frequency response of maximum wave power capture factor, $\eta_{max,c}$, the corresponding optimised PTO damping, $c_{opt,c}\sqrt{gh}/(\rho g a L^4)$, and the amplitude of the relative rotation displacement between the hinged floats, $|X_{opt,c}|$. Four values of the upper limit amplitude of the relative rotation displacement are considered, i.e., $\delta = \pi/12, \pi/6, \pi/4$, and ∞. For the case with $\delta = \infty$, the constraint of the relative rotation is released and hence $\eta_{max,c} = \eta_{max}$, $c_{opt,c} = c_{opt}$, and $|X_{opt,c}| = |X_{opt}|$. Figure 6.22a indicates that in the four upper limits examined, the larger the value of δ, the lager the peak value of $\eta_{max,c}$, and the smaller the kL where the peak of $\eta_{max,c}$ occurs. Meanwhile, the peak of $\eta_{max,c}$ associated $c_{opt,c}$ is smaller for a larger δ (Figure 6.22b). As expected, when $\eta_{max,c} < \eta_{max}$, $|X_{opt,c}| = \delta$ (Figure 6.22c).

6.4.5.2 Maximum power absorption with optimised power take-off damping and stiffness/inertia

For simplicity, it is assumed that either k_{pto} or I_{pto} vanishes. Since $z_{pto} = k_{pto} - \omega^2 I_{pto}$, if $z_{pto} > 0$, there is no inertia needed in the PTO system, $I_{pto} = 0$, and $k_{pto} = z_{pto}$; else if $z_{pto} < 0$, it means that there is no stiffness needed in the PTO system, $k_{pto} = 0$, and $I_{pto} = -z_{pto}/\omega^2$.

Figure 6.23 presents the maximum wave power capture factor when PTO damping and PTO stiffness/inertia are optimised simultaneously, the corresponding optimised PTO parameters, and the amplitude of the relative rotation displacement between

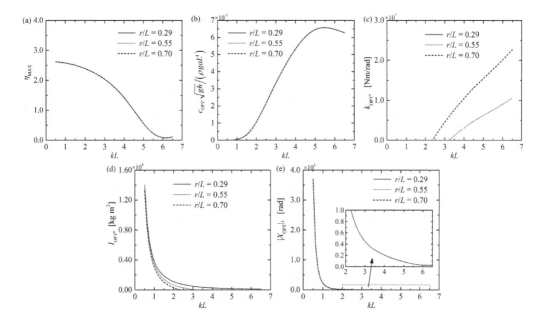

Figure 6.23: Frequency response of the maximum wave power capture factor, the corresponding optimised PTO damping, the corresponding optimised PTO stiffness/inertia, and the amplitude of the relative rotation between the hinged floats: (a) η_{MAX}; (b) $c_{OPT} \sqrt{gh}/(\rho ga L^4)$; (c) k_{OPT}; (d) I_{OPT}; and (e) $|X_{OPT}|$.

the hinged floats. Different curves represent the device with different values of the float rotary inertia radius, $r/L = 0.29$ (where the float mass is uniformly distributed), 0.55 and 0.70.

The float rotary inertia radius r/L is found to have no influence (or rather limited influence) on η_{MAX}, c_{OPT} and $|X_{OPT}|$ (see Figure 6.23a, b, and e). As kL decreases from 5.0 to 0.5, the maximum wave power capture factor of the device with a totally optimised PTO system (Figure 6.23a) shows a slow increase; whereas the relative rotation response (Figure 6.23e) shows a rapid growth.

For $kL < 2.0$, η_{MAX} can be larger than 2.3 and the corresponding $|X_{OPT}|$ is larger than 1.5 rad. As kL increases from 0.5 to 6.5, $c_{OPT} \sqrt{gh}/(\rho ga L^4)$ first increases and then decreases after reaching a peak around $kL = 5.5$. The optimal PTO stiffness and PTO inertia are found to be dependent on the float rotary inertia radius (Figure 6.23c and d). For a smaller float rotary inertia, e.g., $r/L = 0.29$, a PTO system with no stiffness, a proper PTO damper and an optimised PTO inertia dependent on wave frequency is available to maximise the wave power capture factor for all the examined wave conditions. Nevertheless, for a larger float rotary inertia radius, e.g., $r/L = 0.70$, a proper PTO stiffness is needed instead of a PTO inertia for large wave frequencies, e.g., $kL > 2.5$.

A $|X_{OPT}|$ larger than 1.5 rad is required for $kL < 2.0$, as illustrated in Figure 6.23e, but this is obviously unattainable in practice. Figure 6.24 shows the frequency response of $\eta_{MAX,c}$, $c_{OPT,c} \sqrt{gh}/(\rho ga L^4)$, and $|X_{OPT,c}|$ with consideration of the impact of the upper limit of the relative rotation displacement amplitude. The

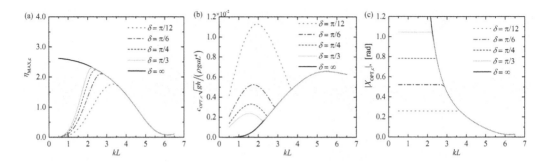

Figure 6.24: Frequency response of the maximum wave power capture factor with consideration of the upper limit impact, the corresponding optimised PTO damping and the amplitude of the relative rotation between the hinged floats with $r/L = 0.70$: (a) $\eta_{MAX,c}$; (b) $c_{OPT,c}\sqrt{gh}/(\rho gaL^4)$; and (c) $|X_{OPT,c}|$.

optimised PTO stiffness/inertia are independent of δ, as proved in Section 6.4.2.2 (see Eq. (6.44)), meaning that they are the same as those illustrated in Figure 6.23c and d with $r/L = 0.70$.

As shown in Figure 6.24a, there are peak values of the maximum wave power capture factor if the motion constraints are considered. The more rigid the motion constraint, the larger the optimal non-dimensional PTO damping coefficient $c_{OPT,c}\sqrt{gh}/(\rho gaL^4)$ for the peak of $\eta_{MAX,c}$ (Figure 6.24b). When the curves of $\eta_{MAX,c}$-kL for any certain upper limit δ disjoint from that without any constraints, $|X_{OPT,c}|$ holds the value of δ (Figure 6.24c).

6.4.5.3 Wave power extraction with different optimisation principles

The performance of the device is significantly influenced by the PTO optimisation principles. Figure 6.25 presents a comparison of the maximum wave power capture width of the device using different optimisation principles, especially for a certain wave condition $kL = 3.2$ (i.e., $T = 5$ s) and $A = 1.0$ m. If the PTO system is allowed to adjust itself to have c_{pto} and z_{pto} optimised simultaneously and to adapt to different wave conditions, then $\eta_{MAX,c}$ and η_{MAX} could be achieved for all kL, respectively, depending on whether the motion constraints are considered or not. If z_{pto} is fixed and only c_{pto} is free to change for various wave conditions, then η_{max} could be achieved for all kL. $r/L = 0.29$ is associated with the case where the float mass is uniformly distributed all over the float. $r/L = 0.55$ denotes the optimal rotary inertia radius that gives the maximum wave power absorption of the device at $kL = 3.2$ (i.e., $T = 5$ s). Although the wave power capture factor is optimised at $kL = 3.2$ for the case of $k_{pto} = I_{pto} = 0$ and $r/L = 0.55$, the peak of the η_{max}-kL curve occurs at $kL = 2.95$.

6.4.5.4 Performance of a device with a fixed optimised power take-off system

It is not practical to design an optimised c_{pto} and/or optimised z_{pto} for various wave conditions and the respective frequency dependent properties of the device as it would significantly increase the system complexity and costs. For simplicity, choosing proper

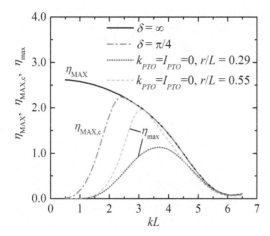

Figure 6.25: Frequency response of the maximum wave power capture factor for different optimisation principles.

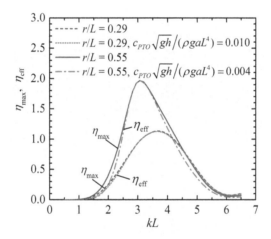

Figure 6.26: Comparison of η_{max} with optimised c_{pto} and η_{eff} with fixed c_{pto} for $k_{pto} = I_{pto} = 0$.

wave frequency independent c_{pto} and z_{pto} to improve the power capture ability of the device may also be welcomed. Figure 6.26 shows the frequency response of η_{max} and η_{eff} for different r/L and $(r/L, c_{pto})$, respectively. The specified c_{pto} is selected in the same value of c_{opt} at $kL = 3.2$. The $\eta_{\text{eff}} - kL$ curves are found to own narrower bandwidths compared with those of $\eta_{max} - kL$. Compared with the case of $r/L = 0.29$, the difference between η_{eff} and η_{max} for $r/L = 0.55$ is more obvious.

6.5 SUMMARY

This chapter is focused on attenuator WECs, which are mainly composed of a series of hinged floats or a submerged flexible tube lying parallel to the wave direction. After delivering an overview of attenuator WECs, the modelling methods of two hinged floats in both the frequency and the time domain are introduced. The numerical models are based on linear potential flow theory and the three-dimensional wave radiation-diffraction theory. A mathematical model is also developed to evaluate the maximum wave power absorption of a device with an optimal PTO damper and a certain or an optimal PTO stiffness/inertia. The effect of the relative rotation constraints is also considered.

Numerical models are employed to examine the effect of multiple parameters, e.g., float length, float rotary inertia, Coulomb damping, and latching control. The maximum wave power absorption of the two hinged floats for different optimisation principles is also evaluated and compared. The following conclusions may be drawn:

- For certain wave conditions, when both the float length and the linear PTO damping coefficient are optimised simultaneously to maximise energy absorption, there still exist obvious phase lags between pitch excitation moments and pitch velocities. The phase lags can be reduced by using a non-uniform distribution of the mass along the floats, leading to a much larger wave power capture factor.

- The latching control can make the floats rotate in phase with the wave excitation moments without changing the radius of the float gyration. For a specified damping, there is a proper latching time to maximise the wave power absorption.

- For long incident waves, the maximum wave power capture factor η_{MAX} of a device with an optimised PTO system under constraints is much smaller than that without any motion constraints. Nevertheless, whatever the upper limit amplitude of the relative rotation displacement of the hinged floats, the corresponding optimal PTO stiffness/inertia is the same as that without any motion constraints.

Although the numerical models presented in this chapter are mainly about devices consisting of two-hinged floats, they can be extended to study the hydrodynamic characteristics of hinged-float-based attenuator WECs with three or even more floats. The models can be further applied to the cases of flexible-tube-based attenuator devices should generalised modes be included to represent structural deflections and tube deformation.

The numerical models are proposed in the framework of linear potential flow theory, which does not capture viscous effects. Hence the models may not be suitable for extreme wave–structure interactions.

CHAPTER 7

Oscillating water column wave energy converters

Rongquan Wang, Dezhi Ning, Robert Mayon
Dalian University of Technology, rqwang@dlut.edu.cn

7.1 INTRODUCTION

This chapter presents the current state of the art design method for the oscillating water column (OWC) type of wave energy converters. As one of the most promising types of wave energy device, the OWC WEC has been extensively studied and a number power of plants have been constructed due to its mechanical and structural simplicity [168, 842] and reliability. Furthermore, several full sized OWC prototypes have been installed and tested around the world. However, at present, OWC technology has not been fully commercialised [371]. The main reason is that the hydrodynamic behaviour at the OWC devices has not been fully understood. Thus, further hydrodynamic investigations into OWC devices still need to be carried out through theoretical, numerical, and experimental approaches.

According to the previous research, it is recommended that OWC devices operate at near-resonance conditions to achieve high-efficiency performance [166]. As a result, the effective frequency bandwidth of the device is usually narrow. To improve its hydrodynamic performance, the concept of a dual-chamber OWC device was proposed. Along with the efficiency, the survivability of the OWC is another important factor that needs to be considered at the design stage. Hydrodynamic loads have remarkable effects on the subsurface structure of marine renewable energy converters [458] and determine the device's survivability. Thus, in the present study, the hydrodynamic performance of both single and dual-chamber OWC devices are investigated experimentally and numerically. The main foci of the study are the hydrodynamic efficiency of the device and the effect of wave loads on the OWC structure. The effects of wave conditions, front wall draughts, chamber widths and the opening ratios of the duct on hydrodynamic performance are discussed.

DOI: 10.1201/9781003198956-7

7.2   WORKING PRINCIPLE

There are various types of OWC device, yet the main working principle is similar, i.e., the trapped air inside the chamber is cyclically compressed and decompressed by wave action. The air is consequently expelled from the chamber and drawn into the chamber through an orifice. A turbine is located at this orifice and the dynamic action of the air through the turbine transfers the wave energy into pneumatic energy and finally into electrical energy. In terms of the standard classification of WECs, the OWC device could be categorised as a fixed or a floating device [42].

7.2.1   Fixed oscillating water column

Conventional OWCs. The most common OWC device structure is shown in Figure 7.1a. Under the action of the incident wave, the oscillating vertical motion of the water column inside the chamber forces the trapped air to flow in and out of the duct periodically, thus driving a turbine which is coupled to a generator to generate electricity. Therefore, the energy conversion of the OWC device is composed of three stages: at the primary conversion stage, wave energy is converted to pneumatic energy, at the secondary conversion stage, the pneumatic energy is transformed in to mechanical energy, and at the third conversion stage, the mechanical energy is transformed to electric energy.

   U-OWCs. The vertical duct significantly changes the excitation form of the U-OWC device (see Figure 7.1b) with respect to conventional OWC devices. The incident wave trains cannot propagate into the inner chamber, but the hydrodynamic force exerted at the upper opening of the vertical duct induces an up and down motion of the water column and then a reciprocating air-flow through the orifice. Compared with the conventional OWC devices, the Eigen period of a U-OWC device and the air pressure inside the U-OWC chamber are larger [77].

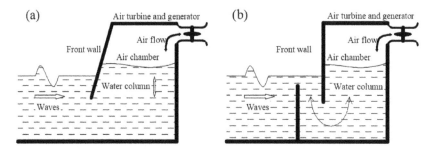

Figure 7.1: Schematic of a (a) conventional OWC and (b) U-OWC.

7.2.2   Floating oscillating water column

Backward bent duct buoys (BBDB). The concept of a Backward Bent Duct Buoy (BBDB) OWC was first proposed by Masuda in 1986 [841]. The device consists of an L-shaped duct, a buoyancy module, an air chamber, an air turbine, and a generator (see Figure 7.2a). The L-shaped duct has its opening facing the leeward side of the device. The oscillating water in the duct creates an air flow in the air chamber, which drives the air turbine coupled to the generator to generate electricity. The draught of the BBDB device is usually small and thus it can be deployed into relatively shallow water, which may be an important advantage in some coastal regions.

U-Gen OWCs. A U-Gen OWC consists of an asymmetric floater with an interior U shaped tank partially filled with water and two lateral air chambers connected by a duct (see Figure 7.2b). The oscillating water column is enclosed in the floating structure and is not connected to the outer sea water. The pitching motion of the device causes a periodic rise and fall of the water column inside the U-tank. This leads to a bi-directional air flow through the duct, where a self-rectifying air turbine is installed to convert the energy.

Spar-buoy OWCs. The spar-buoy OWC (see Figure 7.2C) is an axisymmetric device (and hence insensitive to wave direction) basically consisting of a submerged vertical tail tube open at both ends, fixed to a floater that moves essentially in heave. The length of the tube determines the resonance frequency of the inner water column. The air-flow displaced by the motion of the OWC's inner free surface, relative to the buoy, drives an air turbine [221].

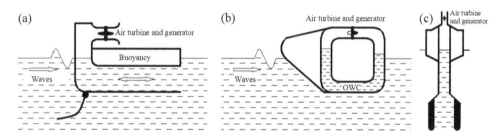

Figure 7.2: Schematic of (a) Backward Bent Duct Buoy (BBDB) , (b) U-Gen, and (c) spar-buoy OWCs.

7.3 EXISTING PROTOTYPES

This section introduces some typical prototypes of OWC devices, with the fixed and floating prototypes detailed in Sections 7.3.1 and 7.3.2, respectively.

7.3.1 Fixed oscillating water column prototypes

The fixed OWC devices are mainly constructed in onshore locations, or in nearshore conditions. The notable ones are shown in Figure 7.3, including the Pico power plant (see Figure 7.3a), the Shanwei power plant (see Figure 7.3b), the Yongsoo power plant (see Figure 7.3c), and the Oceanlinx bottom-standing OWC (see Figure 7.3d), which are detailed below.

Pico power plant. The plant is on the island of Pico, Azores, Portugal. Construction began in 1996 and was completed in 1999. The plant was rated 400 kW. The concrete structure of the chamber (square planform with inside dimensions of 12 m × 12 m at the mean water level) was built in-situ on a rocky bottom (about

8 m water depth), spanning a small natural harbour [222]. The submerged structure of the plant was significantly damaged by waves [549] and the structural foundation of the plant eventually collapsed due to a strong storm in 2018 [222].

Shanwei power plant. The plant is at Zhelang Town, Shanwei City, Guangdong Province, China. The construction of the station started in June 1998 and was completed in February 2001. This device converts wave power into pneumatic power using a bidirectional air flow and a special unidirectional turbine. An asynchronous generator is used to convert the pneumatic power into electricity. The power system is designed to deliver a peak power of 100 kW at significant wave height of 1.5 m [864].

Yongsoo power plant. The plant was completed in 2015, about 1 km away from the south coast of Jeju Island, South Korea. Two horizontal-axis impulse turbines connect to a synchronous generator (250 kW, 150–400 rpm) and an induction generator (250 kW, 450 rpm), respectively. The total nominal rated power of Yongsoo plant is 500 kW. The plant is 37 m long and 31.2 m wide [108].

Oceanlinx bottom-standing OWC plant. The plant was developed by Oceanlinx, an Australian company. The plant was tested at Port Kembla, Australia in 2005. The structure of the plant was made of steel [223].

Figure 7.3: Typical prototypes of fixed OWC devices, with (a) the Pico power plant [222], (b) the Shanwei power plant [864], (c) the Yongsoo power plant [108], and (d) the Oceanlinx bottom-standing OWC [463].

Breakwater-integrated OWCs. The integration of the OWC chamber structure into a breakwater for dual functionality coastal or harbour protection coupled with energy generation has several advantages, for example, the construction costs are shared, ease of construction, and operational and maintenance tasks for the wave energy plant are improved. The breakwater-integrated OWC concept was achieved successfully for the first time in the harbour of Sakata, Japan, in 1990, rated 60 kW (see Figure 7.4a). One of the caissons making up the breakwater had a purpose designed shape to accommodate the OWC and the ancillary mechanical and electrical equipment. Boccotti [76] proposed a U-shaped OWC embedded into a breakwater, with the device's outer opening facing upwards. This design allows the total length of the water column to be increased without placing the opening too far below the sea's surface. This type of OWC breakwater was constructed at the harbor of Civitavecchia (near Rome), Italy, in 2014 (see Figure 7.4b). The structure is composed of a total of 17 caissons and 136 OWCs [223].

Figure 7.4: Breakwater-integrated OWCs: (a) OWC plant integrated into a breakwater (Sakata Harbour, Japan) [223] and (b) U-OWCs integrated into a breakwater (Civitavecchia Harbour, Italy) [220].

7.3.2 Floating oscillating water column prototypes

The floating type OWCs can be located either onshore or in deeper waters offshore. As one example of a floating type OWC, the BBDB concept was model tested in Japan in the mid-1980s. Between 2008 and 2011, a 1:4 scale BBDB OWC (the OE Buoy) (see Figure 7.5a) was tested in Galway Bay, Ireland. In 2010, the 1:3 scale Oceanlinx Mk3 multi-chamber floating OWC device (see Figure 7.5b) was deployed by the Australian company, Oceanlinx. The Mk3 is a floating platform proposed with 8 OWC chambers, with each chamber having its own dedicated air turbine. However, during the testing phase only two turbines (of different types) were installed. In 2012, a 1:16 scale spar-buoy OWC was tested at NAREC, in teh UK (see Figure 7.5c). From a report prepared for the British Department of Trade and Industry in 2005, the OWC spar-buoy was considered to be the lowest risk and most economic option for further development after being compared with several types of floating OWCs for electricity generation in an Atlantic environment [223].

Figure 7.5: Prototypes of fixed OWC devices, with (a) the Backward Bent Duct Buoy (BBDB) (image courtesy of Ocean Energy), (b) the Oceanlinx Mk3 [223], and (c) the Spar-buoy OWC [223].

7.4 CASE STUDY – OWC WEC

In this section, a conventional OWC WEC, as shown in Figure 7.1a, is taken as an example to demonstrate some basic time-domain modelling and experimental approaches for OWC WECs. By considering the effects of wave conditions and geometric parameters, the hydrodynamic performance of two different types of OWC device (i.e., single chamber and dual-chamber OWC device) are considered. The focus of the case study is on the optimisation of the devices' hydrodynamic efficiency and wave loads on the devices.

7.4.1 Modelling

7.4.1.1 Boundary Integral Equation

Based on potential flow theory and the time-domain higher-order boundary elemental method (HOBEM), a two-dimensional fully nonlinear numerical model was built to simulate the hydrodynamic performance of an OWC device. A schematic diagram of the numerical wave flume with a fixed OWC located at its right-hand end is shown in Figure 7.6. A Cartesian coordinate system is defined with the origin on the plane of the undisturbed free surface $z = 0$, with the z-axis positive upwards. As shown in Figure 7.6, h denotes the calm water depth, B_c the chamber width, C the thickness of the front wall, d_w the width of the duct, d the immergence of the front wall, and

h_c the height of the air chamber (i.e., the distance between the still water surface and the chamber ceiling), and L the length of the damping zone.

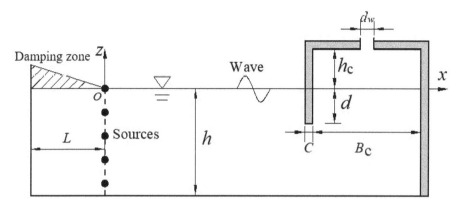

Figure 7.6: Sketch of the numerical wave flume.

It is assumed that the water is incompressible, inviscid, and the fluid flow is irrotational. The fluid motion can therefore be described by a velocity potential ϕ. Generally, the velocity potential satisfies the Laplace equation. However, the incident waves are generated using the inner sources in the computational domain, the governing equation is then described with the Poisson equation:

$$\nabla^2 \phi = q^*(x_s, z, t), \tag{7.1}$$

where $\nabla^2 = \frac{\partial^2}{\partial x^2} + \frac{\partial^2}{\partial z^2}$ is the 2D Laplacian operator, and $q^*(x_s, z, t)$ is the pulsating volume flux density of the internal source distribution described below. Following Brorsen and Larsen [92], the incident wave is specified by writing the flux density as follows:

$$q^*(x, z, t) = 2v\delta(x - x_s), \tag{7.2}$$

where v is the horizontal fluid speed corresponding to the wave to be generated; x_s is the horizontal position of the vertical source, and $\delta(x - x_s)$ is the Dirac delta function. In the present study, the horizontal velocity v is given by the second-order Stokes analytical solution.

On the instantaneous free surface, both the fully nonlinear kinematic and dynamic boundary conditions are satisfied. In the present work, the mixed Eulerian-Lagrangian method is used to describe the time-dependent free surface with moving nodes in both the horizontal and the vertical directions. A damping layer with coefficient $\mu_1(x)$ at the left end of the numerical flume is implemented to absorb the reflected wave from the OWC device. To incorporate the viscous effect due to water viscosity and flow separation at the chamber skirt, a constant damping layer with coefficient μ_2 is implemented on the free surface in the chamber. Then, the free surface boundary conditions can be written as follows:

$$\begin{cases} \frac{dX(x,z)}{dt} = \nabla\phi - \mu_1(x)(X - X_0), \\ \frac{d\phi}{dt} = -g\eta + \frac{1}{2}|\nabla\phi|^2 - \frac{p_{air}}{\rho} - \mu_1(x)\phi - \mu_2\frac{\partial\phi}{\partial n}, \end{cases} \tag{7.3}$$

where $X_0 = (x_0, 0)$ denotes the initial static position of the fluid particle, η the vertical elevation of the free surface, p_{air} the air pressure inside the chamber, g the gravity acceleration, ρ the water density and t the time. The damping coefficient $\mu_1(x)$ is defined by:

$$
\mu_1(x) = \begin{cases} \omega\left(\frac{x-x_1}{L}\right)^2 &, \quad x_1 - L < x < x_1, \\ \\ 0 &, \quad x > x_1, \end{cases} \tag{7.4}
$$

where x_1 is the starting position of the damping zone, L is the length of the damping zone at the left flume-end, given to be 1.5 times the incident wavelength in the present study. The artificial viscous damping coefficient μ_2 is determined by trial and error and only implemented inside the chamber [584]. ω is the angular frequency and satisfies the following dispersion relation in Eq. (2.21).

The boundaries at the bottom, back wall and front wall are considered impermeable. Therefore, the zero normal velocity condition (Eq. (2.9)) is satisfied.

To solve the above boundary value problem in the time domain, the initial conditions are required as follows:

$$
\phi|_{t=0} = \eta|_{t=0} = 0. \tag{7.5}
$$

By applying Green's second identity to the fluid domain Ω, the above boundary value problem can be converted in the usual manner into the following boundary integral equation:

$$
\alpha(p_0)\phi(p_0) = \int_\Gamma \left[\phi(q_0)\frac{\partial G(p_0, q_0)}{\partial n} - G(p_0, q_0)\frac{\partial \phi(q_0)}{\partial n}\right] d\Gamma + \int_\Omega q^* G(p_0, q_0) d\Omega, \tag{7.6}
$$

where Γ represents the entire computational boundary, $p_0 = (x_0, z_0)$ and $q_0 = (x, z)$ are the source and the field points, respectively; and α is the solid angle coefficient determined by the surface geometry of a source point position. G is a simple Green function and can be written as $G(p_0, q_0) = \ln r/2\pi$, where $r = [(x - x_0)^2 + (z - z_0)^2]^{1/2}$. The above boundary integral equation is solved by the boundary element method using a set of collocation nodes on the boundary and higher-order elements. More details regarding the numerical model can be found in [581].

7.4.1.2 *Pneumatic Model*

Inside the chamber, the air and water are strongly coupled. By assuming a quadratic relationship between the chamber pressure and the air duct velocity, the air pressure in the chamber can be expressed as follows [415]:

$$
p_{air}(t) = C_{dm}U_d(t) + D_{dm}|U_d(t)|U_d(t), \tag{7.7}
$$

where C_{dm} and D_{dm} are the linear and quadratic pneumatic damping coefficients, respectively; and $U_d(t)$ is the air flow velocity through the duct. Based on the assumption of a negligible spring-like effect of air compressibility inside the chamber, the air flow velocity $U_d(t)$ can be expressed as follows:

$$U_d(t) = \frac{\Delta V}{S_0 \Delta t}, \tag{7.8}$$

where $\Delta V = V_{t+\Delta t} + V_t$ represents the change of air volume in the chamber within each time step Δt, which can be calculated with the variation of the free surface. S_0 is the cross-sectional area of the air duct.

7.4.1.3 Hydrodynamic Efficiency

In this study, the hydrodynamic efficiency of the OWC device is calculated using the relationship between the pneumatic power and the power of the corresponding incident wave. The average pneumatic power, i.e., the power absorbed from the waves by the OWC, during a wave period T, can be expressed as the time-averaged flow rate multiplied by the air pressure variation:

$$\overline{P}_{pto} = \frac{1}{T} \int_t^{t+T} Q(t)p(t)dt, \tag{7.9}$$

where $Q(t) = S_0 U_d(t)$ is the flow rate.

The average energy transported by the per-unit frontage of the incident wave J can be calculated from Eqs. (2.24), (2.25), and (2.26).

Thus, the hydrodynamic efficiency is given by:

$$\eta_{\text{eff}} = \frac{\overline{P}_{pto}}{J}. \tag{7.10}$$

The hydrodynamic efficiency η_{eff} is in the range of $(0, 1)$. $\eta_{\text{eff}} = 1$ denotes that, effectively, all the incident wave energy is captured by the OWC device. However, this is not feasible due to the radiated waves generated by the oscillatory motion of the water column and the scattered waves by the device and various sources of viscous damping.

7.4.1.4 Wave loads

The velocity potential on the OWC structure surface is obtained once Eq. (7.6) is solved. Then, based on the Bernoulli equation, the wave-induced pressure on the OWC device can be obtained. Finally, the wave-induced force $\boldsymbol{F}$ and bending moment $\boldsymbol{M}$ can be calculated by integrating the wave pressure over the instantaneous wetted OWC device surface Γ_b as:

$$\boldsymbol{F} = \int_{\Gamma_b} p_{dyn}\boldsymbol{n}d\Gamma = -\rho \int_{\Gamma_b} \left[\frac{\partial \phi}{\partial t} + g\eta + \frac{1}{2}(\nabla\phi \cdot \nabla\phi) + \frac{p_{air}}{\rho} - \mu_2 \frac{\partial \phi}{\partial t} \right] \boldsymbol{n}d\Gamma, \tag{7.11}$$

$$M = \int_{\Gamma_b} p_{dyn}(\mathbf{r} \times \mathbf{n})d\Gamma = -\rho \int_{\Gamma_b} \left[\frac{\partial \phi}{\partial t} + g\eta + \frac{1}{2}(\nabla\phi \cdot \nabla\phi) + \frac{p_{air}}{\rho} - \mu_2 \frac{\partial \phi}{\partial t} \right] (\mathbf{r} \times \mathbf{n})d\Gamma,$$

(7.12)

in which p_{dyn} denotes the wave induced pressure on the OWC surface; $\mathbf{r}$ denotes the distance vector and $\mathbf{r} \times \mathbf{n}$ is the length of the moment arm. The pressure term p_{air}/ρ represents the pneumatic pressure in the chamber. The linearised Bernoulli equation $p = \partial\phi/\partial t$ is used here to account for the pressure drop due to turbulence phenomena. Therefore, the last term $\mu_2 \frac{\partial \phi}{\partial t}$ in Eqs. (7.11) and (7.12) is introduced to account for these turbulence effects on the wave force. The acceleration-potential method is used here to calculate the time derivative of the velocity potential $\partial\phi/\partial t$ [817].

7.4.2 Experimental investigation

The experimental tests described in this case study were carried out in the wave-current flume at the State Key Laboratory of Coastal and Offshore Engineering, Dalian University of Technology, China. A piston-type unidirectional wave-maker and a wave-absorbing beach are located at the two ends of a flume to generate the desired incident waves and absorb the outgoing waves, respectively. The wave maker is able to generate regular and irregular waves with periods ranging from 0.5 s to 5.0 s. The glass-walled wave flume is 69 m long, 2 m wide and 1.8 m deep. The test section of the flume was divided into two parts along the longitudinal direction, which were measured as 1.2 m and 0.8 m in width, respectively (see Figure 7.7). The OWC model was installed in the 0.8 m wide part, 50 m away from the wave maker. To avoid wave energy transfer through the device, the model was designed to span across the width and depth of the flume. To have a clear view of the water free-surface within the OWC chamber, the main body of the model was made of transparent Perspex sheets.

It is well known that the power take-off (PTO) system plays an important role in power absorption [584, 343]. However, it is not practical to simulate the PTO system in the present scaled model tests [666]. Instead, the PTO effects are usually simulated by the incorporation of a duct located on the air chamber. In the present study, one circular duct is positioned on the roof of each sub-chamber.

Here we investigate the hydrodynamic performance of both a single chamber and a dual-chamber OWC device. The schematics of the experimental setup are shown in Figure 7.7 (single chamber) and Figure 7.9 (dual-chamber). In each of these figures, D denotes the diameter of the duct, and the other symbols in the figures are identical to those in Figure 7.6. In the experiments, resistance-type wave gauges with a resolution of 0.01 cm were used to measure the instantaneous free surface elevations at various locations along the flume length, pressure sensors were used to measure the air pressure inside the chamber and the hydrodynamic pressure on the OWC chamber structure's surface. As shown in Figure 7.8, six pressure sensors were fixed symmetrically on both sides of the front wall of the single chamber OWC device to record the wave pressure. One pair of sensors (S_{O1} and S_{I1}) was situated at 1.5 cm above the bottom edge of the wall, one pair (S_{O3} and S_{I2}) was fixed at the mid

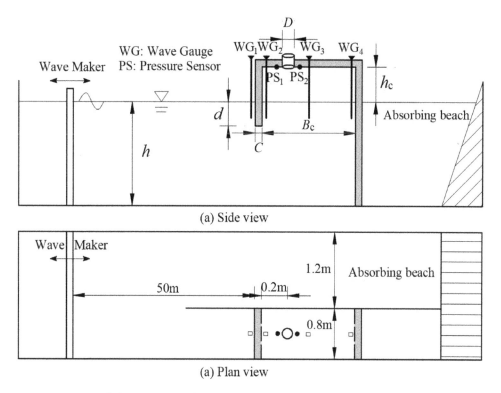

(a) Side view

(a) Plan view

Figure 7.7: Schematic of the single chamber OWC experimental setup.

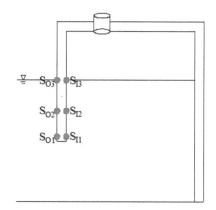

Figure 7.8: Locations of the pressure sensors on the front wall.

depth of the submerged part of the front wall and the third pair of sensors (S_{O3} and S_{I3}) was fixed at the height of the still water surface.

7.4.3 Comparisons between measured and simulated results

To validate the numerical model, the simulation results are compared with the experimentally-generated measured data. The OWC device equipped with a single chamber is taken as an example test case.

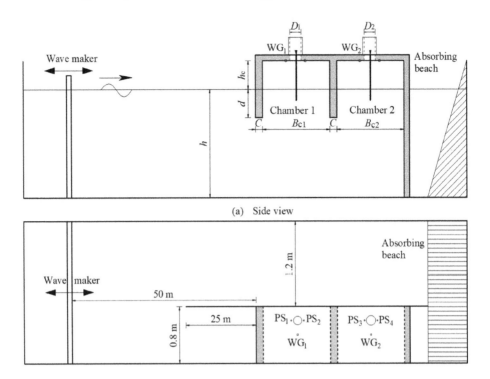

(a) Side view

Figure 7.9: Schematic of the dual-chamber OWC experimental setup.

The geometric parameters of the single chamber OWC device are chosen as follows: chamber width $B_c = 0.55$ m, front wall thickness $C = 0.04$ m, and front wall draught $d = 0.14$ m. In the experiment setup, the duct diameter $D = 0.06$ m, and in the numerical model, an air duct width $d_w = 0.0036$ m, provides the the same opening area as the circular air duct in the experiment.

The water depth is $h = 0.8$ m and the incident wave amplitude is set to $A_i = 0.03$ m. The viscous coefficient and the pneumatic damping coefficient in Eqs. (7.3) and (7.7) are set as $\mu_2 = 0.2$ and $C_{dm} = 9.5$, and $D_{dm} = 0$, respectively. The length of the numerical flume is set to 5λ, in which 1.5λ at the left side is used as the damping layer. After the convergence check, the size of the boundary elements Δx is set to $\lambda/30$, and time step Δt is set to $T/80$.

Figures 7.10 and 7.11 show the time series of the surface elevation at the chamber centre and the air pressure inside the chamber for $T=1.366$ s and $T=1.610$ s, respectively. Figure 7.12 shows the time series of the hydrodynamic pressure on the front wall of the single chamber OWC device. Overall, the experiment measurements and numerical simulation surface elevations, chamber internal air pressures, and the hydrodynamic pressures on the front wall match well with each other as shown in Figures 7.10, 7.11, and 7.12. Figure 7.13 illustrates the variation of the hydrodynamic efficiency with the dimensionless wave number kh. It can be seen that the present potential model results with a specified damping term agree well with the experimental data.

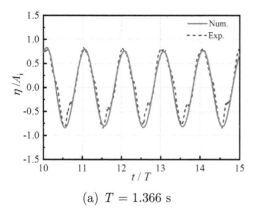
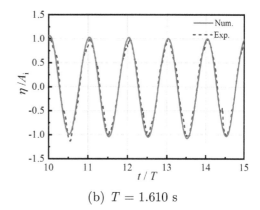

(a) $T = 1.366$ s (b) $T = 1.610$ s

Figure 7.10: Time series of the simulated and measured surface elevation at the chamber centre at $T = 1.366$ s and 1.610 s.

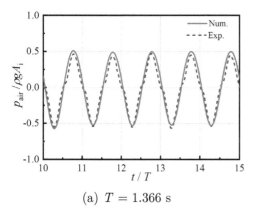
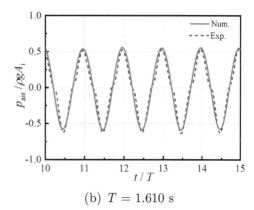

(a) $T = 1.366$ s (b) $T = 1.610$ s

Figure 7.11: Time series of the simulated and measured air pressure inside the chamber at $T = 1.366$ s and 1.610 s.

The aforementioned comparisons indicate that the present numerical model is capable of simulating the processes of wave interactions with the OWC device accurately. The wave-induced dynamic pressure on the front wall is also well captured by taking the air pressure inside the chambers and the turbulence effects into account.

7.4.4 Single chamber oscillating water column device

In this sub-section the hydrodynamic performance of a single chamber OWC device will be discussed. The effect of the wave conditions and chamber geometry on the hydrodynamic efficiency and wave loads are investigated.

7.4.4.1 Hydrodynamic Efficiency

Effect of chamber width. By keeping front wall thickness $C = 0.04$ m, front wall draught $d = 0.14$ m, duct diameter $D = 0.06$ m, water depth $h = 0.8$ m and incident wave amplitude $A_i = 0.03$ m constant, Figure 7.14 shows the hydrodynamic efficiency

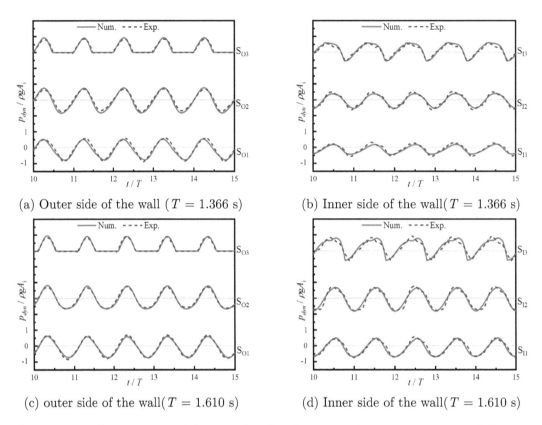

(a) Outer side of the wall ($T = 1.366$ s)

(b) Inner side of the wall($T = 1.366$ s)

(c) outer side of the wall($T = 1.610$ s)

(d) Inner side of the wall($T = 1.610$ s)

Figure 7.12: Comparisons of the simulated and measured wave pressures at different measuring points.

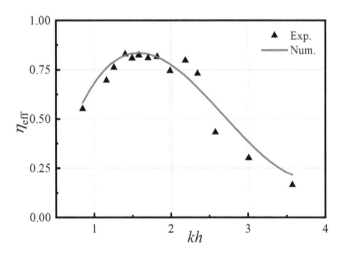

Figure 7.13: Variation of the simulated and measured hydrodynamic efficiency with kh.

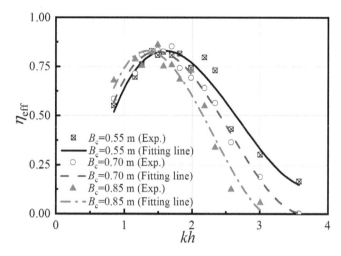

Figure 7.14: Hydrodynamic efficiency versus dimensionless wave number for different chamber widths.

of the OWC device for three different chamber widths: $B_c = 0.55$ m, 0.70 m and 0.85 m. The hydrodynamic efficiency increases with the increase of chamber width B_c in the low-frequency region (about $kh < 1.5$), but follows a completely opposite trend in the high-frequency region, as shown in Figure 7.14. Furthermore, the resonant frequency decreases with the increase of B_c due to the increasing inertia of the OWC water column. The optimal points are around $kh = 1.58$ ($B_c = 0.55$ m), $kh = 1.50$ ($B_c = 0.70$ m) and $kh = 1.36$ ($B_c = 0.85$ m) resulting in a hydrodynamic efficiency of approximately 0.83 for each of the three cases. The dependence of the natural frequency on the width of the chamber can be clearly seen in $\omega_n = (g/(d+0.41 B_c B_t)^{\frac{1}{2}}))^{\frac{1}{2}}$ (where $B_c B_t$ is the horizontal cross sectional area of the air chamber, and B_t the transverse width of the chamber), which is used to predict the natural piston frequency of a water mass oscillating in a moonpool [796]. It should be noted that the coefficient 0.41 in the above formula is empirical and hence does not necessarily provide accurate results in the case of all OWC devices. However, the dependence of the natural frequency on the width of the chamber can be seen clearly.

It should be noted that the hydrodynamic efficiency is near to zero for $kh = 3.57$, $B_c = 0.70$ m. In this case, the incident wave period T is 0.950 s with a corresponding wavelength $\lambda = 1.406$ m. This wavelength is approximately twice the chamber width B_c, i.e., $\lambda/B_c = 2.01$. Figure 7.15 shows the snapshots of the surface elevations profiles in the chamber for this case. The water surface in the chamber is rising at one wall and falling at the other wall and the intersection node of the two lines lies at the chamber centre. This is the so-called seiching phenomenon, which is a characteristic that occurs in typical standing wave conditions. Furthermore, the total mass of water inside the chamber has not changed and the air pressure also remains constant at close to the external atmospheric pressure. In this condition no energy is extracted from the incident waves. Therefore, such seiching phenomena should be avoided in the OWC design.

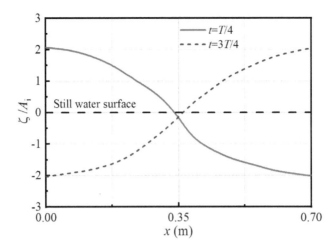

Figure 7.15: Snapshots of surface elevations profiles in the chamber for $kh = 3.57$ ($T = 0.95$ s and $B_c = 0.70$ m).

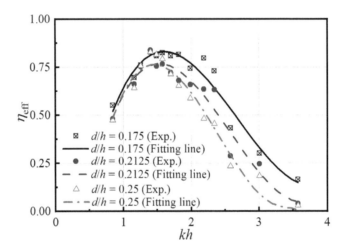

Figure 7.16: Hydrodynamic efficiency versus dimensionless wave number kh for different draughts d/h.

Effect of front wall draught. Figure 7.16 illustrates the hydrodynamic efficiency of the OWC device for different front wall draughts, $d = 0.14$ m ($d/h = 0.175$), 0.17 m ($d/h = 0.2125$), and 0.20 m ($d/h = 0.25$) with the other dimensional parameters remaining the same as those in the previous study investigating the effect of chamber width. It can be observed that both the resonant frequency and the peak efficiency decrease with the increase of the submerged depth d/h due to the increased mass of the water column in the chamber. The resonant frequency occurs at $kh = 1.59$ ($d/h = 0.175$), 1.50 ($d/h = 0.2125$), and 1.41 ($d/h = 0.25$) with corresponding hydrodynamic efficiencies of 0.83, 0.77 and 0.76, respectively. Additionally, as shown in Figure 7.14, the hydrodynamic efficiency is not particularly sensitive to the change of draught d/h in the low-frequency zone (in the region $kh < 1.0$) due to the small

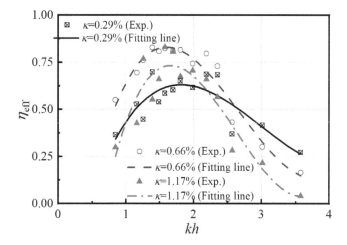

Figure 7.17: Variations of hydrodynamic efficiency for different duct opening ratios.

relative length of the draught to wavelength in the low-frequency, long wave region. However, in the high-frequency short wave region, the draught of the front wall is not insignificant relative to the wavelength and the reflected wave energy increases with the front wall draught. As a result, the hydrodynamic efficiency decreases with an increase in the front wall draught.

Effect of the duct scale. The effect of the duct scale on the hydrodynamic efficiency is usually described by the opening area ratio $\kappa = S_0/S$, where S_0 and S are the cross-sectional areas of the duct and the air chamber respectively. In this study, three duct diameters ranging from $D = 0.04$ m, 0.06 m, and 0.08 m, corresponding to opening ratios of 0.29%, 0.66%, and 1.17%, respectively, are experimentally tested, whilst the other parameters were kept the same as those in the previous study investigating the effect of chamber width. The optimal hydrodynamic efficiency η_{eff} is highly sensitive to the opening ratio, with a maximum $\eta_{\text{eff}} = 0.63$ ($\kappa = 0.29\%$), 0.83 ($\kappa = 0.66\%$) and 0.74 ($\kappa = 1.17\%$) as shown in Figure 7.17. Moreover, $\kappa = 0.66\%$ exhibits the largest the hydrodynamic efficiency η_{eff} amongst the three opening ratios, except $\kappa = 0.29\%$ in the high-frequency zone (about $kh > 2.6$). To further investigate this behaviour, Figures 7.18a and 7.18b present the maximum air pressure in the chamber and the maximum water surface elevation at the chamber centre for different opening ratios. It can be seen that the internal air pressure decreases with the increase in opening ratio, whilst the maximum surface elevations vary with an opposite trend. For the largest opening ratio $\kappa = 1.17\%$, the air pressure variation in the chamber is minimal; however, this opening ratio produces the largest surface elevation. Considering the smallest opening ratio $\kappa = 0.29\%$, it is shown that the largest pressure fluctuation in the chamber leads to the smallest oscillation amplitude of the water column. According to Eq. (7.9), the wave energy extracted by an OWC is attributed to the product of air pressure and air volume variations in the chamber. In the opening ratio cases considered here, the optimal opening ratio κ is 0.66% from Figures 7.17 and 7.18. The present analysis may help to determine the turbine damping by the OWC device to achieve optimal wave energy extraction.

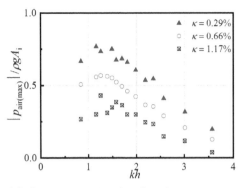

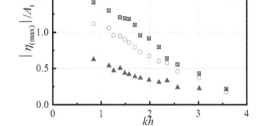

(a) Air pressure in the chamber

(b) Surface elevations at the chamber centre

Figure 7.18: Variations of the (a) air pressure in the chamber and (b) surface elevation at the chamber centre for different opening ratios.

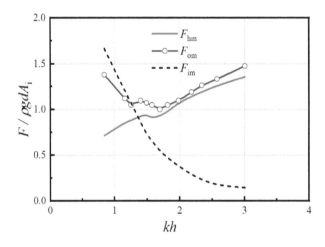

Figure 7.19: Maximum horizontal wave forces versus kh for $A_i = 0.03$ m.

7.4.4.2 Wave Loads

The chamber geometrical parameters are chosen to be the same as those in Section 7.4.3. Figure 7.19 shows the variations in the dimensionless maximum horizontal wave forces on the front wall with respect to variations in kh. F_{hm} denotes the maximum total horizontal wave force, F_{om} and F_{im} denote the maximum horizontal wave forces on the outer side and inner side of the front wall respectively, where the subscript "m" stands for the maximum force. It can be seen that F_{hm} increases with an increase in kh. In addition, an opposite trend is observed for F_{im}. F_{om} initially decreases to its minimum and then increases with the increase in kh. The variation of the wave forces with kh can be explained using the relationship between the wave transmission ability and wavelength as follows.

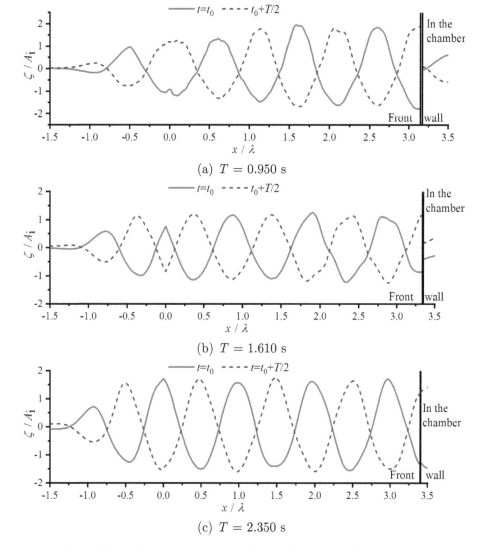

Figure 7.20: Snapshot of wave elevations along the wave flume at $t = t_0$ and $t = t_0 + T/2$ $A_i = 0.03$ m.

Figures 7.20a, 7.20b, and 7.20c present snapshots of wave elevation along the wave flume for different wave periods, $T = 1.037$ s, 1.610 s, and 2.350 s respectively. From Figure 7.20a it can be seen that the free surface fluctuation on the inner surface of the front wall is very small. However, the amplitude of the wave elevation outside the chamber is nearly twice the incident wave amplitude. Almost full wave reflection at the external face of the front wall occurs because of the weak transmission ability of the short wave. Thus, the maximum wave force on the outer side of the wall F_{om} is large and the maximum wave force on the inner side of the wall F_{im} is small.

As the wave period increases to $T = 1.610$ s, the transmission ability of the wave is enhanced as its wavelength increases. Then, from Figure 7.20b, it can be seen that the fluctuation of the inner surface is much greater than that of the $T = 1.037$ s wave and the outside wave elevation amplitude is reduced. Consequently, the wave force on the outer surface of the front wall F_{om} decreases and the wave force on the inner surface of the front wall F_{im} is increased.

With the wave period further increased to $T = 2.350$ s, the transmission ability of the wave is further enhanced. From Figure 7.20b and 7.20c it can be seen that the surface elevation in the wave flume is greater than that for $T = 1.610$ s. This is because most of the wave energy is transmitted into the chamber through the front wall due to the wave's strong transmission ability. Then the wave energy is totally reflected by the rear wall of the chamber. The reflected waves from the rear wall also pass through the front wall with very little reflection. Furthermore, it can be noticed that the wave surface motions on the inner and the outer surfaces of the front wall are almost in phase with each other and have similar motion amplitudes. As a result, both F_{im} and F_{om} increase and there is a $T/2$ phase difference between them. The crest of F_{im} and the trough of F_{om} exhibit similar values, and the opposite is observed with the trough of F_{im} and the crest of F_{om}. Therefore, the value of F_{hm} is relatively small in the low frequency region, as shown on Figure 7.19.

7.4.5 Dual-chamber oscillating water column device

In this sub-section the hydrodynamic performance of a dual-chamber OWC device will be discussed. The effects of the wave conditions and chamber geometry on hydrodynamic efficiency and wave loads are presented. The water depth is set to $h = 1.0$ m, and the incident wave amplitude is $A_i = 0.03$ m. The draughts of the two barrier walls are always maintained with identical values in the present study.

7.4.5.1 Hydrodynamic Efficiency

Comparison with a typical single-chamber device. First, the hydrodynamic performance of a dual-chamber OWC is compared with that of a typical single-chamber OWC device in terms of the overall energy conversion efficiency. The overall power absorbed/energy efficiency of the dual-chamber is the sum of the two sub-chambers, due to the fact that efficiency is a nondimensional, scalar parameter, i.e., $\overline{P}_{pto} = \overline{P}_{pto1} + \overline{P}_{pto2}$, $\eta_{eff} = \eta_{eff1} + \eta_{eff2}$. The width of the single-chamber is 0.7 m, equivalent to the overall width of the two chambers in the dual-chamber OWC device, i.e., $B_{c1} + B_{c2} + C = 0.7$ m. The width ratio of the two sub-chambers is set to be $B_{c1} : B_{c2} = 1 : 1$.

Figure 7.21 shows the comparison of the instantaneous power absorbed by the dual-chamber and single-chamber OWC devices for $kh = 1.40$. It can be seen from Figure 7.22 that the overall power extracted from the dual-chamber OWC device is larger than that of the single-chamber OWC device. It can also be seen from Figure 7.22 that the overall hydrodynamic efficiency is increased by $\sim 10\%$ by adopting the dual-chamber design.

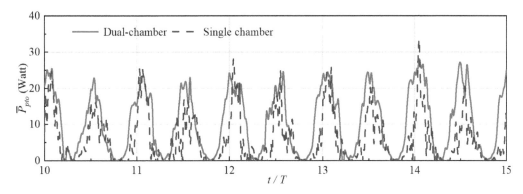

Figure 7.21: The instantaneous wave power absorbed by the dual-chamber and single-chamber OWC devices for $kh = 1.40$.

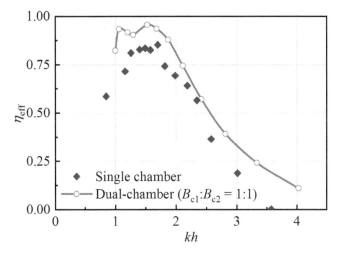

Figure 7.22: Variations of the hydrodynamic efficiency of both the dual-chamber and single-chamber OWC devices with kh.

Additionally, it can be observed that the effective frequency bandwidth, i.e., the frequency range of surrounding sea-states resulting in a larger wave energy conversion efficiency, is greater for the dual-chamber OWC device than that of a typical equivalent single-chamber OWC. The associated hydrodynamic efficiency curve for the dual-chamber OWC is less sensitive to variations in kh (when kh is in the range of 1~2). This may be due to the fact that the dual-chamber OWC system has more than one resonant frequency. It is worth mentioning that a similar phenomenon was presented by Rezanejad et al., [662], in which this increase in the frequency bandwidth was attributed to three resonant modes of the system. The basic resonance frequency of the whole dual-chamber OWC system is expected to be close to the resonance frequency of the equivalent single chamber OWC device, and the existence of the internal barrier wall would introduce two additional resonance frequencies for the two water columns trapped inside the two sub-chambers. The three resonant frequencies are close to each other and together they result in the higher performance

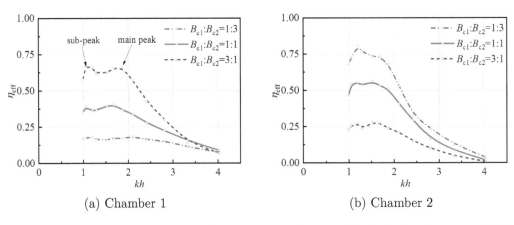

(a) Chamber 1   (b) Chamber 2

Figure 7.23: Variations of the hydrodynamic efficiency for (a) chamber 1 and (b) chamber 2 with kh for all three chamber width ratios.

and larger effective frequency bandwidth of the dual-chamber OWC device. It should also be noted that the frequency bandwidth at maximum efficiency is significantly wider than the bandwidth of maximum efficiency in the single chamber device.

Effect of chamber width. Three chamber width ratios (i.e., $B_{c1} : B_{c2} = 3{:}1$, 1:1 and 1:3) are considered with the overall chamber width kept constant, i.e., $B_{c1} + B_{c2} + C = 0.7$ m. The barrier wall draught d is also kept constant at 0.2 m. The variations in the hydrodynamic efficiency for chamber 1 and chamber 2 with respect to kh are shown in Figure 7.23a and 7.23b. It can be seen that the efficiency increases with the increase of kh until it reaches its maximum value, and then it decreases with a further increase of kh, for both chambers, in all three chamber width ratio cases. The hydrodynamic efficiency of each chamber increases with the increase of chamber width for a certain wave condition (kh). This is not surprising because the energy that can be extracted via the system is proportional to the cross-sectional area of the trapped water surface, thus, the chamber width, as indicated in Eq. (7.9), is a significant factor in determining the hydrodynamic efficiency. The overall conversion efficiency of the system is observed to be insensitive to the sub-chamber width variation when the overall chamber width and the barrier walls draughts are maintained constant [582]. This can be explained as being due to the fact that the length and the overall mass of the trapped water column in the system do not change with the variation of the sub-chamber widths when the two barrier wall draughts are kept constant.

It is notable that there are double peaks in the efficiency curves for both chambers (1 and 2). These double peaks most obvious for chamber 1, with a width ratio of 1:3 (blue dashed line in Figure 7.23a). This phenomenon may result from the interaction between the two water columns in the two sub-chambers and the aforementioned diverse resonant modes of the system. For convenience, we refer to the larger peak as the main peak, and the slightly smaller peak as the sub-peak. More detailed discussion on this can be found in the following sub-section.

Effect of barrier wall draught. Three sets of barrier wall draughts (i.e., $d = 0.125$, 0.2 and 0.25 m) are considered. The chamber breadth ratio is kept constant at

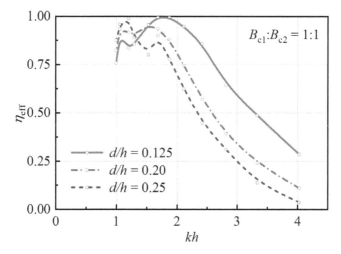

Figure 7.24: Variations of the overall hydrodynamic efficiency of the dual-chamber system with kh for all three barrier wall draughts.

$B_{c1} : B_{c2} = 1{:}1$ in this analysis. Figure 7.24 shows the variation in the overall efficiency of the dual-chamber OWC system for the three sets of barrier wall draughts (i.e., d/h = 0.125, 0.2 and 0.25). It can be observed that the dual-chamber device performs consistently better for the relative barrier wall draught of 0.125 over the majority of the frequency range considered in this study. This observation is in line with the single-chamber OWC device, which also exhibits better performance with a smaller front wall draught [584, 343, 552, 760]. Other effects, such as wave transformations and tides near the installation site should also be considered. It is emphasised that the maximum efficiency of a dual-chamber OWC system (which is notably close to 1) is much larger than that of a typical single-chamber OWC device.

7.4.5.2 Wave Loads

Diagrams of the wave force, bending moment and 4 rotation centres for the dual-chamber OWC device are illustrated in Figure 7.25. The dual-chamber OWC device can be divided into three cantilever structures; therefore, the inflection points 1, 2, 3 and 4 shown in Figure 7.25 are the positions on the device that are subjected to local maximum loads. These four points are taken as the rotational centres of wave bending moments. The wave force and bending moment can be obtained from Eqs. 7.11 and 7.12. $F_{i\mathrm{H}}$ and $F_{i\mathrm{V}}$ ($i = 1, 2$) denote the horizontal and vertical component of the wave force respectively, on i# barrier wall.

The total wave loading vector on the i# barrier wall $\mathbf{F}_i$ ($i = 1, 2$) consists of the total horizontal and vertical components:

$$\mathbf{F}_i = F_{i\mathrm{H}} \cdot \boldsymbol{i} + F_{i\mathrm{V}} \cdot \boldsymbol{k} \quad (i = 1, 2) \tag{7.13}$$

in which $\boldsymbol{i}$ and $\boldsymbol{k}$ represent the unit normal vector in the x and z direction, respectively.

Wave force vectors $\mathbf{F}_1$ and $\mathbf{F}_2$ denote the total wave forces on the 1# and 2# barrier wall respectively. $\mathbf{F}_4$ denotes the total wave force on the back wall of the

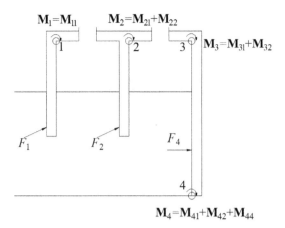

Figure 7.25: Schematic of wave force and bending moment and 4 rotation centres on the dual-chamber OWC device.

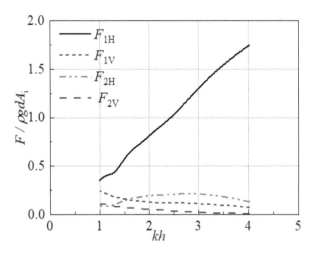

Figure 7.26: Maximum horizontal and vertical wave forces on the two barrier walls versus kh.

dual-chamber OWC device, which is equal to the horizontal component of the wave force because there is zero vertical component on the back wall. The moment component $\mathbf{M}_{qj}$ represents the moment about rotation centre q ($q = 1, 2, 3, 4$) by the wave force $\mathbf{F}_j$ ($j = 1, 2, 4$). For example, $\mathbf{M}_{42}$ represents the moment component about rotation centre 4 due to the wave force $\mathbf{F}_2$ on 2# barrier wall. $\mathbf{M}_q$ denotes the total moment about rotation centre q and is the sum of all moment components at the rotation centres q. For example, $\mathbf{M}_4 = \mathbf{M}_{41} + \mathbf{M}_{42} + \mathbf{M}_{44}$.

Wave force. The peak horizontal and vertical wave forces on 1# and 2# barrier walls are shown in Figure 7.26. The peak wave forces F are normalised by $\rho g d A_i$. It can be seen that F_{1H} increases with the increase in dimensionless wave frequency kh, while F_{2H} increases very gradually to a peak value and then decreases with the increase in kh. Furthermore, as shown in Fig. 7.26, F_{1H} is much larger than F_{2H}. This indicates that the incident wave side barrier wall, i.e., 1# barrier wall, was subjected

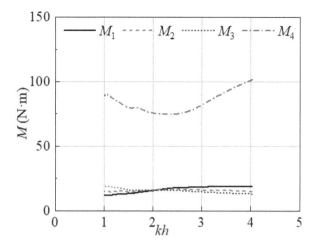

Figure 7.27: Wave moments about the four rotation centres versus kh.

to the majority of the horizontal wave load. Thus, the strength of 1# barrier wall needs to be reinforced to ensure the structural durability of the device. The strength requirements for 2# barrier wall do not need to be as large as 1# barrier wall and less expensive material could be used to reduce construction costs. Additionally, as illustrated in Figure 7.26, the vertical components of the wave forces are much smaller than the horizontal components, i.e., $F_{1V} < F_{1H}$ and $F_{2V} < F_{2H}$. This is due to the fact that the force components of the vertical forces are much smaller than those of the corresponding horizontal forces. Thus, for simplicity, only the horizontal wave forces are examined in detail in the present study, and the vertical wave forces are only considered in the wave moment calculations.

Wave moment. Figure 7.27 shows the wave moments about the four rotation centres, indicated in Figure 7.25, as a function of kh. It is evident that the moment about rotation centre 4, M_4, is much larger than the moment loading about the other three rotation centres. This suggests that the junction of back wall and seabed is the most vulnerable part of OWC system to moment loading failure. Thus, the connection between the back wall and seabed should be strengthened to prevent overturning of the device. The largest moment occurs at rotation centre 4 due to the fact that this position has the greatest lever arm distance for all the applied wave loads, as can be seen from Figure 7.25. Additionally, it can be seen that M_4 firstly decreases with increasing kh to its minimum and then increases with kh. This indicates that the rotation centre 4 experiences a relatively larger moment at both low-frequency, long waves and also at high-frequency, short waves. More details regarding the moment loading can be found in Wang et al. [817]

7.5 SUMMARY

In this chapter, the hydrodynamic performance of the OWC device was experimentally and numerically investigated. The numerical model is based on potential flow theory and the time-domain higher order boundary elemental method (HOBEM).

Wave forces and bending moments were calculated based on the Bernoulli equation, which was modified by taking the air pressure inside the chamber and turbulence effects into account. This equation was then solved with an acceleration-potential method. Overall good agreements in the surface elevation, air pressure in the chamber and the hydrodynamic pressure on the barrier wall between simulations and experiments are observed. This suggests that the hydrodynamic performance of the fixed OWC device can be simulated successfully by the time-domain HOBEM model with a properly calibrated pneumatic damping coefficient and artificial viscous damping coefficient.

It is found that the opening ratio κ has a significant influence on the maximum hydrodynamic efficiency of the OWC device. Optimal efficiency occurs with an opening ratio of $\kappa = 0.66\%$. The water surface motion in the chamber is highly dependent on the relative wavelength λ/B_c. The Seiching phenomenon, which can lead to an absence of energy extracted from the waves, can be instigated when the relative wavelength is $\lambda/B_c = 2$. This phenomenon should be avoided in the design of an OWC device.

The total horizontal wave force on the front wall of the OWC device increases with the increase of kh. For a given chamber geometry, under the action of short waves, the total horizontal force on the front wall is mainly due to the large amplitude water surface oscillation on the outer surface of the front wall, while the very small force amplitude on the inner surface makes little contribution to the total force.

To enhance the hydrodynamic performance of the OWC device, the land-based, dual-chamber type OWC is designed and investigated. By conducting a comparison analysis with the single chamber OWC device, it is shown that both the maximum efficiency and the range of wave frequency that leads to a higher rate of wave energy absorption, i.e., the effective frequency bandwidth, are significantly augmented in the dual-chamber OWC system. It can be seen that the width ratios of the individual chambers play a negligible effect on the overall extraction efficiency when the given total chamber width is kept constant. It can also be seen that a smaller barrier wall draught is favourable in terms of the overall extraction efficiency.

The horizontal components of the wave forces on the two barrier walls of a dual-chamber OWC are much larger than the corresponding vertical components for a small thickness of the barrier wall. Due to the sheltering effect by the wave-side barrier wall (i.e., 1# barrier wall), the horizontal wave force on the inner barrier wall (i.e., 2# barrier wall) is much smaller than that on the former, i.e., $F_{1H} > F_{2H}$. The joint between the device and seabed was subjected the largest wave moment M_4 loading. Furthermore, this position experienced a relatively larger moment at both low-frequency long waves and high-frequency short waves. It should be noted that this bending moment may cause an overturning of the device, and thus, should be one of the major concerns for structural design and construction of the device.

IV

Wave Energy Converter Arrays/Farms

CHAPTER 8

Large-scale computation of wave energy converter arrays

Yingyi Liu

Kyushu University, liuyingyi@riam.kyushu-u.ac.jp

8.1 INTRODUCTION

This chapter is dedicated to the numerical modelling of wave energy converter (WEC) arrays. Deployment of WEC devices into arrays becomes a trend when the technology moves into the large-scale commercial exploitation stages, creating so-called "WEC farms". The arrays can be deployed in nearshore zones or those further offshore, depending on the type of the devices. One big advantage of "WEC farms" is that the devices can share some common infrastructures such as power substations, mooring systems and cables, which can significantly reduce the cost of construction and maintenance, as shown in Figure 8.1. In addition, the electricity generated by arrays of WECs can be far more stable than that generated by a single individual device. Wave farms of this nature are thereby beneficial to reduce the overall cost. The objective of optimising wave farm performance will be discussed extensively in Chapter 9.

Due to the fact that a WEC device in an array is not only subject to ambient incident waves, but also to those that have been reflected and radiated from the other WECs [124], an important issue that must be taken into account is the hydrodynamic interactions between the devices. In this chapter, we will first introduce the existing knowledge of the WEC array modelling methods in Section 8.2, and then present a hybrid methodology combining interaction theory (IT) and the boundary element method (BEM) in Sections 8.3–8.5. Evaluation of array properties are illustrated in Section 8.6 where a case study is given. The method described in this chapter is expected to be promising for a WEC farm of generic device geometries.

DOI: 10.1201/9781003198956-8

(a)

(b)

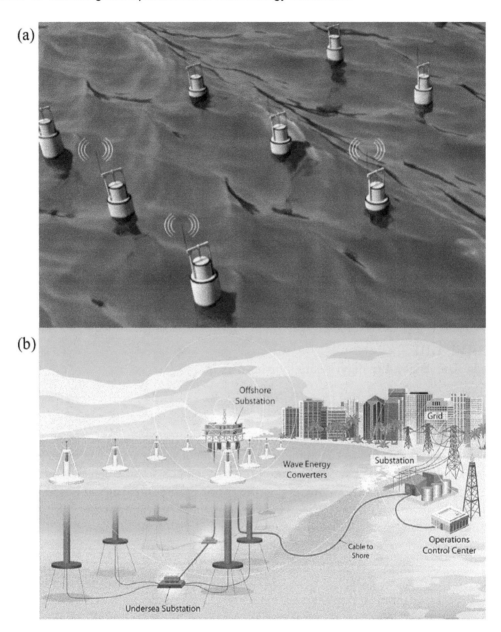

Figure 8.1: Artist's impression of a wave energy farm: (a) an array of point absorbers (artwork credit and copyright, 2018: Lu Wang, Ph.D.). This picture is taken from Ref. [876], under the Creative Commons Attribution 4.0 International License (`http://creativecommons.org/licenses/by/4.0/`); (b) the conceptual design of a wave farm. Illustration by Alfred Hicks, NREL.

8.2 REVIEW OF EXISTING ARRAY MODELLING METHODS

The theory for evaluating the power absorption of WEC arrays traces back to Refs. [97, 209, 224]. D. V. Evans [209] and J. Falnes [224] independently derived a general power absorption theory for arrays of oscillating bodies around 1979–1980. In the earliest works, simplified theories were proposed to reduce the complexity of array interactions due to the limitations of computational technologies. Two of the representative simplified theories are: *point-absorber approximation* [97, 208, 209, 210, 224, 524], assuming that the device's characteristic length are small enough in comparison with the incident wave length, and *plane-wave approximation* [718, 523, 525, 525], assuming that the devices are widely spaced relative to the wave length. Refs. [524, 510] performed comparative studies on these approximations.

Beside quick approximations, exact theories were also developed, such as the multiple scattering method [591, 507, 508, 717] and the direct matrix method [389, 295, 393, 124, 530, 245]. The limitation of the first approach is that only expansions with explicit analytical expressions can be applicable; while in the latter approach, the advantages of the multiple scattering method and the numerical BEM can be combined to account for devices with complex geometries.

Kagemoto and Yue (1986) [389] developed an exact interaction theory based on multiple-scattering interaction theory [591] with the direct matrix method [740, 718, 527, 523], and applied it to axisymmetric bodies using a hybrid element method [859]. Goo and Yoshida (1990) [295] extended this method to bodies of complex shape by using the source panel method and the Green function in polar coordinates [71, 239, 362]. Further extensions were made to interaction theory in infinite-depth water by Peter and Meylan (2004) [632], and to hierarchical interaction theory with multiple layers by Kashiwagi (2000) [393]. Child and Venugopal (2010) [124] applied semi-analytical techniques to study optimal configurations of WEC arrays. McNatt et al. (2015) [530] developed a simplified method that can derive the diffraction transfer matrix (DTM) and the radiation characteristics (RC) from the standard output of wave potentials of a BEM solver. However, their method considers only the progressive mode and ignores the evanescent modes. Later, Flavia et al. (2018) [245] implemented the method of [295] on the open-source BEM code Nemoh [46] and derived general identities to water-wave multiple-scattering problems [246]. Liu et al. (2021) [457] derived the formulations for evaluating the DTM and RC using the hybrid source and dipole method, increasing accuracy over the methods described in Refs. [295] and [245].

Exact theories on multiple bodies have been applied to various fields, such as the interconnected multi-moduled floating offshore structure [111], ice-floes in the marginal ice zone [632, 69], very-large floating structures [394], and, recently, arrays of wave energy converters [298, 877, 873], in which analytical approaches or semi-analytical approaches applying multiple-scattering interaction theory have been used extensively.

8.3 INTERACTION THEORY

8.3.1 The concept of partial waves

Prior to presenting the interaction theory of multiple floating bodies, for convenience, the concept of "partial waves" needs to be introduced. In a finite-sized array of floating bodies, the wave velocity potential can be expressed as a scalar product between a vector of complex coefficients and a vector of partial cylindrical wave components [529, 245]. As discussed in Section 2.2.1, the incident wave potential to body j involves the incoming wave solutions to the Laplace equation, subjecting it to a set of boundary conditions in polar coordinates

$$\phi_j^I(r_j, \theta_j, z_j) = \sum_{q=-\infty}^{\infty} \left[\left(A_j^I\right)_{0q} \frac{\cosh k\,(z_j + d)}{\cosh k_0 d} J_q(kr_j) \right. $$
$$\left. + \sum_{l=1}^{\infty} \left(A_j^I\right)_{lq} \cos k_l\,(z_j + d)\, I_q(k_l r_j) \right] e^{iq\theta_j}, \qquad (8.1)$$

where J_q is the Bessel function of the first kind of order q, and I_q is the modified Bessel function of the first kind of order q. The scattered and the radiated wave potentials involve the outgoing wave solutions

$$\phi_j^S(r_j, \theta_j, z_j) = \sum_{m=-\infty}^{\infty} \left[\left(A_j^S\right)_{0m} \frac{\cosh k\,(z_j + d)}{\cosh k_0 d} H_m^{(1)}(kr_j) \right. $$
$$\left. + \sum_{n=1}^{\infty} \left(A_j^S\right)_{nm} \cos k_n\,(z_j + d)\, K_m(k_n r_j) \right] e^{im\theta_j}, \qquad (8.2)$$

$$\phi_j^{R,p}(r_j, \theta_j, z_j) = \sum_{m=-\infty}^{\infty} \left[\left(R_j^p\right)_{0m} \frac{\cosh k\,(z_j + d)}{\cosh k_0 d} H_m^{(1)}(kr_j) \right. $$
$$\left. + \sum_{n=1}^{\infty} \left(R_j^p\right)_{nm} \cos k_n\,(z_j + d)\, K_m(k_n r_j) \right] e^{im\theta_j}. \qquad (8.3)$$

where $\boldsymbol{x}_j = (r_j, \theta_j, z_j)$ represents the polar coordinates of the field point $\boldsymbol{x}_j$ in the fluid domain; $H_m^{(1)}$ is the Hankel function of the first kind of order m; K_m is the modified Bessel function of the second kind of order m; and p stands for the pth body rigid mode. Note that the wavenumber k is the positive root of the water wave dispersion equation Eq. (2.54); k_n $(n = 1, 2, ...)$ satisfies Eq. (2.55), characterising the evanescent modes of the eigenfunction expansion. Eqs. (8.1), (8.2), and (8.3) can be written in a compact vector form

$$\phi_j^I = \{A_j^I\}^T \{\psi_j^I\}; \quad \phi_j^S = \{A_j^S\}^T \{\psi_j^S\}; \quad \phi_j^{R,p} = \{A_j^{R,p}\}^T \{\psi_j^S\}, \qquad (8.4)$$

where the superscript T represents the transpose operator for a matrix or vector, the curly bracket stands for a vector; and A_j^I, A_j^S and $A_j^{R,p}$ are the complex incident, scattered and radiated vectors of partial wave coefficients. Herein, indexes (l, q) are associated with incident waves and (n, m) with outgoing waves. The vectors of the incident and scattered cylindrical functions are respectively

$$\left\{\psi_j^I\right\}_{lq} = \begin{cases} \frac{\cosh k(z_j + h)}{\cosh kh} J_q(kr_j) e^{iq\theta_j}, & l = 0 \\ \cos k_l\,(z_j + h)\, I_q(k_l r_j) e^{iq\theta_j}, & l \geq 1 \end{cases}, \qquad (8.5)$$

$$\left\{\psi_j^S\right\}_{nm} = \begin{cases} \frac{\cosh k(z_j + h)}{\cosh kh} H_m^{(1)}(kr_j) e^{im\theta_j}, & n = 0 \\ \cos k_n\,(z_j + h)\, K_m(k_n r_j) e^{im\theta_j}, & n \geq 1 \end{cases}. \qquad (8.6)$$

It is noted that the first terms of the incident and scattered cylindrical functions represent the propagating mode, while the rest of the terms are associated with evanescent modes.

8.3.2 Ambient incident plane waves

Let us consider a long-crested incident wave propagating to the positive x-direction, transmitting with a small amplitude A, a heading angle β measured from the positive x-axis, and a wave number k, in water with a finite depth of h. The ambient wave potential incident to body j can be written as

$$\phi_j^A(x_j, y_j, z_j) = -\frac{igA}{\omega} \frac{\cosh k(z_j + h)}{\cosh kh} e^{ik(x_j \cos\beta + y_j \sin\beta)}. \tag{8.7}$$

By means of the polar coordinates, Eq. (8.7) can be expressed relative to body j

$$\phi_j^A(r_j, \theta_j, z_j) = -\frac{igA}{\omega} \frac{\cosh k(z + h)}{\cosh kh} I_j e^{ikr_j \cos(\theta_j - \beta)}. \tag{8.8}$$

where $I_j = e^{ik(x_j \cos\beta + y_j \sin\beta)}$ is a phase factor dependent on body j, and r_j is the radial coordinate of the point $x_j = (r_j, \theta_j, z_j)$ in the local coordinate system of body j. Using an identity [8], Eq. (8.8) can be further expanded as a summation of partial cylindrical waves incident on the body j

$$\phi_j^A(r_j, \theta_j, z_j) = -\frac{igA}{\omega} \frac{\cosh k(z + h)}{\cosh kh} I_j \sum_{q=-\infty}^{\infty} J_q(kr_j) e^{iq(\pi/2 + \theta_j - \beta)}. \tag{8.9}$$

In comparison with Eq. (8.1), Eq. (8.9) reduces to

$$\phi_j^I(x_j, y_j, z_j) = \{a_j^I\}^T \{\psi_j^I\}, \tag{8.10}$$

where the expansion coefficients are

$$\{a_j^l\}_{lq} = \begin{cases} -i\frac{gA}{\omega} e^{ik_0(x_j \cos\beta + y_j \sin\beta)} e^{iq(\pi/2 - \beta)}, & l = 0 \\ 0 & l \geq 1 \end{cases}. \tag{8.11}$$

8.3.3 Solution of the partial wave coefficients

The complex expansion coefficients A_j^I, A_j^S, and $A_j^{R,p}$ are the only unknowns to be solved for an array of bodies. The primary task in this subsection is to find a relationship between these unknown coefficients and the existing information. We deal with the diffraction and radiation problems separately, as in Refs. [530, 245], rather than treating them simultaneously as a whole (e.g., [124]).

In Eq. (8.4), the scattered wave potential of body j amongst an array of bodies is expressed as the scalar product of scattered coefficients and outgoing partial wave components. Using Graf's addition theorem [8], it is straightforward to obtain

$$H_m(k_0 r_i) e^{im\theta_i} = \sum_{q=-\infty}^{\infty} H_{m+q}(k_0 L_{ij}) J_q(k_0 r_j) e^{i[\alpha_{ij}(m+q) + q(\pi - \theta_j)]}, \tag{8.12}$$

$$K_m(k_n r_i) e^{im\theta_i} = \sum_{q=-\infty}^{\infty} K_{m+q}(k_n L_{ij}) I_q(k_n r_j) e^{i[\alpha_{ij}(m+q) + q(\pi - \theta_j)]}. \tag{8.13}$$

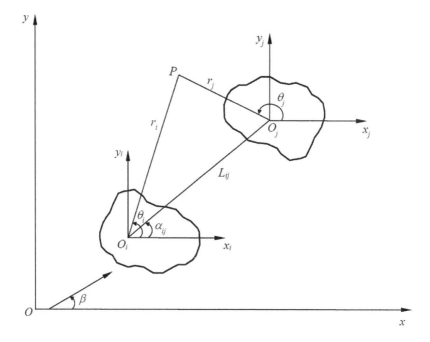

Figure 8.2: Schematic of the local and global coordinate systems.

where L_{ij} is the distance between the centres (origins of the local coordinate systems) of body i and j; α_{ij} is the angle at body i between the positive x-direction and the line joining the centre of i to that of j in an anti-clockwise direction (see Figure 8.2) [121]. In order to apply Graf's addition theorem, it is necessary to assume here that the circumscribing cylinder of each body does not overlap vertically with the other's. Eqs. (8.12) and (8.13) hold true for any integer m, and any non-negative integer n, when $r_j \leq L_{ij}$. The two formulae are the basis to derive a coordinate transformation matrix $[T_{ij}]$, for every pair of i, j except $i = j$:

$$[T_{ij}]^{mq}_{nn} = \begin{cases} H_{m-q}(k_0 L_{ij})e^{i\alpha_{ij}(m-q)}, & n = 0 \\ K_{m-q}(k_n L_{ij})e^{i\alpha_{ij}(m-q)}(-1)^q, & n \geq 1 \end{cases}. \qquad (8.14)$$

Note that the square bracket in this chapter indicates a matrix. Eq. (8.14) can be used to express the scattered partial wave components in terms of the incident partial wave components:

$$\{\psi_i^S\} = [T_{ij}]\{\psi_j^I\}. \qquad (8.15)$$

According to Eq. (8.4), the scattered waves from body i can be expressed as the incident waves to body j:

$$\phi_i^S(r_i, \theta_i, z_i) = \{A_i^S\}^T [T_{ij}]\{\psi_j^I\}. \qquad (8.16)$$

In this way, the total wave potentials incident to body j can be written as a summation of the ambient incident wave and all scattered waves from the other bodies

$$\phi_j^I(r_j,\theta_j,z_j) = \phi_j^A(r_j,\theta_j,z_j) + \sum_{\substack{i=1 \\ i\neq j}}^{N_B} \{A_i^S\}^T [T_{ij}]\{\psi_j^I\}$$

$$= (\{a_j^I\}^T + \sum_{\substack{i=1 \\ i\neq j}}^{N_B} \{A_i^S\}^T [T_{ij}]) \cdot \{\psi_j^I\}, \quad (j=1,2,...,N_B), \tag{8.17}$$

where N_B stands for the number of bodies in the array.

The incident and scattered partial waves can be related by a linear operator, termed DTM $[D_j]$. It transfers the incident partial waves to the corresponding scattered partial waves. The element $[D_j]_{nl}^{mq}$ is defined as the amplitude of the $[n(2M+1)+m+1]^{\text{th}}$ scattered partial wave potential due to a single unit-amplitude incidence of the $[l(2Q+1)+q+1]^{\text{th}}$ mode on body j, where M and Q represent the number of truncation terms in m and q, respectively. Therefore, it is straightforward to write the scattered wave potential from body j as

$$\phi_j^S(r_j,\theta_j,z_j) = (\{a_j^I\}^T + \sum_{\substack{i=1 \\ i\neq j}}^{N_B} \{A_i^S\}^T [T_{ij}]) \cdot [D_j]^T \{\psi_j^S\}, \quad (j=1,2,...,N_B). \tag{8.18}$$

The combination of Eqs. (8.4) and (8.18) yields a new equation. Cancelling the common vector of scattered partial waves and transposing both of the left- and right-hand sides of the equation results in

$$\{A_j^S\} = [D_j] \cdot (\{a_j^I\} + \sum_{\substack{i=1 \\ i\neq j}}^{N_B} [T_{ij}]^T \{A_i^S\}), \quad (j=1,2,...,N_B). \tag{8.19}$$

Following a similar method, the equation for the radiation problem is

$$\{A_j^{R,i,p}\} = [D_j] \cdot (\{a_j^{R,i,p}\} + \sum_{\substack{t=1 \\ t\neq j}}^{N_B} [T_{tj}]^T \{A_t^{R,i,p}\}), \quad (j=1,2,...,N_B), \tag{8.20}$$

where $\{a_j^{R,i,p}\}$ are the expansion coefficients of the radiated wave incident on body j, generated by the unitary motion of body i in its pth degree of freedom:

$$\{a_j^{R,i,p}\} = \begin{cases} 0, & i=j \\ [T_{ij}]^T \cdot \{R_i^p\}, & i\neq j \end{cases}, \tag{8.21}$$

where $\{R_i^p\}$ is termed as RC. Note that Eqs. (8.19) and (8.20) are expressed for each body. It is thereby possible to assemble a large linear algebraic system involving all the bodies in the array to solve the scattered partial wave coefficients.

8.3.4   Wave excitation forces and hydrodynamic quantities

Wave forces can be calculated using matrix manipulation after the scattered partial wave coefficients are obtained. Based on Eqs. (8.17) and (8.18), the wave excitation force can be calculated by integrating the hydrodynamic pressure over the immersed body surface:

$$
F_{j,p}^E = \mathrm{i}\omega\rho \iint_{S_B^j} \left(\phi_j^I + \phi_j^S \right) n_{j,p} \mathrm{d}S = \mathrm{i}\omega\rho \left(\left\{ a_j^I \right\}^T + \sum_{\substack{i=1 \\ i\neq j}}^{N_B} \left\{ A_i^S \right\}^T [T_{ij}] \right)
$$
$$
\times \iint_{S_B^j} \left(\left\{ \psi_j^I \right\} + [D_j]^T \left\{ \psi_j^S \right\} \right) n_{j,p} \mathrm{d}S, \quad (p = 1,2,...,6, j = 1,2,...,N_B)
$$
(8.22)

where $F_{j,p}^E$ is interpreted as the excitation force in the pth DoF of body j, and $n_{j,p}$ is the pth component of the normal vector on the immersed body surface. By defining a new linear operator-force transfer matrix as

$$
\left\{ G_{j,p}^E \right\} - \mathrm{i}\omega\rho \iint_{S_B^j} \left(\left\{ \psi_j^I \right\} + [D_j]^T \left\{ \psi_j^S \right\} \right) n_{j,p} \mathrm{d}S,
$$
(8.23)

and the overall expansion coefficients of the total waves incident to body j, which consists of the ambient incident wave and all the scattered waves from neighbouring bodies as

$$
\left\{ \eta_j^E \right\} = \left\{ a_j^I \right\}^T + \sum_{\substack{i=1 \\ i\neq j}}^{N_B} \left\{ A_i^S \right\}^T [T_{ij}],
$$
(8.24)

calculation of the wave excitation force on body j can be simplified in the following matrix form:

$$
F_{j,p}^E = \left\{ \eta_j^E \right\} \left\{ G_{j,p}^E \right\}.
$$
(8.25)

Following a similar method, the wave radiation force on body j can be evaluated as

$$
F_{j,t}^{R,i,p} = \begin{cases} \left\{ \eta_j^{R,i,p} \right\} \left\{ G_{j,t}^E \right\}, & i \neq j \\ \mathrm{i}\rho(a_{j,p} + \omega b_{j,p}) + \left\{ \eta_j^{R,i,t} \right\} \left\{ G_{j,t}^E \right\}, & i = j \end{cases}
$$
(8.26)

where $a_{j,p}$ and $b_{j,p}$ are the added mass and the radiation damping of body j in the pth DoF due to its own unitary motion in the same mode when the body is in isolation; the overall expansion coefficient in association with the force transfer matrix is

$$
\left\{ \eta_j^{R,i,p} \right\} = \left\{ a_j^{R,i,p} \right\}^T + \sum_{\substack{t=1 \\ t\neq j}}^{N_B} \left\{ A_t^{R,i,p} \right\}^T [T_{tj}].
$$
(8.27)

Note that $F_{j,t}^{R,i,p}$ is interpreted as the radiation force of body j in the tth DoF due to the pth DoF motion of body i. Correspondingly, the added mass and the radiation damping of a body can then be obtained by decomposition of the complex radiation force into real and imaginary parts.

Figure 8.3 gives a comparison between different methods in calculating the wave forces on arrays of bodies, which shows a high-degree match between them. This numerical case consists of an array of 4 truncated cylinders and the details of the layout

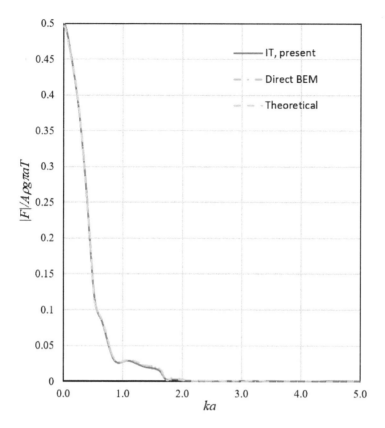

Figure 8.3: Variations of the heave excitation force on Cylinder 1 of an array of 4 truncated vertical cylinders with radius a, draught $T = 2a$, in a water depth of $h = 4a$ with each cylinder placed at the vertex of a square of side length $4a$ for an incident wave heading angle of $\beta = \pi/4$.

are given in Ref. [717]. The "Direct BEM" denotes the complete BEM method based on mixed sources and dipoles as described in Section 2.2.2 of Chapter 2. "Theoretical" denotes the result calculated by Ref. [717]. Although "Direct BEM" is not limited to regular geometries, the computational cost is much more expensive than the "Theoretical" method. In contrast with the other two methods, the present method, based on interaction theory, provides a compromise option for arrays of complex geometries.

8.3.5 Motion responses

Given the body specifications, the power-take-off (PTO) characteristics, the mooring system properties and the wave loads, the motion equation of body j can be constructed as

$$\left\{ -\omega^2 \left([M] + [A_{\mathrm{m}}] \right) - \mathrm{i}\omega \left([B_{\mathrm{rad}}] + [B_{\mathrm{pto}}] \right) + \left([K_{\mathrm{s}}] + [K_{\mathrm{pto}}] + [K_{\mathrm{moor}}] \right) \right\} \{X\} = \{F^E\},$$
(8.28)

where $[M]$ is the device mass matrix, $[A_{\mathrm{m}}]$ and $[B_{\mathrm{rad}}]$ the device added mass and the radiation damping matrices, $[B_{\mathrm{pto}}]$ and $[K_{\mathrm{pto}}]$ the mechanical damping and the stiffness matrices related to the PTO system, $[K_{\mathrm{s}}]$ the hydrostatic restoring matrix, $[K_{\mathrm{moor}}]$ the mooring stiffness matrix, $\{X\}$ the displacement vector, and $\{F^E\}$ the wave excitation force vector. Note that all the above matrices or vectors are of body j. For brevity, the subscript j is omitted in Eq. (8.28).

8.4 LINEAR OPERATOR MATRICES

A linear operator matrix refers to a matrix that relates two different physical variables in interaction theory within the framework of the linear water wave (see Section 2.1.1.2 of Chapter 2). To simulate the wave interactions between an array of bodies, an essential characteristic is that the linear operators are only determined by a single body in isolation. The existence of linear operators facilitates multi-body computations via the matrix form based on the direct matrix method.

8.4.1 Diffraction transfer matrix

A diffraction transfer matrix (DTM) represents the scattering properties of a body and is solved from the boundary value problem (Eq. 8.43 or Eq. 8.45) for the body in isolation. Technically speaking, DTM transforms a vector of incident, cylindrical, partial-wave coefficients into a vector of outgoing, partial-wave coefficients representing waves scattered by the body [389, 530].

8.4.1.1 Alternative method I

The 1st linear operator, the DTM of a specific floating body, as mentioned in Eq. (8.18), can be constructed by considering the wave diffraction when it is in isolation. The scattered potential of a single floating body in a partial incident wave of mode (l, q) without the presence of other bodies can be expressed as

$$
\begin{aligned}
\left\{\varphi_j^S\left(r_j, \theta_j, z_j\right)\right\}_{l,q} = & \frac{\cosh k\left(z_j+h\right)}{\cosh kh} \sum_{m=-\infty}^{\infty}\left[D_j\right]_{0,m}^{l,q} H_m^{(1)}\left(k r_j\right) e^{\mathrm{i} m \theta_j} \\
& + \sum_{n=1}^{\infty} \cos k_n\left(z_j+h\right) \sum_{m=-\infty}^{\infty}\left[D_j\right]_{n,m}^{l,q} K_m\left(k_n r_j\right) e^{\mathrm{i} m \theta_j}
\end{aligned}
\tag{8.29}
$$

where $D_{0m}^{j,lq}$ and $D_{nm}^{j,lq}$ are scattered complex coefficients as well as the DTM elements. The scattered potential at a field point in the fluid domain (other than the body surface) can be determined by the following equation:

$$
\begin{aligned}
\left\{\varphi_j^S\left(\boldsymbol{x}_j\right)\right\}_{l,q} = -\frac{1}{4\pi} & \left\{\iint_{S_B^j}\left\{\varphi_j^S\left(\boldsymbol{\xi}_j\right)\right\}_{l,q} \frac{\partial G\left(\boldsymbol{x}_j ; \boldsymbol{\xi}_j\right)}{\partial n\left(\boldsymbol{\xi}_j\right)} \mathrm{d} S_{\boldsymbol{\xi}_j}\right. \\
& \left. + \iint_{S_B^j} G\left(\boldsymbol{x}_j ; \boldsymbol{\xi}_j\right) \frac{\partial\left\{\psi_j^I\left(\boldsymbol{\xi}_j\right)\right\}_{l,q}}{\partial n_{\xi}} \mathrm{d} S_{\boldsymbol{\xi}_j}\right\}.
\end{aligned}
\tag{8.30}
$$

Substituting Eq. (8.46) into Eq. (8.30) and comparing Eq. (8.29) with Eq. (8.30) yields

$$[D_j]^{l,q}_{0,m} = -\frac{i}{2} C_0 \cosh kh \iint_{S_B^j} \left[\{\varphi_j^S\}_{l,q} \frac{\partial}{\partial n} + \frac{\partial \{\psi_j^I\}_{l,q}}{\partial n} \right] \left[J_m(kR_j) \cosh k(\zeta_j + h) e^{-im\Theta_j} \right] dS_{\xi_j},$$

(8.31)

$$[D_j]^{l,q}_{n,m} = -\frac{1}{\pi} C_n \iint_{S_B^j} \left[\{\varphi_j^S\}_{l,q} \frac{\partial}{\partial n} + \frac{\partial \{\psi_j^I\}_{l,q}}{\partial n} \right] \left[I_m(k_n R_j) \cos k_n(\zeta_j + h) e^{-im\Theta_j} \right] dS_{\xi_j}.$$

(8.32)

The unknown wave scattering potential φ_{jlq}^S in Eq. (8.15) and Eq. (8.16) on the body surface can be solved from the following boundary integral equation:

$$2\pi \{\varphi_j^S(\boldsymbol{x}_j)\}_{l,q} + \iint_{S_B^j} \{\varphi_j^S(\boldsymbol{\xi}_j)\}_{l,q} \frac{\partial G(\boldsymbol{x}_j; \boldsymbol{\xi}_j)}{\partial n(\boldsymbol{\xi}_j)} dS_{\xi_j} = -\iint_{S_B^j} \frac{\partial \{\psi_j^I(\boldsymbol{\xi}_j)\}_{l,q}}{\partial n_\xi} G(\boldsymbol{x}_j; \boldsymbol{\xi}_j) dS_{\xi_j}.$$

(8.33)

8.4.1.2 Alternative method II

Refs. [395, 393] employed a different boundary integral equation to solve the total wave diffraction potential φ_j^D, i.e.,

$$2\pi \{\varphi_j^D(\boldsymbol{x}_j)\}_{l,q} + \iint_{S_B^j} \{\varphi_j^D(\boldsymbol{\xi}_j)\}_{l,q} \frac{\partial G(\boldsymbol{x}_j; \boldsymbol{\xi}_j)}{\partial n(\boldsymbol{\xi}_j)} dS_{\xi_j} = 4\pi \{\psi_j^I(\boldsymbol{x}_j)\}_{l,q},$$

(8.34)

from which the wave scattering potential φ_j^S at a field point in the fluid domain can easily be obtained by subtracting the incident partial wave component

$$\{\varphi_j^S(\boldsymbol{x}_j)\}_{l,q} = \{\varphi_j^D(\boldsymbol{x}_j)\}_{l,q} - \{\psi_j^I(\boldsymbol{x}_j)\}_{l,q} = -\frac{1}{4\pi} \iint_{S_B^j} \{\varphi_j^D(\boldsymbol{\xi}_j)\}_{l,q} \frac{\partial G(\boldsymbol{x}_j; \boldsymbol{\xi}_j)}{\partial n(\boldsymbol{\xi}_j)} dS_{\xi_j}.$$

(8.35)

Given the unknown wave scattering potential, the DTM elements can be found from Eqs. (8.29), (8.35) and (8.46):

$$[D_j]^{l,q}_{0,m} = -\frac{i}{2} C_0 \cosh kh \iint_{S_B^j} [\{\varphi_j^S\}_{l,q} + \{\varphi_j^I\}_{l,q}] \frac{\partial}{\partial n} [J_m(kR_j) \cosh k(\zeta_j + h) e^{-im\Theta_j}] dS,$$

(8.36)

$$[D_j]^{l,q}_{0,m} = -\frac{1}{\pi} C_n \iint_{S_B^j} [\{\varphi_j^S\}_{l,q} + \{\varphi_j^I\}_{l,q}] \frac{\partial}{\partial n} [I_m(k_n R_j) \cosh k_n(\zeta_j + h) e^{-im\Theta_j}] dS.$$

(8.37)

8.4.1.3 Comparison of accuracy and efficiency

In general, the accuracy of the two alternative methods is similar. However, there are cases (e.g., geometries with sharp corners) when Method II performs better than Method I. This is because at the right-hand side of the boundary integral equations, the integration of the normal derivative of the incident wave potential over the body surface in Method I might have a larger cumulative numerical error than simple evaluation of the incident partial wave potential at a single field point in Method II. In addition, Method II is also superior to its counterpart in terms of computational efficiency as the right-hand side is faster to evaluate. This advantage may not be noticeable in single-body computations, but when it comes to a multi-body problem, the difference is remarkable. Ref. [457] compares the computational time for per-frequency DTM computation, showing that Method II is far more efficient than Method I when the truncation number of modes increases.

As an example, Figure 8.4 shows a comparison between the two alternative methods in calculating the DTM terms. McNatt et al. (2015) denotes the results given in Ref. [530]. Generally, there is a good level of agreement between the two methods in calculating both the real and the imaginary parts. However, it can be noticed that Method II performs slightly better than Method I in approaching the results of McNatt et al. (2015). Furthermore, the computation time of Method I in this numerical case is *more than ten times* that of Method II.

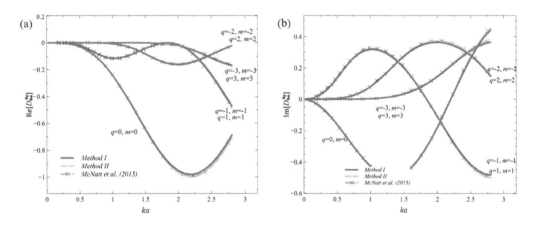

Figure 8.4: DTM progressive terms for a circular cylinder of 3 m radius, 6 m draft in a 10 m water depth, a comparison between Method I and Method II: (a) real part; (b) imaginary part. This figure is adapted from Ref. [457].

8.4.2 Radiation characteristics

Let us derive the expressions for the 2nd linear operator, the RC, as mentioned in Eq. (8.21). Physically, RC characterises the way in which a floating body radiates waves. The wave radiation potential $\{\varphi_j^{R,k}\}$ away from a single floating body of mode (n, m) without the influence of other bodies can be constructed as

$$
\varphi_j^{R,p}(r_j, \theta_j, z_j) = \frac{\cosh k(z_j + h)}{\cosh kh} \sum_{m=-\infty}^{\infty} \{R_j^p\}_{0,m} H_m^{(1)}(kr_j) e^{im\theta_j}
$$
$$
+ \sum_{n=1}^{\infty} \cos k_n(z_j + h) \sum_{m=-\infty}^{\infty} \{R_j^p\}_{n,m} K_m(k_n r_j) e^{im\theta_j},
$$

$$(8.38)$$

The radiation potential at a field point in the fluid domain (other than the body surface) can be determined by the following equation:

$$\varphi_j^{R,p}(\boldsymbol{x}_j) = -\frac{1}{4\pi}\left\{\iint_{S_B^j}\varphi_j^{R,p}(\boldsymbol{\xi}_j)\frac{\partial G(\boldsymbol{x}_j;\boldsymbol{\xi}_j)}{\partial n(\boldsymbol{\xi}_j)}\mathrm{d}S_{\boldsymbol{\xi}_j}\right.$$
$$\left. -\iint_{S_B^j}n_{j,p}G(\boldsymbol{x}_j;\boldsymbol{\xi}_j)\mathrm{d}S_{\boldsymbol{\xi}_j}\right\}. \tag{8.39}$$

combining Eqs. (8.38), (8.39) and (8.46) leads to the following expressions of the RC elements:

$$\{R_j^p\}_{0,m} = -\frac{\mathrm{i}}{2}C_0\cosh kh\iint_{S_B^j}(\varphi_j^{R,p}\frac{\partial}{\partial n} - n_{j,p})[J_m(kR_j)\cosh k(\zeta_j+h)\mathrm{e}^{-\mathrm{i}m\Theta_j}]\mathrm{d}S, \tag{8.40}$$

$$\{R_j^p\}_{n,m} = -\frac{1}{\pi}C_n\iint_{S_B^j}(\varphi_j^{R,p}\frac{\partial}{\partial n} - n_{j,p})[I_m(k_nR_j)\cosh k_n(\zeta_j+h)\mathrm{e}^{-\mathrm{i}m\Theta_j}]\mathrm{d}S. \tag{8.41}$$

In Eq. (8.22) and Eq. (8.23), the unknown wave radiation potential $\varphi_j^{R,k}$ is solved from the following boundary integral equation:

$$2\pi\varphi_j^{R,p}(\boldsymbol{x}_j) + \iint_{S_B^j}\varphi_j^{R,p}(\boldsymbol{\xi}_j)\frac{\partial G(\boldsymbol{x}_j;\boldsymbol{\xi}_j)}{\partial n_\xi}\mathrm{d}S = \iint_{S_B^j}n_{j,p}G(\boldsymbol{x}_j;\boldsymbol{\xi}_j)\mathrm{d}S. \tag{8.42}$$

Figure 8.5 verifies the present method for calculating the RC progressive terms, by comparing the results given respectively in Refs. [245] and [530]. Good agreement is found between the three methods in all the RC terms shown. Furthermore, the values of RC terms of two additive inverse modes (e.g., $m = -1$ and $m = 1$) are exactly the opposite; whereas the values of DTM terms of two such modes are exactly the same (see Figure 8.4).

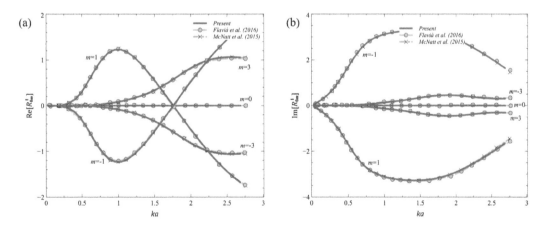

Figure 8.5: RC progressive terms for a cube box of 6 m side, 6 m draft moving in surge in a 10 m water depth: (a) real part and (b) imaginary part. This figure is taken from Ref. [457], under the Creative Commons Attribution 4.0 International License (http://creativecommons.org/licenses/by/4.0/).

8.4.3 Force transfer matrix

The 3rd linear operator, the force transfer matrix (FTM) $\{G_{j,p}^E\}$, given in Eq. (8.23), transforms a vector of incident partial cylindrical wave coefficients into forces (either diffraction or radiation) on a floating body. The idea of FTM was first introduced by McNatt et al. (2015) [530]. The elements of an FTM can be determined using the BEM (as introduced in Section 8.5) by integrating the hydrodynamic pressure (indeed, the diffraction or radiation potential) due to a partial wave.

8.5 BOUNDARY INTEGRAL EQUATIONS FOR PARTIAL WAVES

Wave radiation and diffraction of multiple floating bodies can be solved within the limits of potential flow theory, as discussed extensively in Section 2.2.2 of Chapter 2. Based on the assumption that the fluid is inviscid, incompressible, and with an irrotational motion, the fluid flow can be described by an ideal velocity potential satisfying the Laplace equation. However, the difference with Chapter 2 is that, in this section, partial wave potentials are considered instead of the incident plane wave.

8.5.1 Indirect approach

The indirect approach is based on the source formulation, which means that only sources are distributed on the immersed body surface. An isolated floating body is considered by adopting the polar coordinate system, the partial scattered wave potential should satisfy the following boundary integral equation:

$$2\pi\{\sigma_j^S(\boldsymbol{x}_j)\}_{l,q} + \iint_{S_B^j} \{\sigma_j^S(\boldsymbol{\xi}_j)\}_{l,q} \frac{\partial G(\boldsymbol{x}_j; \boldsymbol{\xi}_j)}{\partial n(\boldsymbol{x}_j)} \mathrm{d}S_{\boldsymbol{\xi}_j} = V_{\mathrm{n}}(\boldsymbol{x}_j), \qquad (8.43)$$

where σ_j^S is the source strength and $G(\boldsymbol{x}_j; \boldsymbol{\xi}_j)$ is the free-surface Green's function. At the right-hand side of Eq. (8.43), the Neumann boundary condition prescribes the normal velocity of the fluid on the immersed and impermeable body surface:

$$V_{\mathrm{n}}(\boldsymbol{x}_j) = \begin{cases} n_{j,p}(\boldsymbol{x}_j), & p = 1, 2, ..., 6 \\ -\dfrac{\partial\{\psi_j^I(\boldsymbol{x}_j)\}}{\partial n(\boldsymbol{x}_j)}, & p = 7 \end{cases} \qquad (8.44)$$

in which $n_{j,p}$ is defined in Eq. (8.22). Note that in Eq. (8.43) and Eq. (8.44), the body boundary condition and the normal derivative are applied at the field point $\boldsymbol{x}_j = (r_j, \theta_j, z_j)$ rather than at the source point $\boldsymbol{\xi}_j = (R_j, \Theta_j, \zeta_j)$.

8.5.2 Direct approach

The direct approach is based on the potential formulation, involving both the sources and dipoles. Unlike the indirect approach, the wave potential can be directly solved from the boundary integral equation. Applying Green's second identity, a Fredholm integral equation of the second kind can be constructed as the follows:

$$2\pi \left\{ \varphi_j^S \left(\boldsymbol{x}_j \right) \right\}_{l,q} + \iint_{S_B^j} \left\{ \varphi_j^S \left(\boldsymbol{\xi}_j \right) \right\}_{l,q} \frac{\partial G \left(\boldsymbol{x}_j; \boldsymbol{\xi}_j \right)}{\partial n(\boldsymbol{\xi}_j)} dS_{\boldsymbol{\xi}_j} = \iint_{S_B^j} V_{\mathrm{n}}(\boldsymbol{\xi}_j) G \left(\boldsymbol{x}_j; \boldsymbol{\xi}_j \right) dS_{\boldsymbol{\xi}_j}.$$

(8.45)

Following Refs. [71, 239, 362], by applying Graf's addition theorem [820], it is straightforward to expand the free-surface Green function in the eigenfunction expansion and express it in polar coordinates in the form of

$$G \left(\boldsymbol{x}_j; \boldsymbol{\xi}_j \right) =$$

$$2\pi i C_0 \cosh k \left(z_j + h \right) \cosh k \left(\zeta_j + h \right) \sum_{m=-\infty}^{\infty} \left\{ \begin{array}{c} H_m^{(1)} \left(k r_j \right) J_m \left(k R_j \right) \\ H_m^{(1)} \left(k R_j \right) J_m \left(k r_j \right) \end{array} \right\} e^{im(\theta_j - \Theta_j)} +$$

$$4 \sum_{n=1}^{\infty} C_n \cos k_n \left(z_j + h \right) \cos k_n \left(\zeta_j + h \right) \sum_{m=-\infty}^{\infty} \left\{ \begin{array}{c} K_m \left(k_n r_j \right) I_m \left(k_n R_j \right) \\ K_m \left(k_n R_j \right) I_m \left(k_n r_j \right) \end{array} \right\} e^{im(\theta_j - \Theta_j)},$$

(8.46)

where the expansion coefficients are

$$C_0 == \frac{k^2 - K^2}{\left(k^2 - K^2 \right) h + K} = \frac{2k}{2kh + \sinh 2kh},$$

(8.47)

$$C_n == \frac{k_n^2 + K^2}{\left(k_n^2 + K^2 \right) h - K} = \frac{2k_n}{2k_n h + \sin 2k_n h},$$

(8.48)

where $K = \omega^2 / g$, and $k_n (n = 0, 1, 2 \dots)$ are the roots of the wave dispersion equation in finite-depth water. In Eq. (8.46), the upper terms in the brackets are used when $r_j \geq R_j$ (the region outside of a circular cylinder that circumscribes the body or bodies) and the lower terms when $r_j < R_j$. The Green function in Eq. (8.46) was also named as the "ring source" by Ref. [362].

8.5.3 Removal of irregular frequencies

Similar to the wave interaction with a single floating body, directly solving Eq. (8.43) or Eq. (8.45) can lead to some unphysical numerical distortions in the computation results around the eigen-frequencies of the sloshing modes inside the floating body, which is normally termed the "irregular frequencies" phenomenon. For the source formulation, the "extended integral equation method" is recommended to remove the irregular frequencies, by adding a "rigid lid" at the interior waterplane section of the floating body. The boundary integral equations that need to be solved together are as below:

$$2\pi \left\{ \sigma_j^S \left(\boldsymbol{x}_j \right) \right\}_{l,q} + \iint_{S_B^j} \left\{ \sigma_j^S \left(\boldsymbol{\xi}_j \right) \right\}_{l,q} \frac{\partial G \left(\boldsymbol{x}_j; \boldsymbol{\xi}_j \right)}{\partial n \left(\boldsymbol{x}_j \right)} dS_{\boldsymbol{\xi}_j} +$$

$$\iint_{S_F^j} \left\{ \sigma_j^W \left(\boldsymbol{\xi}_j \right) \right\}_{l,q} \frac{\partial G \left(\boldsymbol{x}_j; \boldsymbol{\xi}_j \right)}{\partial n \left(\boldsymbol{x}_j \right)} dS_{\boldsymbol{\xi}_j} = V_{\mathrm{n}} \left(\boldsymbol{x}_j \right),$$

(8.49)

$$-4\pi \left\{ \sigma_j^S (\boldsymbol{x}_j) \right\}_{lq} + \iint_{S_B^j} \left\{ \sigma_j^S (\boldsymbol{\xi}_j) \right\}_{l,q} \frac{\partial G(\boldsymbol{x}_j; \boldsymbol{\xi}_j)}{\partial n(\boldsymbol{x}_j)} \mathrm{d}S_{\boldsymbol{\xi}_j} +$$

$$\iint_{S_F^j} \left\{ \sigma_j^W (\boldsymbol{\xi}_j) \right\}_{l,q} \frac{\partial G(\boldsymbol{x}_j; \boldsymbol{\xi}_j)}{\partial n(\boldsymbol{x}_j)} \mathrm{d}S_{\boldsymbol{\xi}_j} = V_{\mathrm{n}}'(\boldsymbol{x}_j). \tag{8.50}$$

In Eq. (8.49) and Eq. (8.50), σ_j^W is the source strength on the waterplane area. The proper condition of the function V_{n}' has been discussed in Ref. [432]. Note that using this method, the logarithmic singularity should be subtracted from the Green function and then integrated analytically. For the potential formulation, it is recommended to use the "overdetermined integral equation method", as described in Refs. [592, 429, 454, 446], as it is not necessary to integrate the logarithmic singularity and sufficient accuracy can be achieved with only a few discrete points on the waterplane [446]. Using this method, the following additional equation needs to be solved together with Eq. (8.43):

$$\iint_{S_B^j} \left\{ \varphi_j^S (\boldsymbol{\xi}_j) \right\}_{l,q} \frac{\partial G(\boldsymbol{x}_j; \boldsymbol{\xi}_j)}{\partial n(\boldsymbol{\xi}_j)} \mathrm{d}S_{\boldsymbol{\xi}_j} = \iint_{S_B^j} V_{\mathrm{n}}(\boldsymbol{\xi}_j) G(\boldsymbol{x}_j; \boldsymbol{\xi}_j) \mathrm{d}S_{\boldsymbol{\xi}_j}. \tag{8.51}$$

It should be noted in Eq. (8.51) that, the field point is taken from the discrete points on the interior waterplane area rather than those on the immersed body surface. Since the field point and the source point can never be coincident with each other, the diagonal terms with the solid angle coefficient diminish in Eq. (8.51), and the "irregular frequencies" can be effectively removed.

8.6 EVALUATION OF THE ARRAY PROPERTIES

8.6.1 Interaction factor and directionality

The interaction factor is a key metric to assess the performance of wave energy arrays. It is defined as

$$q(\beta) = \frac{\overline{P}_{\mathrm{array,max}}}{N_B \overline{P}_{\mathrm{isolated,max}}}, \tag{8.52}$$

where $\overline{P}_{\mathrm{array,max}}$ represents the maximum power absorbed by an array of N_B identical devices, $\overline{P}_{\mathrm{isolated,max}}$ represents the maximum power absorbed by a single such device in isolation, and β is the incident wave direction [839]. Eq. (8.52) means that if $q < 1$, the average power per WEC in the array is less than the power of an isolated WEC [41]. Hence, wave interactions have a destructive effect on the power absorption of the wave farm. Conversely, if $q > 1$, the park effect is constructive. Evans [209] and Falnes [224] independently derived the time-averaged power that can be absorbed by an array of oscillators in response to a regular wave train

$$\overline{P}_{\mathrm{array}} = \frac{1}{4} \left(\{U\}^* \{F^E\} + \{F^E\}^* \{U\} \right) - \frac{1}{2} \{U\}^* [B_{\mathrm{rad}}] \{U\}\}, \tag{8.53}$$

where $\{U\}$ and $\{F^E\}$ are the $N_B \times 1$ vectors of complex amplitudes of the body velocities and the wave excitation forces respectively; the asterisk $*$ denotes the complex conjugate transpose; and $[B_{\mathrm{rad}}]$ represents the $N_B \times N_B$ radiation damping matrix. The first term in Eq. (8.53) represents the total absorbed power from the incident waves; whereas the second term is the power radiated back to the sea due to the motion of the bodies [508]. Provided $[B_{\text{rad}}]$ is positive definite, the maximum total absorbed power of the array can be derived as [209, 224]

$$\overline{P}_{\text{array,max}} = \frac{1}{8} \{F^E\}^* [B_{\text{rad}}]^{-1} \{F^E\}, \tag{8.54}$$

which occurs at the optimum condition

$$\{U\}_{\text{opt}} = \frac{1}{2} [B_{\text{rad}}]^{-1} \{F^E\}. \tag{8.55}$$

Under the assumption of point absorber theory, Ref. [243] proved the following variation relationship of q with respect to β for a fixed wave frequency:

$$\frac{1}{2\pi} \int_0^{2\pi} q(\beta) \mathrm{d}\beta = 1. \tag{8.56}$$

When neither point absorber theory nor optimised individual power take-off characteristics is used, Ref. [124] proposed an analogous consistency constant c:

$$c = \frac{1}{2\pi} \int_0^{2\pi} q(\beta) \mathrm{d}\beta. \tag{8.57}$$

Eq. (8.57) shows that when $c = 1$, the q-factor obeys the consistency condition.

By using the reciprocity relationship between the wave radiation damping and the wave excitation force [571], Ref. [839] proved that Eq. (8.56) holds for not only arrays of heaving axisymmetric devices but also for arrays of axisymmetric devices moving in uncoupled heave and surge or pitch degrees of freedom. For bodies with a vertical axis of symmetry, this leads to a result relating the capture width to the interaction factor [637, 243]:

$$\eta_{\text{array,max}} = \frac{\lambda}{2\pi} N_B q(\beta). \tag{8.58}$$

Equations (3.65) and (3.67) in Section 3.2.3.2, and Eq. (8.58) determine that the following relationship exists for an array of heaving point absorbers

$$q(\beta) = \frac{1}{N_B} \frac{\eta_{\text{array,max}}}{\eta_{\text{isolated,max}}}. \tag{8.59}$$

where $\eta_{\text{isolated,max}}$ is the maximum capture width of an isolated WEC. Eq. (8.59) illustrates that the interaction factor of an array of heave point absorbers can be evaluated as the ratio of the averaged capture width of the array to that of an individual device. Furthermore, it can be shown that a symmetry in the interaction factor with respect to the incident wave angle exists and is [522]

$$q(\beta) = q(\beta + \pi). \tag{8.60}$$

8.6.2 Overall energy production of the WEC array

Eq. (8.54) gives the maximum power of a WEC array at the optimum condition, i.e., Eq. (8.55). However, in many cases such an optimum condition is not satisfied. Assuming sinusoidal waves, the mean absorbed power of a generic WEC device over a wave period can be calculated as

$$\overline{P}_{\text{isolated}} = \frac{1}{T}\int_0^T B_{\text{pto}}U^2 \mathrm{d}t = \frac{1}{2}\omega^2 B_{\text{pto}}|U|^2, \tag{8.61}$$

The formulation can be extended to an array of multi-DoF WECs using the following matrix manipulation:

$$\overline{P}_{\text{array}} = \frac{1}{2}\omega^2 \sum_{j=1}^{N} \{U_j\}^T [B_{\text{pto},j}]\{U_j\}^*, \tag{8.62}$$

where j represents the jth WEC converter; $\{U_j\}$ is an $M \times 1$ vector and $[B_{\text{pto},j}]$ is an $M \times M$ matrix; and M is the number of the total modes of each individual device. In many cases, when there is no coupling between the PTO systems of different DoFs in each individual WEC device, it is possible to formulate Eq. (8.62) as a summation of the non-zero terms given the majority of the elements in the PTO damping matrix are zeros, except for the diagonal terms. In such cases, Eq. (8.62) can be simplified as

$$\overline{P}_{\text{array}} = \frac{1}{2}\omega^2 \sum_{j=1}^{N}\sum_{i=1}^{M} B_{\text{pto}}^{j,i}|U_{j,i}|^2. \tag{8.63}$$

where i stands for the ith mode in each individual device.

8.6.3 Case study

An idealised example is given here to illustrate how to evaluate the interaction factor $q(\beta)$ for a WEC array and verify the accuracy using the numerical method presented in this chapter. As shown in Figure 8.6, three heaving hemispherical point absorbers

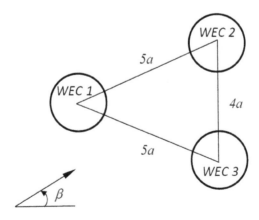

Figure 8.6: Plan-view layout of the three-device array configuration [839].

are displayed in an isosceles-triangle layout (Configuration B in Ref. [839]). The radius of each individual device is a and the water depth is $10a$. Half of each device is meshed by 2058 panels on its immersed hull and 331 panels on its waterplane area (taking advantage of the xoz symmetry plane). For heaving point absorbers, it is straightforward to derive the following expression from Eq. (8.59)

$$q(\beta) = \frac{2\pi}{\lambda N_B} \frac{\overline{P}_{\text{array,max}}(\beta)}{J},$$

(8.64)

where $\overline{P}_{\text{array,max}}(\beta)$ can be evaluated by Eq. (8.54), and the average energy flux J can be evaluated by Eqs. (2.24) and Eq. (2.25) in Section 2.1.1.3, respectively.

The relationship between the interaction factor and the wave incident angle is plotted in Figure 8.7. The results are obtained using sufficient truncation numbers for the angular modes, as well as the wave number modes (see Eqs. 8.1–8.3, herein 10 terms are used for all the modes). It is found that, given a high wave number ka, this relationship varies dramatically with respect to the wave heading. The maximum value of q in the figure reaches 1.5, while the minimum approaches 0.6: both are at the extremes of the range. The results evaluated by the present method, based on interaction theory, exhibit good agreement with the results of Ref. [839]), which was evaluated by a complete boundary element method (quadratic polynomial approximation) for multiple-body interactions. Note that although a simple small array

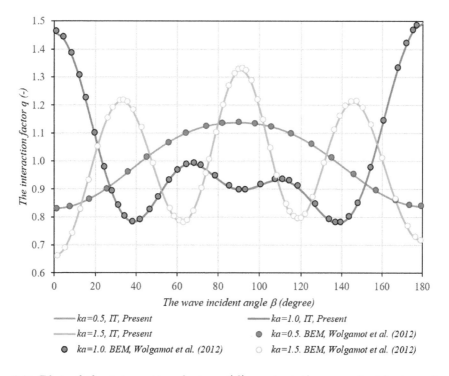

Figure 8.7: Plot of the interaction factor $q(\beta)$ against the wave incident angle β for an array of heaving hemispherical point absorbers in an isosceles-triangle layout (as indicated in Figure 8.6).

of three hemispherical point absorber WECs is shown for the case study here, the method itself can, in principle, be applied to large arrays of various WEC types with generic geometries.

8.7 SUMMARY

This chapter presents a sophisticated hybrid methodology combining the interaction theory of Kagemoto and Yue (1986) and the boundary element method. This new methodology successfully avoids evaluating the interactions between different bodies numerically, hence it is superior to the conventional boundary element method for wave interactions between multiple bodies in terms of computational efficiency. The general process of implementing the present methodology can be summarised as below. The diffraction transfer matrix, the radiation characteristics and the force transfer matrix are calculated by the BEM in advance for a single device in isolation. The total wave potential incident on each body is expressed as the summation of the ambient incident plane wave and all the scattered waves from other bodies. The wave potentials can be expanded as a Fourier series, in which the expansion coefficients are solved from the resultant linear algebraic system. Wave excitation forces, as well as added masses and radiation dampings, are then obtained by the product of the wave elevation and the force transfer matrix. Given the values of these quantities, the averaged capture width and the interaction factor of an array of WEC devices can finally be evaluated in a very straightforward manner. The case study of a simple array of hemispherical point absorbers provided at the end of the chapter illustrates the application in detail, and verifies the accuracy of the present methodology.

CHAPTER 9

Optimisation of wave farms

Malin Göteman

Uppsala University, malin.goteman@angstrom.uu.se

9.1 INTRODUCTION

For wave energy to reach commercial viability, most concepts will require that the WECs are deployed in arrays, or parks, or farms, as in Figures 9.1–9.3. This will reduce the cost of the requisite infrastructure for electrical subsystems (e.g. cables and substations with transformers and other power electronics), moorings and foundations, wave measurement instrumentation, maintenance and repair (vessels, cranes, and replacement components), and access to personnel with the required expertise. The costs per WEC will be reduced when they are constructed as part of a larger installation and the produced energy per ocean area will increase when devices are installed in a farm. In addition, a few WECs can be under maintenance while the majority are still operational, and this redundancy increases the reliability of the produced electricity.

Depending on the WEC technology, a farm can consists of anything between a few devices, up to several hundred components. Each WEC will modify the wave field within and outside the farm, and the resulting wave field will be a complex superposition of all scattered and radiated waves from all the devices, which again affects the dynamics of each WEC. As the waves are scattered and propagate in all horizontal directions, the WECs in the back of a farm (in the direction of incident waves) can affect WECs in the leeward region, making the interactions in wave farms more complex than the analogous situation for wind farms. Thus, to understand the dynamics and performance of a wave farm, and the resulting wave conditions outside the farm, the hydrodynamic interactions must be well understood. As these will depend on many parameters such as the layout of the farm, the separation distance between WECs, the mooring and PTO configurations, the WEC dimensions and properties, the wave conditions and directions, the bathymetry, etc, the complexity of the problem is significant and grows with the number of interacting devices.

Since the far-field effects of wave farms may affect wave height and sediment transports, with positive or negative implications for the local environment at the

DOI: 10.1201/9781003198956-9

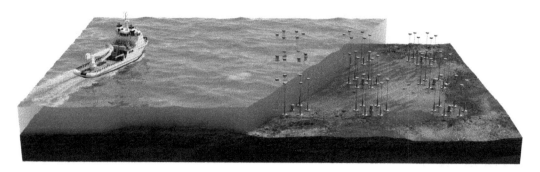

Figure 9.1: Illustration of a wave power farm consisting of point-absorbing wave energy converters with linear generators installed at the seabed. Illustration by Dan Hambe, Hambe Illustration AB. Copyright: Malin Göteman.

coast as well as for commercial and recreational activities, the coastal impact of wave farms has been studied by many authors [683, 2, 1, 536, 294, 602, 588, 746, 668]. Another application for wave energy farms is in conjunction with wind farms, either co-located or integrated, using the same structural foundations. Such hybrid systems have been shown to increase energy extraction per square km (capacity density), reduce power fluctuations, improve power system reliability, and reduce costs due to shared infrastructure. In addition, an efficient layout of the wave farm can provide a sheltered environment for the wind farm, reducing harmful wave loads on the structures. Hybrid systems have been studied in a number of works, using numerical models [631, 37, 861, 361], and experiments [542, 391].

However, despite multiple applications for wave farms, such as protection for coasts or other offshore structures, the main focus and purpose of wave farms is electricity production, and the vast majority of research on wave farms has been

Figure 9.2: (a) An Eco Wave Power array, installed at a former World War II Ammunition Jetty in Gibraltar and providing 100 kW to the electric grid. (b) A birds-eye view of the Mutriku wave power plant, consisting of 16 oscillating water column chambers installed at a breakwater structure [108]. Pictures republished under the Creative Commons CC-BY-SA-4.0 license.

Figure 9.3: Two of the very few examples of wave farms deployed offshore. (a) Three buoys of full-scale point-absorber WECs, as well as one dummy buoy [649]. (b) The WaveStar prototype in 1:2 scale with two floats [241].

dedicated to modelling and optimising the energy uptake or a related performance measure, such as the levelised cost of energy. This chapter aims to review these works and identify some directions for future research. The interested reader is directed to other review papers on modelling and optimisation of wave farms, including [249, 126, 303, 293].

In Section 9.2, the analytical and numerical methods used to model wave farms will be reviewed. Most of the work on wave farm analysis is based on numerical modelling, although a few wave tank experiments and even fewer offshore experiments for wave farms have been presented, as will be discussed in Section 9.2.3. The different approaches to modelling full wave energy systems including mooring and PTO dynamics will also be discussed.

Optimisation is the procedure for identifying the best solution from a solution space, under given constraints. In the simplest case, an optimisation problem consists of maximising or minimising a real function by systematically choosing input values from an allowed set and computing the value of the function. In a more complex situation, optimisation includes finding the best available values for some objective functions given a defined parameter space. Many research groups have approached wave farm optimisation, and an overview of different optimisation methods will be given in Section 9.3, including both simple parameter sweeps and more advanced heuristic optimisation algorithms. A particular focus will be put on different optimisation objectives, relating to Section 1.2.3 on performance measures. In section 9.4, examples of wave farm optimisation results will be discussed and compared, and some general conclusions will be drawn.

9.2 MODELLING METHODS OF WAVE FARMS

As discussed in Chapter 2, ocean waves and fluid-structure interactions can be modelled using different analytical and numerical methods, and to different levels of complexity, ranging from low-fidelity methods based on linear potential theory, to high-fidelity CFD methods.

For the study of wave farms, most modelling and optimisation works have been based on linear potential flow theory, sometimes with additional approximations. Some of these methods were discussed in detail in Chapter 8. Even if there are some advances in modelling the hydrodynamic interactions between WECs using CFD methods, the computational cost involved is too high to model even one configuration of a large wave farm, let alone carry out an optimisation study across many configurations. The analytical and numerical methods used to model the hydrodynamic interactions in wave farms are reviewed in Section 9.2.1. In addition, there are many methods of varying complexity available for modelling the dynamics and performance of the wave farms, including power take-off systems, mooring dynamics, and electrical subsystems, as will be discussed in Section 9.2.2. As for hydrodynamics, the complexity and computational cost will increase with the number of units in the park, which often implies that approximations must be made. To validate the simplified numerical and analytical models, physical experiments should be carried out, as reviewed in section 9.2.3.

9.2.1 Modelling of hydrodynamic interactions

The first step in analysing wave farm performance is to model the hydrodynamic interactions between the wave energy converters. Here, the different analytical and numerical methods to model the hydrodynamic interactions as well as the incident waves will be reviewed.

9.2.1.1 Analytical modelling

The early works on wave farm performance mainly relied on analytical methods based on linear potential flow theory. Several approximations were used to facilitate models of large farms, and their validity and limits were investigated. As was discussed in Chapter 8, the point-absorber approximation relies on the assumption that the floats are small enough that the interaction due to scattered waves can be neglected; whereas the plane-wave approximation assumes that the floats are sufficiently far apart that they interact only by plane waves, and the evanescent modes can be neglected. The point-absorber approximation has been found to perform well at low frequencies and for large separating distances, whereas the plane-wave approximation works better at high frequencies. However, the approximation error is expected to grow with an increasing number of interacting units [524, 510]. Simplifications such as the point-absorber approximation are still used to reduce computational costs in wave farm optimisation works [301, 522].

As reviewed in Section 8.2, the iterative multiple scattering method was developed in [591] to study offshore platforms. This was achieved by applying an acoustic multi-body diffraction theory to water waves, which was later extended for wave farm systems [509, 507]. To obtain the full wave field within a system of structures, diffraction and radiation properties of isolated bodies are used and the reflected waves within an array are added iteratively until convergence is achieved.

This iterative method was combined with the direct matrix method of [718] by [389] to obtain a non-iterative multiple scattering method (also denoted as the direct matrix method as in Section 8.2). In the non-iterative multiple scattering method, the wave amplitude around each body is computed simultaneously, by solving a diffraction equation with a large number of unknowns, involving all hydrodynamic interaction terms for the array. The resulting potentials are exact within the assumptions of linear potential flow theory. The infinite matrices in the solution must be truncated for floating bodies, and the method is semi-exact. The method was extended to include the single-body diffraction solution of [275] by [854], and later to include independent radiation by [717].

Several other analytical methods have been developed and used for wave farm studies, including matched asymptotic expansions [526], multipole expansions [449], and Bragg scattering [445].

These analytical methods have been used to study wave farms in many works [376, 659, 660, 586, 414, 298, 680, 283, 231, 873, 871, 453], along with being extended in various directions. For instance, the analytical multiple scattering method was coupled to a numerical method to allow for arbitrary body geometry [530], and was further used to study wave farm interactions and the existence of trapped wave modes [245]. To reduce the computational cost and allow for the optimisation of large wave farms, methods using interaction distance cut-offs [300], nearest neighbor approaches [689], resonant modes [164, 837], Haskind's relation [244], and multiple cluster scattering [299] have been introduced, connecting the iterative and non-iterative versions of multiple scattering.

9.2.1.2 Numerical modelling

The majority of wave farm studies are based on numerical modelling using boundary element methods (BEM). As reviewed in Section 2.2.2 and extended to arrays in Chapter 8, in BEM the boundaries of the fluid domain are discretised and the integral representation of the fluid velocity potential is used. The boundary conditions on the body and free surface are applied and the fluid potential can then be determined using Green's functions anywhere in the fluid domain. Several commercial and open-source BEM software are available, such as WAMIT [431], Ansys AQWA [23] and NEMOH [46]. In recent works by [800, 798, 799] and [62], wave propagation models have been coupled to BEM software, in the latter case to study the wave-structure interaction in a wave farm consisting of two clusters, or subgroups. Examples of wave farm studies based on BEM for resolving hydrodynamic interactions can be found in [559, 719, 755, 680, 62, 80, 470].

Other numerical methods to study wave farm hydrodynamics exist. In [559], a finite element model based on linear potential flow theory was developed to study scattered waves around both single OWCs and arrays of OWCs, and it was found that the performance of the array could be enhanced by up to 30% by optimal positioning of the devices. Interaction between a few WECs has been studied using high-fidelity CFD methods [173, 174], but the computational cost is still too high to allow for larger wave farm optimisations.

As will be seen in Section 9.2.3, there are only a few experimental works on wave farms, that can be used to validate the numerical simulations. Many analytical and numerical codes have also been compared with each other, such as in [392] where the numerical code ITU-WAVE was validated against analytical results and used to model the wave-structure interactions in arrays of truncated cylinders.

9.2.1.3 Incident waves

The hydrodynamic forces can be obtained by different methods, as discussed above. As the excitation force is attributed to the incident waves, the waves incident on the wave farm must be modelled and used as input. Many works have studied the performance of wave farms as a function of incident wave conditions, such as wave direction and wave climate.

Even if a single point-absorber WEC absorbs wave energy independently of wave direction, the performance of an array of WECs is strongly dependent on the direction of the incident waves. Array layouts with rotational symmetry are often less affected by the wave direction, than corresponding parks with rectangular or linear layouts [162, 300], and the array layout can be more easily optimised if they are installed at sites with a narrow wave direction range [522]. According to [563, 564, 565], an optimal wave farm layout would align perpendicular to the predominant incident wave direction at sites with a narrow wave direction range. In contrast, similar layout configurations were not obtained at sites with more diverse wave directionality.

As irregular waves are composed of many regular waves, the waves may travel in many directions simultaneously, i.e., they are short-crested. Whilst most wave farm optimisation studies have covered only long-crested waves, some works have also considered short-crested waves. In [748], an experimental wave tank investigation on a large wave farm was presented, and it was concluded that the wave height reduction due to the farm occurred earlier in short-crested than in long-crested waves. In [376], a numerical and experimental study was conducted on an array of bottom-mounted cylinders, and it was observed that unidirectional wave modelling would overpredict the normal force and underpredict the transversal force during wave run-up on the cylinders. Wave farms in short-crested waves were studied numerically in [755] and their performance was found to be slightly lower than that in long-crested waves. A somewhat conflicting conclusion was reached in [304], where short- and long-crested waves resulted in a similar energy absorption, although the power fluctuations were considerably lower in the short-crested waves.

An alternative approach to solving the hydrodynamics and dynamics of the WECs explicitly as a response to the incident waves, is to use spectral wave models. In these models, the effect of the WEC on the wave field is obtained by modelling the WEC as an energy sink. Spectral models are often used to model the far-field effect on the wave field by wave farms. Several codes are available for spectral wave modelling, perhaps the best known being the open-source third-generation wave model SWAN developed by Delft University of Technology [78]. In [528], the wave field obtained by the spectral model SNL-SWAN (an extension of SWAN by Sandia National Laboratories, USA [679]) was compared with the wave field obtained by an explicit BEM computation using WAMIT. Farms of different WEC types were compared, and as expected, the two methods produced very different wave fields within the farm, but agreed well in the far-field area, away from the farm.

9.2.2 Modelling full farms and power absorption

Wave energy converters are complex systems, as has been discussed in earlier chapters. As shown in Eq. (2.46), the system is subjected to hydrodynamic forces, power take-off and control forces, mooring forces, and mechanical forces, which may all be non-linear. When extending the analysis from single devices to wave farms, the complexity of the problem increases. As for the hydrodynamic wave-structure interaction, most wave farm models rely on simplifying assumptions to enable feasible simulations.

9.2.2.1 *Power take-off and control in wave farms*

The purpose of a PTO is to extract energy from the system, which will affect the device's motion and the hydrodynamics. As shown in Figure 3.6, the PTO can be a hydraulic system or an air turbine driving a rotating generator, or a direct-driven linear generator. Many different PTO systems and methods of modelling them exist in the literature (see [249, 255, 161], and [622]). Nevertheless, as was discussed in Section 3.3, it is common when studying WECs to model the PTO as a spring-damper system, $\mathbf{F}_{pto} = -B_{pto}(t)\dot{\mathbf{x}}(t) + K_{pto}\mathbf{x}(t)$, even though this is a gross simplification of a complex PTO system. The damping, $B_{pto}(t)$, can be a pressure differential of a hydraulic system or the generator damping of a direct-driven generator, and is often modelled as a constant. This simplification is even more common in optimisation studies of wave farms. The dynamic part of the PTO, $\mathbf{x}(t)$ can be restricted to some degrees of freedom, e.g. for linear generators where the moving part of the generator is restricted to one degree of freedom.

To gain confidence in the accuracy of simplified PTO models, validation with non-linear models for single WECs and with experimental data has been carried out. A reasonable agreement between linear and non-linear PTO models for the Wavepiston WEC was obtained in [654], both with and without viscosity effects, and in the time and frequency domains. Similarly, numerical simulations based on both frequency and time domain models were compared with experiments for multi-body WEC systems with eddy current brake PTOs in [427] and [286]. In both works, the time domain model was able to predict non-linear dynamics that the frequency model could not capture.

In the wave tank model tests, simplified physical models are used to replicate generators. Very few validations of PTO models in wave farms have been carried out with data from realistic electrical generators, although a few exceptions exist. In [205], experimental data from a full-scale WEC deployed offshore was used to validate a PTO circuit time-domain model. The generator damping in the model was time-dependent and accorded an excellent agreement with the actual generator. Onshore experiments on a similar WEC were carried out in [787], and it was concluded that the generator damping coefficient was approximately constant when the translator was entirely within the stator, and reduced with a decreasing active area.

With the expanding field of control algorithms to steer the WEC dynamics, the complexity of PTO modelling is growing with advanced time-domain models and wave-to-wire models, as discussed in Section 3.3.

9.2.2.2 Mooring systems

For most wave energy concepts, the primary purpose of the mooring system is station-keeping. For several WECs, however, the mooring system is integrated into the PTO. In both cases, for single WECs, it is well established that mooring dynamics may be highly non-linear and may significantly affect the dynamics, energy absorption, and survivability of the system [70, 606]. Due to the high computational demand required for non-linear mooring simulations, moorings in wave farm simulations are often approximated as linear springs, or simply neglected.

Some exceptions exist. The mooring dynamics of nine WECs were modelled in the time domain by [269], and it was demonstrated that considerable mooring line tension could occur in certain situations. Similarly, in [803] and [804], it was found that the mooring system could significantly alter the performance of an array. In [423], the mooring system of an OWC array was investigated experimentally, and it was found that mooring peak loads may be considerably higher in an array than for a single device configuration. A system of three floating OWC was studied by [413], and it was concluded that the power absorption was higher when mooring was considered, than for freely floating OWCs. Mooring in WEC arrays was studied using a finite element model in [59]. In [851], the mooring cables of a WaveEL WEC system were modelled for arrays of 2 and 10 WECs, and it was seen that the hydrodynamic interaction might affect the fatigue in the mooring lines up to tenfold.

9.2.2.3 Solving the equations of motion

After the forces affecting the wave farm have been established, the system of equations must be solved for each WEC to obtain the dynamics and performance of the wave farm. As explained in Section 3.1, time-varying forces and control methods require solving of the equations in the time domain. However, with the approximations applied to PTO systems, mooring dynamics and hydrodynamic interactions discussed above, the standard approach when optimising wave energy parks is to solve the equations of motion in the frequency domain. A few examples of real-time array simulations in the time-domain exist but are restricted to small arrays due to the high computational cost [770] or rely on approximations such as replacing the radiation convolution with a state-space model [255]. A linear time-domain model was used in [783, 748] and was seen to overpredict the measured total power for single WECs when compared with the experimental data [783]. Results from the numerical software WEC-Sim were compared with experimental data from five pitching WaveStar WECs in a staggered layout in [491]. An array of five OWCs was studied in the time domain by [80] using the open-source software WEC-Sim, and the same software was used by [491] to study an array of five pitching WaveStar devices in the time-domain. Similarly, WEC-Sim was coupled to NEMOH and used to study a 5-WEC array by [61]. Layout optimisation of a 6-body array was performed employing real-time simulations by [60], and control models have been introduced both in experiments and numerical modelling of arrays in [55, 56, 441, 537, 558, 770].

9.2.3 Experimental modelling

In the previous subsections, analytical and numerical modelling of wave farms have been discussed. However, to gain confidence in the results and quantify the uncertainties, simulations must be validated with experimental data. Although much experimental work has been carried out within wave energy research, there have been very few experiments with wave farms. This is despite the fact that wave energy converters will probably have to be deployed in array configurations and share infrastructure such as submarine cables or substations in order to achieve economic viability.

Only a handful of wave farms have been installed and tested offshore. The Agucadoura Wave Farm consisted of three Pelamis attenuators and was installed off the Portuguese coast in 2008. Due to technical issues the farm had to be de-commissioned, and no data were ever published. The Mutriku wave power plant, commissioned in 2011 and still operating, consists of 16 OWCs installed along a breakwater [779], see Figure 9.2. Its average annual energy output during 2014-2016 was 246 MWh, resulting in a capacity factor of around 0.11 [363]. In 2009, three full-scale point-absorbers were installed off the west coast of Sweden, see Figure 9.3(a). By analysing the aggregated output power, it could be seen that the array effectively reduced the power fluctuations, and technical solutions for grid connections were presented [650, 194]. A 1:2 scale prototype of the WaveStar WEC was installed at Hanstholm outside Denmark in 2009, and began operating in 2010, as shown in Figure 9.3(b). Two floats were operated by hydraulic cylinders from a platform and constituted a small array of point-absorbers. The power produced whilst operating in both single and dual float configurations was measured and analysed as a function of the wave climate, and it was concluded that the performance was in agreement with the predictions [420]. Three point-absorber CETO devices were deployed by Carnegie in the Perth Wave Energy Project in 2015 [411], but very little performance data has been published.

The number of experimental analyses on wave farms tested on a small scale in wave tank laboratories is more extensive; however, these are often constrained to small arrays of reduced-scaled WECs with simplified complexity.

Resonance and (near)-trapped wave modes within arrays have been observed experimentally in several tests. An array of 50 truncated cylinders was studied in regular waves by [388], and it was found that resonance modes were present, although they were smaller than those predicted by numerical simulations. Trapped and near-trapped modes were also observed within an array of eight fixed truncated cylinders [838], and in short-crested waves [376].

In [772], an array of five closely-spaced heaving floats was studied experimentally, and the results were compared with numerical simulations. The study was extended to 12 devices in a rectangular grid layout and irregular waves in [828], and it was seen that the interaction q-factor was close to 1 when compared with an isolated device. Wave tank experiments on five small floats at 1:67 scale were carried out in [12] to measure the change in wave characteristics due to the array. Arrays of three and five point-absorber multi-body WECs at 1:33 scale in a staggered layout were studied in [325, 602], and it was found that wave shadowing was the predominant phenomenon affecting the array performance. Three Spar buoy WECs in a triangular array layout were studied experimentally at 1:32 scale by [142], and it was found that, in wave climates with large energy periods, the array performed better than three isolated devices, i.e. a q-factor larger than one was observed. Wake and array effects were studied experimentally in large arrays of up to 25 heaving point-absorber WECs [123, 783, 748]. In [123], it was found that the hydrodynamic array interactions could cause losses of up to 26% in energy yield, compared with WECs in isolation. Similarly, large hydrodynamic array effects were observed in [748, 783], where the wave height was reduced downstream by up to 18%, while it increased upstream of the array. A WaveStar model at 1:20 scale with five point-absorber floats connected to linear control PTO systems was studied experimentally by [537], and the data were used to validate the hydrodynamics tool implemented in the DTOcean software.

Arrays of up to six point-absorber WECs moving in six degrees of freedom were carried out by different research groups in [558] and [771, 285, 281]. The work of [558] was carried out at the Australian Maritime College and measured the interaction factor for 1-2 floats moving in heave and surge. In [771], a control algorithm based on machine learning and artificial neural networks was used for the array, whereas linear damping was used in [285, 281], see Figure 9.4.

An array of five OWCs at a scale of 1:20 was tested by [35], and an increase in power capture was observed when multiple devices were installed. The interaction factor was studied in an array of onshore OWC wave pumps for water desalination in [479, 483], and it was concluded that better performance was achieved by smaller separation distances between the devices. Both extreme mooring loads and power capture were investigated for an array of up to five floating OWC in [95, 423], see Figure 9.4. The results indicated that the array effects might increase both the energy capture and the peak loads in the mooring lines. Floating or fixed breakwater

Figure 9.4: Pictures of two experimental campaigns that have been carried out on wave farms in wave tank laboratories. (a) Wave farm consisting of five floating OWCs. Experiments carried out at the NTNU Trondheim basin [423]. (b) Wave farm consisting of six point-absorber buoys connected to linear damping PTOs. Carried out at the COAST laboratory OCEAN basin, Plymouth, UK [281].

structures installed with chambers utilised for wave energy extraction have been studied by a number of authors. In [341, 342], it was observed that the pneumatic chambers enhanced the performance of the floating breakwater and that asymmetric chambers could enhance the energy extraction. A 1:50 scale model of 32 OWCs arranged in a V-shaped floating platform was studied experimentally [397]. Five 1:20 scale OWCs integrated with a breakwater structure were investigated experimentally in a shallow water basin by [382], and it was found that under certain conditions, the array performed better than five isolated devices. Similarly, five fixed OWCs were studied with two different array layouts in [80], and it was found that the power in the optimal layout was 12% larger than in the sub-optimal layout. However, when considering the average power, only a modest increase could be established. In [582] and [585], two different stationary OWCs with dual chambers were studied using both physical and numerical modelling, and it was found that the dual-chamber could enhance the energy absorption compared with a single chamber. An experimental investigation into a combined platform consisting of a wind turbine and floating OWCs was presented in [396], with further analysis carried out in [515]. The air pressure inside the WECs, as well as the device dynamics, was compared with the numerical predictions with a good agreement. Five OWCs were studied experimentally in a wave tank in [604], and the wave field variation around the array was measured. It was concluded that radiated waves might account for up to 50% of the effects on the wave climate in the near field in particular operating conditions, highlighting the importance of accounting for the full extent of the WEC behaviour when assessing impacts on the wave field.

Arrays with the primary purpose of protecting the coast and with energy absorption only as the secondary purpose were studied experimentally in [860]. Three floating DEXA type devices were tested in a wave tank at a scale of 1:60 and compared with a single device at a scale of 1:30. By changing the wave conditions and water depth, properties such as wave transmission and hydrodynamic interaction among the devices were examined, and the resulting guidelines for the optimal design concluded that the WECs should be installed nearby and with heavier floats.

In [542], the survivability of a hybrid wind-wave system consisting of a floating wind turbine and three flap-type WECs was studied in extreme wave scenarios, at a scale of 1:50, and the experimental data were compared at numerical predictions with a good agreement. Point-absorber WECs installed around an offshore platform, intended for hybrid wind-wave systems, were investigated experimentally in regular and irregular wave conditions [391]. The study indicated that there were some advantages to the wind turbine foundation's stability by installing the WECs, but that this effect depended on the damping of the WECs.

9.3 OVERVIEW OF OPTIMISATION METHODS

As discussed earlier in this chapter, optimisation is the procedure for finding the best available solution given a defined domain or solution set. The definition of *best* has to be specified by an appropriate objective function, or a multi-objective combination of several objectives. Different constraints may limit the size of the available solution space. In wave farm optimisations, the objective may involve, for example, the maximisation of absorbed annual energy, and minimisation of costs or power fluctuations. Constraints may be given in terms of ocean area or parameters available from a gridded domain. Parameters that can affect the performance of a wave farm include the number of WECs in the farm, farm layout, separation distance between WECs, individual WEC dimensions, mooring configurations and power take-off systems, control algorithms, water depth and bathymetry, wave characteristics and wave directions, etc. Wave farm optimisation differs fundamentally from the analogue for wind farms, due to the coupling between WEC devices. Changing the parameters of one WEC in a farm may affect the performance of all others. The coupling can be hydrodynamic due to scattered and radiated waves throughout the farm, mechanical (shared mooring, breakwater structures, or platforms), electrical (shared sea cables and other electrical infrastructure), and financial (shared capital and operational expenses).

It is clear that wave farm optimisations pose a pervasive challenge, and no study can address all parameters and objectives simultaneously. Whereas early wave farm optimisations primarily compared a few configurations or varied one parameter, many works today apply intelligent optimisation algorithms to identify optimal solutions in a much larger solution space [41, 303].

9.3.1 Grid search over specified solution subsets

Many authors have presented comparisons or parameter sweeps of different wave farm configurations to analyse the performance of the farm as functions of different input parameters and to identify the optima among the given configurations. In [79], different layouts and parameter settings for arrays of heaving cylinders and surging barges were analysed, particularly as a function of the separation distance, and a constructive array interaction was observed. The separation distance between WECs and array layouts was also studied in [802] in wave farms of 12 WECs using different numerical methods, and separation distances, such that constructive array interactions were identified. The energy absorption and power fluctuations were compared for three different layouts of 32 WECs [202], and three array layouts and separation distances were analysed for life-cycle performance in [466], see Figure 9.5. In [162], it was found that triangular arrays of two-body heaving WECs were optimal in multi-directional waves; whereas square arrays performed better in unidirectional wave conditions, and high interaction factors were obtained at certain separation distances. Layout optimisation by sweeping over parameters such as separation distances and angles between devices was performed by [727] and [300]. The performance of five different array layouts consisting of 12 WECs was studied in [719] as a function of different wave directions, and in [850] the performance of four array layouts of 10 WECs was compared. Arrays of heaving cylinders were studied in [625], and constructive interactions were observed for small arrays at certain separation distances. Power production from arrays of 12 OWSCs was studied in regular, long-crested and short-crested waves by [755], and the interaction function was evaluated as a function of the separation distance between devices. A 10-year wave spectra at four Italian

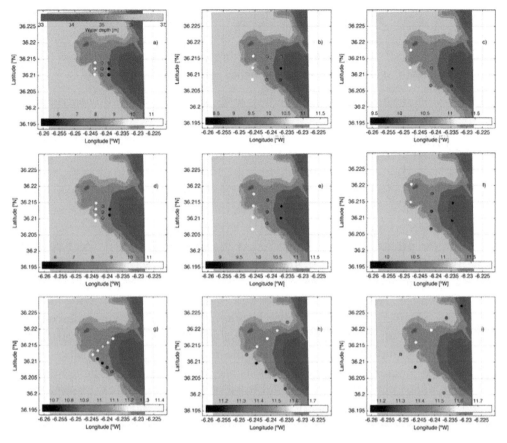

Figure 9.5: Average absorbed power for aligned, staggered, and arrow array layouts with three different separation distances [466].

sites was used in [297] and [83] to compare different array layouts of Oyster, Pelamis and point-absorber WECs.

The performance of wave farms as a function of the number of WECs was studied in [775, 79, 41, 802, 162, 301]. Furthermore, parameter sweeps to determine the impact of wave parameters such as wave number, wave direction, and significant wave height have been conducted in many works, including [414, 719, 537, 755, 558, 846, 245, 873, 466, 522].

9.3.2 Global optimisation algorithms

When searching for the optima among a very confined subset of solutions, as in the parameter sweep or grid searches discussed in the previous subsection, the results will inevitably depend on initial biases in the available solutions. Wave farm performance depends on many parameters including the number of devices, the layout and separation distance between devices, PTO and mooring systems, and wave climate and bathymetry. The solution space is therefore vast and more advanced optimisation methods are required to identify optima. In recent years, many wave farm optimisation works have applied different global optimisation algorithms to resolve this problem.

9.3.2.1 Non-linear optimisation and coordinated control

Non-linear optimisation with motion constraints and other advanced control methods have been applied to the optimisation of arrays, see review papers [622, 213, 670]. For wave farms, the hydrodynamic modelling and wave forecasting is complicated by the radiated and scattered waves from the many devices. However, the motion of each WEC can also be used to provide forecasts for the excitation force applied to the other WECs in the park, the so-called collaborative control. As demonstrated by [619] and discussed in [670], the added complexity to the wave field can be compensated for by the information sharing between the WECs.

Non-linear programming optimisation is useful to handle non-linear constraints on the WEC dynamics [211, 637], and can identify optimal parameters in moderately large solution spaces. In particular, the method can be used to add realistic motion constraints on the dynamic parts of the WECs, e.g. a maximum heaving amplitude.

In [519, 520, 521], linear or circular layouts of spherical heaving point-absorbers were optimised, and, in [522], the array layout was not restricted to a certain geometry. The arrays were optimised upon the mean interaction factor, using a sequential quadratic programming method. It was seen that the WECs aligned differently depending on the motion constraint and wave direction. The PTO damping was optimised to obtain maximal energy absorption for three arrays of 2-3 WECs under motion constraints in [56, 55]. An iterative quasi-Newton optimisation algorithm was applied by [846] to identify the maximum captured power of an array of Duck WECs under motion constraints. Similarly, [812, 813] optimised individual dampings for an array of point-absorbing WECs constrained to heave motion, see Figure 9.6. The efficiency of the model predictive control approach to wave farm optimisation was demonstrated by [441], where increases in power of up to 20% were obtained. In [546], an iterative quadratic programming was used to optimise a farm layout for up to 15 point-absorbers.

Different control strategies applied to a WaveStar array were compared by [561], and the fully-coordinated global array control was found to maximise the energy absorption. In the works of [270] and [712], two different approaches to coupling the optimisation of an array layout to WEC control were presented.

Recent trends in wave farm control focus on developing and employing models with a reduced complexity [624, 670], to eliminate the requirement for wave forecasting or model dependency. In [577], an optimal control law for the maximum absorbed mean power from an array of point absorbers with non-linear buoyancy forces was derived based on a stochastic model of the wave climate, using an infinite time horizon. A real-time centralised control method for a wave farm was developed in [758], with the objective being to meet the power quality requirements of the electric grid. Autonomous operation and control of wave farms was proposed in [340] using a multi-agent system. The method developed in [541] demonstrated how the excitation force

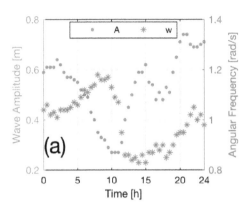 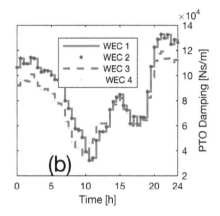

Figure 9.6: In [812], it was observed that the absorbed energy of an array of 4 heaving point-absorbers was maximised if the WECs adopted different individual PTO dampings as functions of the incident waves. The figure shows the wave amplitude in the array and the resulting individual PTO damping applied to optimise the energy absorption [812].

forecast could be provided by measurements of the WECs' motion in the farm, and then be used to control the devices.

Some of the first works to use neural networks in wave energy systems were [67, 793], and machine learning methods have since been used in a variety of different control and optimisation works [17, 19, 496, 442]. Recently these AI approaches have also been employed for the optimisation of wave farms [770, 771, 627].

9.3.2.2 *Metaheuristic optimisation algorithms*

A metaheuristic optimisation algorithm is a search method for sufficiently good solutions to certain optimisation problems. They are useful when the solution space is too large for all grid points to be evaluated, and possibly multi-peaked. Whilst metaheuristics do not guarantee that the found solution is the global optimum, they use advanced convergence tests and other tools to avoid local minima and can optimise the problem rapidly. Metaheuristic optimisation algorithms have been developed and applied to several wave farm problems.

Genetic algorithms Genetic algorithms (GAs) are inspired by the theory of natural evolution, where the fittest individuals in a population survive for reproduction and produce offspring for the next generation, see Figure 9.7. The offspring are created using a crossover mechanism to inherit a combination of properties from their parents. With each generation, the population thus comprises individuals with increasingly better fitness values and, at the end of the iteration, the best individual in the latest generation is chosen as the optimum. To avoid local minima, mutations are programmed to move randomly within the solution space.

Wave farm optimisation utilising a GA was first carried out in [124] using the GA toolbox from MATLAB, and further developed to obtain optimal layouts

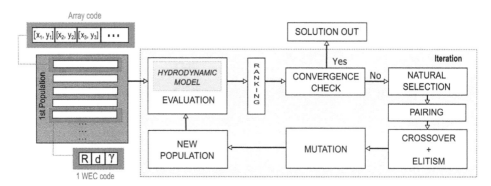

Figure 9.7: Overview of the genetic algorithm applied to wave farms. In each iteration (generation), natural selection, pairing, crossover, elitism, and mutation are carried out [282].

and PTO settings for point-absorber arrays in [122]. In [689], the layout of a large wave farm consisting of 40 OWSCs was optimised with a GA as well as a Monte-Carlo analysis. The hydrodynamic interaction was restricted between nearest neighbours of WECs to allow for fast evaluations, and a machine learning approach was used to identify the optimal 3-WEC clusters, after which the GA was applied to optimise the complete park layout. With the same number of evaluations as in the Monto-Carlo simulation (16 million), the GA was able to find layouts with better performance than those found in the Monte-Carlo simulation. The layout optimisation problem was coupled to active device control in [712]. In [710, 711] the layout of five point-absorber WECs was optimised with a binary GA. An in-house GA was built and connected to an analytical hydrodynamic model of a wave farm in [282] (see Figure 9.7). Validation against

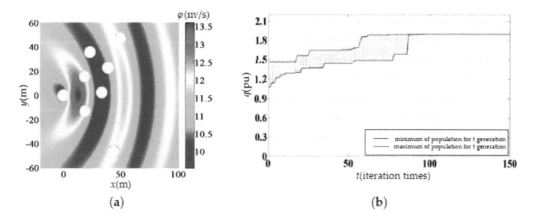

Figure 9.8: Optimal layout of an array with 8 WECs, as obtained by the differential evolution algorithm in [231]. Figure (b) shows how the individual fitness value increases with the generations (iterations). The upper red and lower blue lines show the minimum and maximum of population for generation t, respectively.

parameter sweep for a single WEC showed excellent agreement and fast convergence. It was further used in [284] to find optimal layouts of arrays with WECs of different sizes, and in [283] to obtain optimal wave farm layouts with interaction factors larger than one, and finally in [280] to optimise wave farms dependent upon economic objectives.

Differential evolution Similar to the GAs, differential evolution (DE) algorithms also take their inspiration from natural evolution. The population of the solution space evolves by allowing the individuals to inherit strengths from parents in previous generations. The main difference from GAs is that real-valued vectors are used to represent the individuals, and mutations and crossovers are performed by adding the weighted difference between two solutions to achieve a third solution.

A DE algorithm was used in [72] to optimise the design and control of a single WEC. An adaptive mutation factor was introduced by [231] to improve the convergence rate of the array optimisation using a DE algorithm. In [460], a differential evolution algorithm was used to optimise the layout of an array of OWSCs, based on a wake model validated by a high-fidelity smoothed particle hydrodynamics model. In the study, staggered layouts with narrow and wide OWSCs placed at the front and back of the array were identified as optimal.

Covariance matrix adaptation Covariance matrix adaptation (CMA) evolution strategies also belong to the category of evolutionary algorithms: individuals are ranked based on their fitness and combined to generate new solutions in the next generation. The covariance matrix represents the pairwise dependencies between solutions, and gives a geometric interpretation of the solution distribution in a given generation. Based on its shape, the search space for the next generation of solutions is adaptively increased or decreased; an ample search space is used when the solutions are far from the global minimum, but is automatically reduced when the solutions are close to a minimum and the solution space requires only fine adjustments.

In [844], a CMA method was used to optimise a wave farm of 25 and 50 WECs. The DTOcean tool for wave farm modelling and optimisation is equipped with a CMA tool, which was compared with other metaheuristic optimisation algorithms in [680]. The objective was to maximise the yearly absorbed power as a function of the number of WECs and the park layout, under the constraints of minimal interaction factor $q \geq 0.9$, minimal separation distance 65 m and maximal ocean area $500\,\mathrm{m} \times 500$ m. The CMA was found to be computationally efficient when compared with the other algorithms. However, the solutions were not as effective. The accuracy and computational cost of the CMA algorithm in the DTOcean modelling tool were compared with a meta-model optimisation method by [240], with the conclusion being that the former was the more accurate but with a higher computational cost. A CMA algorithm was used along with other optimisation algorithms in [27] to optimise the layout of wave farms subjected to three objectives: maximising energy absorption, minimising cable length, and minimising ocean area. The work was extended in [28] to more realistic wave scenarios employing a sparse incomplete LU decomposition to reduce the computational cost.

Particle swarm algorithms Particle swarm algorithms do not mimic an evolutionary process, but are nevertheless inspired by biology, in particular the movement of flocks of birds or other animals. In each iteration, the solutions communicate and are guided towards the best-known positions in the search space. Thus, as a collective group, the population moves towards optimal solutions. In the related glowworm optimisation algorithm [421], the particles are equipped with a luciferin level, depending on their fitness values. Each glowworm is attracted by the brighter glow of other neighboring glowworms, bringing the swarm towards the optimal solutions.

In [232], a particle swarm optimisation was compared with a GA for the optimisation of wave-farm separation distances and PTO tuning. The former algorithm was found to outperform the latter for a small array consisting of 2 WECs, but not for larger arrays. In [373], the LCOE for a hybrid wind-wave farm situated on the west coast of Ireland was minimised using a particle swarm optimisation, and reduced costs of up to 73% were obtained from a short-term perspective.

Grey wolf algorithm The grey wolf algorithm is another optimisation algorithm inspired by how animals live in flocks, and mimics the leadership hierarchy and hunting mechanisms of packs of grey wolves. The fittest individuals lead the pack during hunting, and the optima are reached when the pack is converging around the prey.

A grey wolf optimisation algorithm was used in [15] to identify optimal sites for wave energy installations in the Caspian Sea. A large number of optimisation algorithms were applied to arrays of CETO WECs in [567, 563, 564, 565, 566], including grey wolf optimisation, covariance matrix adaptation, differential evolution, and particle swarm optimisation, see Figure 9.9. The first approach in [567, 563, 565] included optimising the layout to achieve maximal absorbed power under the constraints of maximal ocean area and minimal separation distances, considering the wave climates at several Australian sites. The optimisation problem was extended to include the individual PTO parameters in [564, 566].

9.3.3 Optimisation objectives

The aim of wave power farm optimisation is to optimise the performance of wave farms with regards to a specified objective. In most cases, the objective is to maximise the average energy absorbed by the farm, for instance in terms of annual energy production in Eq. (1.4) at a given site. As discussed in Section 1.2.3 however, the performance of a wave energy system can be assessed using many different measures.

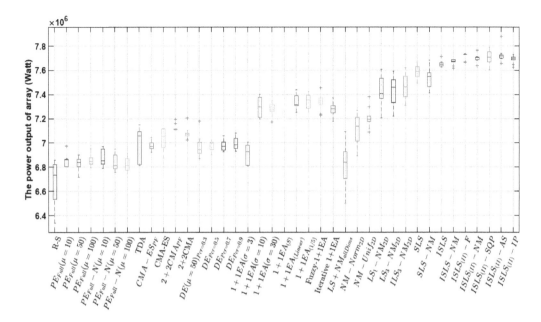

Figure 9.9: Performance comparison of different heuristic optimisation algorithms applied to optimise the array layout of 16 WECs [565]. Partial evaluation (PE), co-variant matrix adaptation (CMA), differential evolution (DE) and variations thereof are some of the algorithms compared. The layout identified by an improved smart local search (ISLC) was found to produce the most power.

Numerous wave farm optimisation works have optimised performance based on the interaction factor, or q-factor (1.5). A factor $q > 1$ indicates that the WECs interact constructively and absorb more energy than they would in isolation. Often, design parameters were identified to give a maximal interaction factor. In [519, 520, 521], the mean of the interaction factor was optimised as it was argued that the optimal layout was often just slightly different from poorly performing layouts, leading to wave farms becoming unstable with minor changes. In [14], the q-factor for small arrays of 3-tethered devices with spherical buoys were optimised at different sites, and average values close to one were obtained.

The wave farm configuration with the maximal energy output or q-factor may not be optimal if the costs are too high (see, for instance, [720]), or, if the power fluctuations are too high to be compatible with grid integration. In recent years, many optimisation studies have developed more advanced objective functions based on analyses from either techno-economics, grid requirements, or other considerations.

9.3.3.1 Economic cost functions

As a simple estimate of the costs of a wave energy system, the mass of the system has sometimes been used, as this is said to represent the costs in terms of steel, concrete and other materials [51]. The unit kWh/kg is commonly used as a measure to compare different wave energy concepts, or kWh/m which is relevant for extended OWSC as in [460]. The objective of energy/mass was compared with other simple performance measures in [284]. Different objective functions generated different optimal configurations. This implies that care must be taken when defining the objective function, such that the optimal solution is the one sought. In [711], another simplified economic cost function was defined as the ratio of the "costs" and the produced power, where the costs were approximated by a simple formula obtained from Sandia National Lab's reference model project [641], $\text{Cost} = 3(10)^7 \cdot N^{0.6735}$, with N being the number of WECs in the park.

For a techno-economic optimisation study of a wave farm, the input values for the economic objective function will greatly affect the results. As there is almost no publicly available data on specific costs for wave energy installations and operations, authors have to base the input values on models and estimations. It can also be assumed that the costs of present-day technology will reduce once the technology reaches a more mature stage [365, 36, 234, 163]. As discussed by [636], the EU Strategic Energy Technology Plan expects the LCOE of wave energy to decrease to 0.15 EUR/kWh by 2030 and 0.10 EUR/kWh by 2035.

The electrical subsystem including sea cables contributes significantly to the cost of the park, both in terms of installed capital costs and costs due to energy losses [350, 708, 709, 203]. The cable dimensions and lengths, distances from the grid connection point, number of substations and related infrastructure all affect the total revenue. In [708, 709], an electrical network cost reduction study was undertaken by optimising spacing and capacity factors of the individual WECs. Optimisation with respect to cable length and costs was carried out in [27, 279].

Operational costs will depend on many factors such as maintenance strategies, failure rates of the WECs, weather conditions, etc. In [759], an economic assessment based on production, location and maintenance operations was carried out for a wave farm consisting of 100 WEC units. In [669], maintenance operations were analysed with respect to annual electricity generation and total gross revenue. A reliability and finance model was developed by [472] and used to study a combined wind and wave power plant with respect to wave farm availability and maintenance.

When the different input values to the capital and operational costs can be computed or estimated, economic performance measures such as the levelised cost of energy (LCOE) (1.8), net present value (NPV) (1.13), and payback period (PBP) (1.14) can be assessed to evaluate the performance of a wave energy system from a more realistic economic long-term perspective. Figure 9.10 shows the LCOE optimisation procedure used by [373] to optimise a hybrid wind-wave farm. An LCOE optimisation in terms of the device dimensions was carried out by [636], and in [232] the LCOE for a system of three WaveSub WECs was minimised. The LCOE of wave energy parks of up to 50 WECs was optimised using a GA by [279]. Based on a study of arrays of WaveEL WECs, in [850], the LCOE values were dependent on realistic modelling of the hydrodynamic and mechanical coupling between the WECs.

9.3.3.2 *Electricity quality*

Most wave energy systems are designed to generate electricity for connection to an electric grid. This puts constraints on the quality of the electricity generated by the

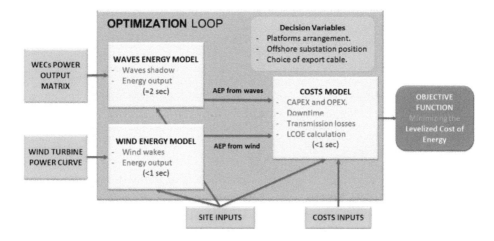

Figure 9.10: LCOE optimisation algorithm of a hybrid wind-wave farm [373].

WECs [758, 622], for example, in terms of maximal power peaks and rapidly varying voltage magnitudes, so-called flicker, as reviewed in Section 1.2.3.

Several numerical and experimental works have concluded that the power fluctuations from a wave farm reduce with an increase in the number of WECs [775, 650, 802, 302, 59] and with strategic array layouts [202, 723]. Voltage fluctuations from wave farms and related grid constraints were studied in [32, 417], and the flicker severity as a function of the grid impedance angle was evaluated in [74]. In [610], a combined battery and supercapacitor system was proposed to minimise the power fluctuations from a wave farm.

9.3.3.3 Survivability and other objectives

In the economic performance measures, the life time duration of the WECs is required as an input. An assumption often used is that the design life time is 20 years. However, the longevity will depend on many factors, which could in themselves also be used as parameters for the optimisation. An example was provided by [851], in which, not only were the LCOE and power output compared, but also the fatigue on the mooring lines for several layouts of 2 and 10 WEC arrays. It was seen that the hydrodynamic interaction in the array could significantly affect fatigue.

In all works discussed in the present sub-section, the optimisation has been carried out using a single objective, such as the annual energy, the LCOE, or the electricity flicker. In a multi-objective optimisation, the performance would be evaluated based on several objectives simultaneously. In [27, 28], several competing objectives (the total energy production, the cable length, and the marine area needed to install the buoys) were optimised in a multi-objective approach. The results showed that the multi-objective method improved the power output of a single-objective optimisation by 3.8%. In [28], the work was extended by considering realistic wave conditions and incident waves from multiple directions. Improvements in terms of power output, minimised cable lengths and marine area were achieved through the optimisation methodology for parks of up to 36 converters.

9.4 CASE STUDIES

In this section, a techno-economic optimisation of a wave farm will be reviewed based on the work in [280], and the results will be compared with similar studies.

9.4.1 Wave farm model

To optimise a wave farm, first an appropriate model of its performance must be defined. As discussed in Section 9.2, this can be done to different levels of complexity and with different numerical and analytical methods. In [280], the hydrodynamic modelling within a farm of PAs connected to bottom-mounted direct-driven linear generators was conducted based on the analytical multiple scattering wave farm model of [298], see Figure 9.11. Large farms of up to 100 WECs were studied. The floats were restricted to a single degree of freedom in heave motion only, the connection line between the buoy and generator was assumed stiff, and all WECs in the array had identical dimensions. Using the multiple scattering approach, the fluid velocity potential representing the radiated and scattered waves throughout the park was obtained and used to compute the hydrodynamic forces on each buoy using Eq. (2.39).

The equations of motion could then be solved individually for each WEC using Eq. (2.46). In [280], the mooring force was integrated into the power take-off force and the viscous force was neglected, such that the resulting equation of motion in

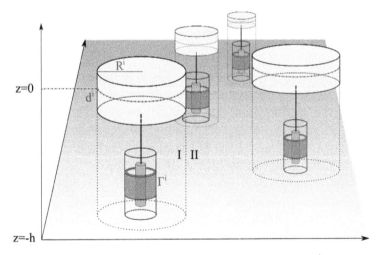

Figure 9.11: Each WEC is characterised by a buoy radius R^i and draft d^i and a generator damping Γ^i, where Γ^i corresponds to notation $\mathbf{B}_{pto}$ used in the text. The fluid domain is divided into interior domains I beneath the buoys and exterior domains II in those regions which are not beneath the buoys, i.e., where $r > R^i$ and r is the radial distance for the axisymmetric centre of the buoy [304].

the frequency domain took the form

$$\left[-\omega^2(\mathbf{M} + \mathbf{A}_m(\omega)) - i\omega(\mathbf{B}_{pto} + \mathbf{B}_{rad}(\omega)) + \rho g\pi \mathbf{R}^2\right]\mathbf{z}(\omega) = \mathbf{F}_e, \quad (9.1)$$

where the added mass and radiation damping are given by the matrices $\mathbf{A}_m(\omega)$ and $\mathbf{B}_{rad}(\omega)$, respectively. The operator acting on the buoy heave position on the left-hand side is the transfer function that is also given in Eq. (3.11).

After solving the equations of motion in the frequency domain, the position of each buoy can be obtained in the time domain by inverse Fourier transform, and the velocity computed by differentiation. Based on an assumption of the stiff connection line between the buoy and the generator, the instant power of the generated electricity is then obtained as $\mathbf{P}(t) = \mathbf{B}_{pto}[\dot{\mathbf{z}}(t)]^2$, corresponding to Eq. (3.57).

As discussed in Section 9.3.3, the wave farm system can be evaluated and optimised based upon many different performance measures, such as the annual absorbed power, q-factor, economic or other objectives. In [280], an advanced economic cost function was built, and the optimisation was carried out to minimise the LCOE and maximise the net present value (NPV), see Section 1.2.3.

The cost function included detailed capital costs (CAPEX) for the WECs, the electrical subsystem including substations and cables within and outside the park, installation and decommissioning. As an example, the costs for the WECs contained the terms

$$\text{CAPEX}_{\text{WEC}} = C_{\text{buoy}} + C_{\text{casing}} + C_{\text{foundation}} + C_{\text{stator}} + C_{\text{translator}} + C_{\text{labour}} + C_{\text{extra}}$$
$$(9.2)$$

and were computed based on choices and costs of materials (steel, copper windings, and concrete), size and rated power of the WECs, and realistic labour costs for manufacturing. The operational expenditures included insurance costs and repair costs due to failures of the WEC. The repair costs for the buoy and translator were computed differently, depending on the different requirements and costs for vessels and offshore operations needed for the repair. Finally, assuming a design life time duration of 20 years for the farm, the LCOE and NPV were computed according to Eqs. (1.8) and (1.13), respectively.

The yearly wave climate at the WaveHub site was used for the simulations, and passive damping corresponding to optimal PTO damping for each sea state was used. The WECs were positioned on a constrained gridded ocean area.

9.4.2 Economic optimisation of a wave farm

The optimisation of the park layout with respect to the economic objective functions was conducted in [280] using a genetic algorithm previously defined in [283]. A k-means clustering algorithm was employed to arrange the WECs in clusters, with a substation positioned at the centre of each cluster.

The results for the normalised economic performance measures (cable costs, annual energy production, LCOE and NPV) as a function of the number of iterations in the optimisation algorithm for a park of 10 WECs are shown in Figure 9.12, together with the corresponding identified optimal park layout. As can be seen from the figure,

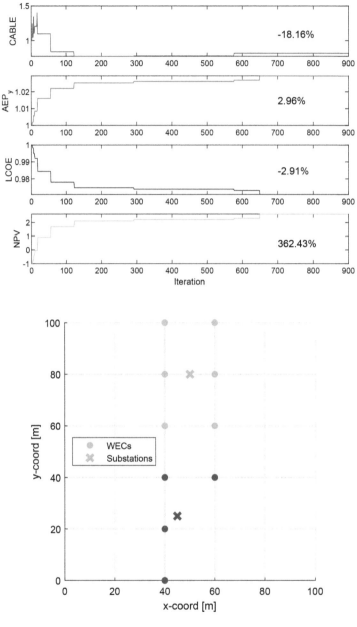

Figure 9.12: Results of the layout optimisation with respect to economic performance measures in [280]. Upper: The cable costs (CABLE) and LCOE reduce while the annual energy (AEP) and NPV increase with the number of iterations in the genetic algorithm optimisation. Lower: layout of the 10 WEC farm corresponding to the configuration obtained in the optimisation process. The crosses indicate the position of the substations in each cluster. The incident waves are propagating along the x-axis.

the LCOE and cable costs reduce with the number of iterations, whereas the annual energy and the NPV increase. In the optimal layout in Figure 9.12, the WECs are positioned in two clusters, each consisting of two rows aligned perpendicular to the wave direction. The positioning in two rows results from the constrained deployment area – if the available ocean area were unlimited, the WECs would instead align in a single row perpendicular to the wave direction. When comparing the impact of the costs and energy output of the farms, it was seen that the variation of the costs between the farms was smaller than the corresponding variation for energy uptake. The optimal layout of Figure 9.12 thus corresponds to the park configuration with the highest energy uptake, rather than the solution with the lowest costs.

9.4.3 Comparison with similar studies

The results obtained in [280] can be considered representative for several similar studies in the field. Two examples of techno-economic wave farm optimisations are shown in Figure 9.13, based on [759] and [232], respectively. The NPV and LCOE were studied and optimised for different wave farm configurations. In [759], it was seen that the NPV would increase with the number of units in the farm; whereas in [232] the LCOE was seen to reduce when employing both a genetic algorithm as well as other optimisation methods.

Even if it has been shown that wave farms can achieve a q-factor larger than one [828, 382, 142], in realistic large farms, the objective is instead to reduce the destructive shadowing effects. Several independent studies have concluded that WECs tend to align perpendicular or in rows at an angle to the incident wave direction in order to minimise shadowing effects and maximise the energy absorption. Similar to the work in [281], several examples of optimal wave farm layouts are shown in Figure 9.14. In all of these studies, the WECs are approximately positioned in rows perpendicular to the predominant incident wave direction in the resulting optimal layout.

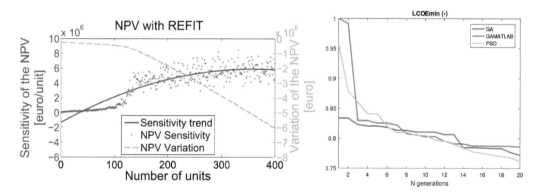

Figure 9.13: Results of two studies on economic performance measures for different wave farm configurations. (a) Variation of the NPV as a function of units in the farm, as well as the sensitivity of the quantity [759]. (b) Genetic algorithm optimisation of a wave farm to minimise the LCOE value [232].

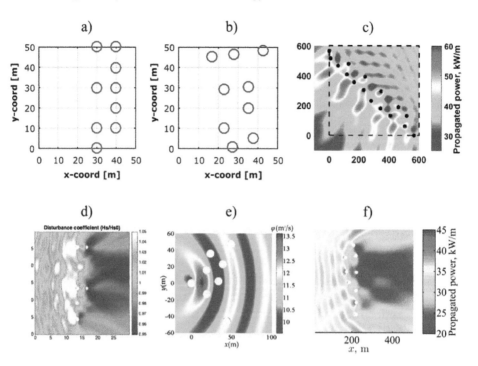

Figure 9.14: Wave farm layout optimisation in constrained ocean areas, with waves propagating along the x-axis in all figures except for c), where the wave direction is a narrow interval around 240°. a-b) GA optimisation in long-crested waves with a gridded and continuous layout, respectively [283]. c) Optimal layout of 16 WECs [564]. d) GA optimisation for the case of 3 m minimum separation distance [711]. e) Optimal 8-WEC array layout obtained in [231] with an evolutionary algorithm and regular waves. f) Multi-objective optimisation of an array with 9 WECs [28]. Figure from [303], based on adapted figures with obtained copyrights from the original sources.

However, this is not always the case and some studies have presented contradicting results. In the works of [124, 122, 231], the trend in the optimal layouts showed that WECs aligned in rows with an angle off the incident wave direction. In locations where the directional spread of the incident waves is large, the optimal layouts are less predictable [564]. Additionally, as discussed above, the energy output or related economic measures are not the only quantities that should perform well. When focusing on electricity quality and low power fluctuations, circular or random layouts may be the best configuration [202, 302, 719].

Most wave farm optimisations rely on numerical simulations and approximations regarding the linearity of the problem, restricted degrees of freedom, simplified PTO models, etc. In a realistic offshore environment, the difference between the layouts might be more negligible, and other considerations such as deployment strategies may influence the choice of layout to a more significant degree. The results presented in [80] indicate that the performance improvement in certain layouts is only marginal when taking the average over different wave conditions and system configurations. In [281], three array layouts, each with 6 WECs, were compared experimentally in a wave tank. The performance difference between the layout with WECs aligned perpendicularly to the wave direction, a staggered layout, and a layout with rows along the wave direction was not large. Even if the latter layout performed worse in most sea states due to the shadowing effect, it was not the case in all wave conditions. This was attributed to the fact that the buoys were moving in 6 degrees of freedom and were not simply confined to heave motion. Nevertheless, even if the differences between layouts were smaller in the experimental work [281] than in the numerical predictions, a q-factor close to or even above 1 was obtained for some layouts and wave conditions. This phenomenon has also been reported by other experimental works on wave farms [828, 142].

9.4.4 General trends

This section has documented a case study on a wave farm optimisation based on [280], and compared it with similar studies to analyse the results and draw general conclusions. A number of authors have carried out attempts to draw general conclusions from the many optimisation studies of wave farms [41, 303, 293]. The WEC designs, wave climates, objective functions, and optimisation methods differ, but some general trends can be observed.

- Although very strong array interaction effects have been found for small arrays by many authors [625], it seems that many of the strong effects disappear when averaging over different wave directions and sea states, as occurs in a realistic long-term scenario, and when studying more complex WECs with multiple degrees of freedom [80, 281].

- At locations where the incident waves have a well-defined predominant wave direction, the optimal layout will tend to position the WECs in rows perpendicular to the incident wave direction to increase the energy uptake [124, 122, 231, 283, 281]. In particular for large parks, shadowing effects can dominate and should be avoided by minimising the number of rows, with a separating distance as large as possible [41].

- The results of the optimisation will depend on the choice of optimisation objective. To account for this, research on wave farm optimisation has transitioned from simple maximisations of the q-factor or energy absorption, to optimisations of more complex economic measures including both the energy absorption in the income functions, and the costs involved with different park configurations. This is a welcome direction, but future work should also try to incorporate multi-dimensional objectives, such as optimising economic measures, electricity quality and survivability requirements simultaneously. In addition, the uncertainty in the economic models is largely due to the early state of WEC technologies and the sparse availability of data for real wave farm operations. Future works will hopefully provide more realistic input values to the models.

- Most wave farm optimisation studies have focused on optimising the array layout; whereas many parameters affect the whole performance. For a comprehensive assessment of a wave farm, the optimisation should be carried out in long-term wave conditions, and covering WEC dimensions, choice of materials, mooring and PTO configurations, array layout, etc. As the computational cost of optimising wave farms is already high, this will further increase the computational load, which will be a challenge. Faster, yet increasingly more accurate models in combination with access to high performing computer clusters are required for this future research direction.

9.5 SUMMARY

Installing WECs in arrays, or farms, is a strategy to reduce capital and operational costs, and simultaneously increase the power reliability and the produced electricity per ocean area. The WECs in the farm will interact in various ways, and the total performance of the farm will depend on the array layout, the individual devices' dimensions and power take-off settings, the wave conditions, and other parameters. Identifying optimal wave farm configurations, with respect to maximum energy absorption, minimal costs, or other objectives, has been a topic of research since the early days of wave energy development. The majority of these works have compared distinct array configurations or performed parameter sweeps of single variables. In recent years, more advanced optimisation methods have been developed and applied, including different meta-heuristic optimisation algorithms based on evolutionary principles, optimisation connected to control methods, and multi-objective approaches. Efforts have also been made to validate the numerical or analytical optimisation results by experimental data.

In the present chapter, these methods have been reviewed, and different optimisation approaches and objective functions have been compared and discussed. A case study of a wave farm layout optimisation using a genetic algorithm has been presented, where the objective was to optimise various economic measures of the farm performance. Based on comparisons of several independent optimisation studies, some general conclusions can be drawn, and some directions for future research have been identified.

Bibliography

[1] J. Abanades, D. Greaves, and G. Iglesias. Coastal defence through wave farms. *Coastal Engineering*, 91:299–307, 2014.

[2] J. Abanades, D. Greaves, and G. Iglesias. Wave farm impact on the beach profile: A case study. *Coastal Engineering*, 86:36–44, 2014.

[3] O. Abdelkhalik, S. Zou, R. Robinett, G. Bacelli, and D. Wilson. Estimation of excitation forces for wave energy converters control using pressure measurements. *International Journal of Control*, 90(8):1793–1805, 2017.

[4] O. Abdelkhalik, S. Zou, R. D. Robinett, G. Bacelli, D. Wilson, R. G. Coe, and U. A. Korde. Multiresonant feedback control of a three-degree-of-freedom wave energy converter. *IEEE Transactions on Sustainable Energy*, 8(4):1518–1527, 2017.

[5] O. Abdelkhalik, S. Zou, R. D. Robinett, and U. A. Korde. Time-varying linear quadratic Gaussian optimal control for three-degree-of-freedom wave energy converter. In *Proceedings of the Twelfth European Wave and Tidal Energy Conference*, pages 728-1–728-8. EWTEC, 2017.

[6] M. Abdelrahman, R. Patton, B. Guo, and J. Lan. Estimation of wave excitation force for wave energy converters. In *Proceedings of the Conference on Control and Fault-Tolerant Systems*, pages 654–659, Barcelona, Spain, 2016. IEEE.

[7] E. Abraham and E. C. Kerrigan. Optimal active control and optimisation of a wave energy converter. *IEEE Transactions on Sustainable Energy*, 4(2): 324–332, 2012.

[8] M. Abramowitz and I. A. Stegun. *Handbook of mathematical functions with formulas, graphs, and mathematical tables*, volume 55. US Government printing office, 1964.

[9] G. Aggidis, A. Bradshaw, M. French, A. McCabe, J. Meadowcroft, and M. Widden. PS frog MK5 WEC developments and design progress. In *Proceedings of the world renewable energy congress*, 2005.

[10] R. Ahamed, K. McKee, and I. Howard. Advancements of wave energy converters based on power take off (PTO) systems: A review. *Ocean Engineering*, 204:107248, 2020.

[11] R. Alcorn and W. Beattie. Power quality assessment from the LIMPET wave-power station. In *Proceedings of the Eleventh International Offshore and Polar Engineering Conference*. OnePetro, 2001.

[12] A. Alexandre, T. Stallard, and P. Stansby. Transformation of wave spectra across a line of wave devices. In *Proceedings of the 8th European Wave and Tidal Energy Conference, Uppsala, Sweden*, volume 710, 2009.

[13] W. Allsop, T. Bruce, J. Alderson, V. Ferrante, V. Russo, D. Vicinanza, and M. Kudella. Large scale tests on a generalised oscillating water column wave energy converter. In *Proceedings of the HYDRALAB IV Joint User Meeting, Lisbon*, 2014.

[14] E. Amini, D. Golbaz, F. Amini, M. Majidi Nezhad, M. Neshat, and D. Astiaso Garcia. A parametric study of wave energy converter layouts in real wave models. *Energies*, 13(22):6095, 2020.

[15] E. Amini, S. T. O. Naeeni, and P. Ghaderi. Investigating wave energy potential in southern coasts of the Caspian Sea and evaluating the application of Gray Wolf Optimizer algorithm. Preprint arXiv:1912.13201, 2019.

[16] M. Ancellin and F. Dias. Capytaine: A Python-based linear potential flow solver. *Journal of Open Source Software*, 4(36):1341, 2019.

[17] E. Anderlini, D. Forehand, E. Bannon, and M. Abusara. Reactive control of a wave energy converter using artificial neural networks. *International Journal of Marine Energy*, 19:207–220, 2017.

[18] E. Anderlini, D. I. Forehand, P. Stansell, Q. Xiao, and M. Abusara. Control of a point absorber using reinforcement learning. *IEEE Transactions on Sustainable Energy*, 7(4):1681–1690, 2016.

[19] E. Anderlini, S. Husain, G. G. Parker, M. Abusara, and G. Thomas. Towards real-time reinforcement learning control of a wave energy converter. *Journal of Marine Science and Engineering*, 8(11):845, 2020.

[20] D. M. Andrade, A. G. Santana, and A. d. l. V. Jaen. Frequency-matching assessment under reactive control on wave energy converters. In *Proceedings of the 4th International Conference on Ocean Energy*, pages 1–6, Dublin, Ireland, 2012.

[21] E. Angelelli, B. Zanuttigh, J. P. Kofoed, and K. Glejbøl. Experiments on the WavePiston, wave energy converter. In *Proceedings of the 9th European Wave and Tidal Conference, Southampton, UK, 5th-9th September 2011*. University of Southampton, 2011.

[22] C. ANSYS. Version 17.0, ansys cfx-solver theory guide. *Canonsburg, PA: Ansys Inc*, 2016.

[23] ANSYS AQWA, Inc. Canonsburg. *Version 15.0; ANSYS*, 2013.

[24] F. d. O. Antonio. Wave energy utilization: A review of the technologies. *Renewable and Sustainable Energy Reviews*, 14(3):899–918, 2010.

[25] M. P. Antonishen. *Harnessing Ocean Wave Energy: The Crest Wing Wave Energy Converter*, 2008.

[26] H. Apostoleris, A. Al Ghaferi, and M. Chiesa. What is going on with Middle Eastern solar prices, and what does it mean for the rest of us? *Progress in Photovoltaics: Research and Applications*, 29(6):638–648, 2021.

[27] D. Arbonès, B. Ding, N. Sergiienko, and M. Wagner. Fast and effective multi-objective optimisation of submerged wave energy converters. In *Proceedings of the Parallel Problem Solving from Nature – PPSN XIV*, pages 675–685. Springer International Publishing, 2016.

[28] D. Arbonès, N. Sergiienko, B. Ding, O. Krause, C. Igel, and M. Wagner. Sparse incomplete LU-decomposition for wave farm designs under realistic conditions. In *Proceedings of the Parallel Problem Solving from Nature – PPSN XV*, pages 512–524. Springer International Publishing, 2018.

[29] F. Arena, V. Fiamma, V. Laface, G. Malara, A. Romolo, A. Viviano, G. Sannino, and A. Carillo. Installing U-OWC devices along Italian coasts. In *Proceedings of the International Conference on Offshore Mechanics and Arctic Engineering*, volume 55423, page V008T09A061. American Society of Mechanical Engineers, 2013.

[30] F. Arena, G. Malara, and A. Romolo. A U-OWC wave energy converter in the Mediterranean Sea: Preliminary results on the monitoring system of the first prototype. *Renewable Energy Offshore; Guedes Soares, C.,* Ed.; Taylor & Francis Group: London, UK, pages 417–421, 2015.

[31] F. Arena, A. Romolo, G. Malara, V. Fiamma, and V. Laface. The first full operative U-OWC plants in the port of civitavecchia. In *Proceedings of the International Conference on Offshore Mechanics and Arctic Engineering*, volume 57786, page V010T09A022. American Society of Mechanical Engineers, 2017.

[32] S. Armstrong, E. Cotilla-Sanchez, and T. Kovaltchouk. Assessing the impact of the grid-connected Pacific marine energy center wave farm. *IEEE Journal of Emerging and Selected Topics in Power Electronics*, 3(4):1011–1020, 2015.

[33] B. Armstrong-Hélouvry, P. Dupont, and C. C. De Wit. A survey of models, analysis tools and compensation methods for the control of machines with friction. *Automatica*, 30(7):1083–1138, 1994.

[34] J. Arnett, L. Schaffer, J. Rumberg, and R. Tolbert. Design, installation and performance of the ARCO solar one-megawatt power plant. In *Proceedings of the 5th Photovoltaic Solar Energy Conference*, pages 314–320, 1984.

[35] I. G. Ashton, L. Johanning, and B. Linfoot. Measurement of the effect of power absorption in the lee of a wave energy converter. In *Proceedings of the International Conference on Offshore Mechanics and Arctic Engineering (OMAE)*, volume 43444, pages 1021–1030, 2009.

[36] S. Astariz and G. Iglesias. The economics of wave energy: A review. *Renewable and Sustainable Energy Reviews*, 45:397–408, 2015.

[37] S. Astariz and G. Iglesias. Enhancing wave energy competitiveness through co-located wind and wave energy farms. a review on the shadow effect. *Energies*, 8(7):7344–7366, 2015.

[38] J. Aubry, H. Ahmed, B. Multon, A. Babarit, and A. Clément. *Marine Renewable Energy Handbook*, chapter Wave energy converters. John Wiley & Sons, 2011.

[39] A. Azzellino, C. Lanfredi, L. Riefolo, V. De Santis, P. Contestabile, and D. Vicinanza. Combined exploitation of offshore wind and wave energy in the Italian seas: a spatial planning approach. *Frontiers in Energy Research*, 7:42, 2019.

[40] A. Babarit. Impact of long separating distances on the energy production of two interacting wave energy converters. *Ocean Engineering*, 37(8):718–729, 2010.

[41] A. Babarit. On the park effect in arrays of oscillating wave energy converters. *Renewable Energy*, 58:68–78, 2013.

[42] A. Babarit. A database of capture width ratio of wave energy converters. *Renewable Energy*, 80:610–628, 2015.

[43] A. Babarit, A. Clément, J. Ruer, and C. Tartivel. SEAREV: A fully integrated wave energy converter. In *Proceedings of the Offshore Wind and other Marine Renewable Energies in Mediterranean and European Seas*, pages 1–11, Rome, Italy, 2006.

[44] A. Babarit and A. H. Clément. Optimal latching control of a wave energy device in regular and irregular waves. *Applied Ocean Research*, 28(2):77–91, 2006.

[45] A. Babarit and A. H. Clement. Shape optimisation of the SEAREV wave energy converter. In *Proceedings of the World Renewable Energy Conference*, pages 1–6, Florence, Italy, 2006.

[46] A. Babarit and G. Delhommeau. Theoretical and numerical aspects of the open source BEM solver NEMOH. In *Proceedings of the 11th European Wave and Tidal Conference (EWTEC)*, Nantes, France, 2015.

[47] A. Babarit, G. Duclos, A. Clément, and J.-C. Gilloteaux. Latching control of a power take off oscillator carried by a wave activated body. In *Proceedings of the 20th International Workshop on Water Waves and Floating Bodies (IWWWFB)*, Longyearbyen, Norway, 2005.

[48] A. Babarit, B. Gendron, J. Singh, C. Mélis, and P. Jean. Hydro-elastic modelling of an electro-active wave energy converter. In *Proceedings of the International Conference on Offshore Mechanics and Arctic Engineering*, volume 55430, page V009T12A033. American Society of Mechanical Engineers, 2013.

[49] A. Babarit, M. Guglielmi, and A. H. Clément. Declutching control of a wave energy converter. *Ocean Engineering*, 36(12-13):1015–1024, 2009.

[50] A. Babarit, J. Hals, M. Muliawan, A. Kurniawan, T. Moan, and J. Krokstad. Numerical estimation of energy delivery from a selection of wave energy converters – final report. Report, Ecole Centrale de Nantes & Norges Teknisk-Naturvitenskapelige Universitet, 2011.

[51] A. Babarit, J. Hals, M. Muliawan, A. Kurniawan, T. Moan, and J. Krokstad. Numerical benchmarking study of a selection of wave energy converters. *Renewable Energy*, 41:44–63, 2012.

[52] A. Babarit, J. Hals, M. Muliawan, A. Kurniawan, T. Moan, and J. Krokstad. Numerical benchmarking study of a selection of wave energy converters. *Renewable Energy*, 41:44–63, 2012.

[53] A. Babarit, J. Singh, C. Mélis, A. Wattez, and P. Jean. A linear numerical model for analysing the hydroelastic response of a flexible electroactive wave energy converter. *Journal of Fluids and Structures*, 74:356–384, 2017.

[54] A. Babarit, J. Singh, C. Mélis, A. Wattez, and P. Jean. A linear numerical model for analysing the hydroelastic response of a flexible electroactive wave energy converter. *Journal of Fluids and Structures*, 74:356–384, 2017.

[55] G. Bacelli, P. Balitsky, and J. Ringwood. Coordinated control of arrays of wave energy devices—benefits over independent control. *IEEE Transactions on Sustainable Energy*, 4(4):1091–1099, 2013.

[56] G. Bacelli and J. Ringwood. Constrained control of arrays of wave energy devices. *International Journal of Marine Energy*, 3:e53–e69, 2013.

[57] G. Bacelli and J. V. Ringwood. A geometric tool for the analysis of position and force constraints in wave energy converters. *Ocean Engineering*, 65:10–18, 2013.

[58] A. Baggini. *Handbook of power quality*. John Wiley & Sons, Hoboken, U.S. 2008.

[59] H. Bailey, B. Robertson, J. Ortiz, and B. Buckham. Stochastic methods to predict WEC array power for grid integration. In *Proceedings of the 11th European Wave and Tidal Energy Conference (EWTEC)*, Nantes, France, 2015.

[60] P. Balitsky, G. Bacelli, and J. Ringwood. Control-influenced layout optimisation of arrays of wave energy converters. In *Proceedings of the ASME 33rd*

International Conference on Ocean, Offshore and Arctic Engineering (OMAE), San Fransisco, USA, 2014.

[61] P. Balitsky, N. Quartier, G. Verao Fernandez, V. Stratigaki, and P. Troch. Analyzing the near-field effects and the power production of an array of heaving cylindrical WECs and osWECs using a coupled hydrodynamic-PTO model. *Energies*, 11(12):3489, 2018.

[62] P. Balitsky, G. Verao Fernandez, V. Stratigaki, and P. Troch. Assessment of the power output of a two-array clustered WEC farm using a BEM solver coupling and a wave-propagation model. *Energies*, 11(11):2907, 2018.

[63] M.-A. Bănică. Energy harvesting from renewable energy sources. In *Proceedings of the International Conference of Mechatronics and Cyber-Mixmechatronics*, pages 247–254. Springer, 2019.

[64] J. Bard and P. Kracht. Linear generator systems for wave energy converters. Technical report, Fraunhofer IWES, 2013.

[65] R. Beharie and J. Side. Acoustic environmental monitoring:–Wello Penguin cooling system noise study. a report commissioned by Aquatera Limited. International Centre for Island Technology. Technical report, Wello Oy, 2012.

[66] S. Behrens, D. Griffin, J. Hayward, M. Hemer, C. Knight, S. McGarry, P. Osman, and J. Wright. Ocean renewable energy: 2015-2050: An analysis of ocean energy in Australia. Report, CSIRO, 2012.

[67] P. Beirao, M. Mendes, D. Valério, and J. S. da Costa. Control of the Archimedes Wave Swing using neural networks. In *Proceedings of the 7th European Wave and Tidal Energy Conference*, 2007.

[68] P. Beirdol, D. Valério, and J. da Costa. Linear model identification of the Archimedes Wave Swing. In *Proceedings of the International Conference on Power Engineering, Energy and Electrical Drives*, pages 660–665. IEEE, 2007.

[69] L. G. Bennetts and V. A. Squire. Wave scattering by multiple rows of circular ice floes. *Journal of Fluid Mechanics*, 639:213–238, 2009.

[70] M. Bhinder, M. Karimirad, S. Weller, Y. Debruyne, M. Guérinel, and W. Sheng. Modelling mooring line non-linearities (material and geometric effects) for a wave energy converter using AQWA, SIMA and Orcaflex. In *Proceedings of the 11th European Wave and Tidal Energy Conference (EWTEC)*, Nantes, France, 2015.

[71] J. L. Black. Wave forces on vertical axisymmetric bodies. *Journal of Fluid Mechanics*, 67(2):369–376, 1975.

[72] M. Blanco, P. Moreno-Torres, M. Lafoz, and D. Ramírez. Design parameters analysis of point absorber WEC via an evolutionary-algorithm-based dimensioning tool. *Energies*, 8(10):11203–11233, 2015.

[73] M. Blanco, J. Torres, M. Santos-Herrán, L. García-Tabarés, G. Navarro, J. Nájera, D. Ramírez, and M. Lafoz. Recent advances in direct-drive power take-off (ddpto) systems for wave energy converters based on switched reluctance machines (srm). *Ocean Wave Energy Systems*, pages 487–532, 2022.

[74] A. Blavette, D. O'Sullivan, R. Alcorn, M. Egan, and T. Lewis. Simplified estimation of the flicker level induced by wave energy farms. *IEEE Transactions on Sustainable Energy*, 7(3):1216–1223, 2016.

[75] C. B. Boake, T. J. Whittaker, M. Folley, and H. Ellen. Overview and initial operational experience of the LIMPET wave energy plant. In *Proceedings of the Twelfth International Offshore and Polar Engineering Conference*. OnePetro, 2002.

[76] P. Boccotti. Caisson breakwaters embodying an OWC with a small opening—part i: Theory. *Ocean Engineering*, 34(5):806–819, 2007.

[77] P. Boccotti. Comparison between a u-owc and a conventional owc. *Ocean Engineering*, 34(5-6):799–805, 2007.

[78] N. Booij, L. Holthuijsen, and R. Ris. The SWAN wave model for shallow water. In *Proceedings of the 25th International Conference on Coastal Engineering*, pages 668–676, 1997.

[79] B. Borgarino, A. Babarit, and P. Ferrant. Impact of wave interactions effects on energy absorption in large arrays of wave energy converters. *Ocean Engineering*, 41:79–88, 2012.

[80] B. Bosma, T. Brekken, P. Lomonaco, B. DuPont, C. Sharp, and B. Batten. Array modeling and testing of fixed OWC type wave energy converters. *International Marine Energy Journal*, 3(3):137–143, 2020.

[81] C. Boström, E. Lejerskog, M. Stålberg, K. Thorburn, and M. Leijon. Experimental results of rectification and filtration from an offshore wave energy system. *Renewable Energy*, 34(5):1381–1387, 2009.

[82] C. Boström, E. Lejerskog, S. Tyrberg, O. Svensson, R. Waters, A. Savin, B. Bolund, M. Eriksson, and M. Leijon. Experimental results from an offshore wave energy converter. *Journal of offshore mechanics and Arctic engineering*, 132(4), 2010.

[83] S. Bozzi, M. Giassi, A. Miquel, A. Antonini, F. Bizzozero, G. Gruosso, R. Archetti, and G. Passoni. Wave energy farm design in real wave climates: the Italian offshore. *Energy*, 122:378–389, 2017.

[84] BP, London, UK. *Statistical Review of World Energy*, 69 edition, 2020.

[85] BP, London, UK. *Statistical review of world energy*, 70 edition, 2021.

[86] T. Brekken, B. Batten, and E. Amon. From blue to green [ask the experts]. *IEEE Control Systems Magazine*, 31(5):18–24, 2011.

[87] J. Bretl, K. Parsa, K. Edwards, J. Montgomery, and M. Mekhiche. The deployment of the PB-3-50-A1 PowerBuoy: Power prediction and measurement comparison. In *Proceedings of the Marine energy Technology Symposium METS*, Washington, D.C., USA, 2016.

[88] C. L. Bretschneider. Wave variability and wave spectra for wind-generated gravity waves. Technical report, DTIC Document, 1959.

[89] A. Brito e Melo, H. Jeffry, and Y. De Roeck. Annual report–an overview of ocean energy activities in 2020. Technical report, The Executive Committee of Ocean Energy Systems, 2021.

[90] A. Brito-Melo, F. Neuman, and A. Sarmento. Full-scale data assessment in OWC Pico plant. *International Journal of Offshore and Polar Engineering*, 18(01), 2008.

[91] H. Brodersen, K. Nielsen, and J. P. Kofoed. Development of the Danish test site DanWEC. In *Proceedings of the European Wave and Tidal Energy Conference*. Technical Committee of the European Wave and Tidal Energy Conference, 2013.

[92] M. Brorsen and J. Larsen. Source generation of nonlinear gravity waves with the boundary integral equation method. *Coastal Engineering*, 11(2):93–113, 1987.

[93] D. Bruschi, J. Fernandes, A. Falcão, and C. Bergmann. Analysis of the degradation in the Wells turbine blades of the Pico oscillating-water-column wave energy plant. *Renewable and Sustainable Energy Reviews*, 115:109368, 2019.

[94] L. Bruzzone, P. Fanghella, and G. Berselli. Reinforcement learning control of an onshore oscillating arm wave energy converter. *Ocean Engineering*, 206:107346, 2020.

[95] I. Bryden and B. Linfoot. Wave and current testing of an array of wave energy converters. In *Proceedings of the HYDRALAB III Joint User Meeting*, 2010.

[96] K. Bubbar and B. Buckham. On establishing generalized analytical phase control conditions in two body self-reacting point absorber wave energy converters. *Ocean Engineering*, 197:106879, 2020.

[97] K. Budal. Theory for absorption of wave power by a system of interacting bodies. *Journal of Ship Research*, 21(04):248–254, 1977.

[98] K. Budal and J. Falnes. A resonant point absorber of ocean-wave power. *Nature*, 256(5517):478–479, 1975.

[99] K. Budal and J. Falnes. Interacting point absorbers with controlled motion. In *Power from Sea Waves*, pages 381–99. Academic Press, Cambridge, U. S. Press, 1980.

[100] K. Budar and J. Falnes. A resonant point absorber of ocean-wave power. *Nature*, 256(5517):478–479, 1975.

[101] T. Cabral, D. Clemente, P. Rosa-Santos, F. Taveira-Pinto, T. Morais, F. Belga, and H. Cestaro. Performance assessment of a hybrid wave energy converter integrated into a harbor breakwater. *Energies*, 13(1):236, 2020.

[102] T. Calheiros-Cabral, D. Clemente, P. Rosa-Santos, F. Taveira-Pinto, V. Ramos, T. Morais, and H. Cestaro. Evaluation of the annual electricity production of a hybrid breakwater-integrated wave energy converter. *Energy*, 213:118845, 2020.

[103] L. Cameron, R. Doherty, A. Henry, K. Doherty, J. Van't Hoff, D. Kaye, D. Naylor, S. Bourdier, and T. Whittaker. Design of the next generation of the Oyster wave energy converter. In *Proceedings of the 3rd International Conference on Ocean Energy*, volume 6, page 1e12. ICOE Bilbao, Spain, 2010.

[104] J. Candido, A. Sarmento, F. Gardner, L. Gato, M. Fontana, and K. Collins. The WETFEET project-a disruptive approach to wave energy. In *Proceedings of the 13th European Wave and Tidal Energy Conference (EWTEC)*, Napoli, Italy, 2019.

[105] C. J. Cargo, A. R. Plummer, A. J. Hillis, and M. Schlotter. Determination of optimal parameters for a hydraulic power take-off unit of a wave energy converter in regular waves. *Proceedings of the Institution of Mechanical Engineers, Part A: Journal of Power and Energy*, 226(1):98–111, 2012.

[106] E. Carpintero Moreno and P. Stansby. The 6-float wave energy converter M4: Ocean basin tests giving capture width, response and energy yield for several sites. *Renewable and Sustainable Energy Reviews*, 104:307–318, 2019.

[107] A. Carrelhas, L. Gato, A. Falcão, and J. Henriques. Control law design for the air-turbine-generator set of a fully submerged 1.5 MW mWave prototype. part 2: Experimental validation. *Renewable Energy*, 171:1002–1013, 2021.

[108] R. Cascajo, E. García, E. Quiles, A. Correcher, and F. Morant. Integration of marine wave energy converters into seaports: A case study in the port of Valencia. *Energies*, 12(5):787, 2019.

[109] L. Castro-Santos, E. Martins, and C. G. Soares. Cost assessment methodology for combined wind and wave floating offshore renewable energy systems. *Renewable Energy*, 97:866–880, 2016.

[110] F. Cerveira, N. Fonseca, and R. Pascoal. Mooring system influence on the efficiency of wave energy converters. *International Journal of Marine Energy*, 3:65–81, 2013.

[111] S. K. Chakrabarti. Response due to moored multiple structure interaction. *Marine Structures*, 14(1-2):231–258, 2001.

[112] J. Chaplin, V. Heller, F. Farley, G. Hearn, and R. Rainey. Laboratory testing the Anaconda. *Philosophical Transactions of the Royal Society A: Mathematical, Physical and Engineering Sciences*, 370(1959):403–424, 2012.

[113] J. Chapman, A. Brask, J.-B. LeDreff, G. Foster, and G. Stockman. Improving energy capture by varying the geometry of a novel wave energy converter. In A. Lewis, editor, *Proceedings of the 12th European Wave and Tidal Energy Conference*, pages 813-1–813-8, University College Cork, Ireland, 2017. EWTEC.

[114] R. H. Charlier and J. R. Justus. *Ocean energies: environmental, economic and technological aspects of alternative power sources*. Elsevier, 1993.

[115] M. A. Chatzigiannakou, I. Dolguntseva, and M. Leijon. Offshore deployments of wave energy converters by Seabased Industry AB. *Journal of Marine Science and Engineering*, 5(2):15, 2017.

[116] Checkmate. `https://www.checkmateukseaenergy.com/anaconda-left-languishing-by-wave-energy-scotland/`Anaconda left languishing by Wave Energy Scotland, 2020. Access date: April 20, 2022.

[117] W. Chehaze, D. Chamoun, C. Bou-Mosleh, and P. Rahme. Wave Roller device for power generation. *Procedia Engineering*, 145:144–150, 2016.

[118] F. Chen, D. Duan, Q. Han, X. Yang, and F. Zhao. Study on force and wave energy conversion efficiency of buoys in low wave energy density seas. *Energy Conversion and Management*, 182:191–200, 2019.

[119] Q. Chen, J. Zang, J. Birchall, D. Ning, X. Zhao, and J. Gao. On the hydrodynamic performance of a vertical pile-restrained wec-type floating breakwater. *Renewable Energy*, 146:414–425, 2020.

[120] X. Chen. Evaluation de la fonction de Green du probleme de diffraction/radiation en profondeur d'eau finie-une nouvelle méthode rapide et précise. *Actes des 4e Journées de l'Hydrodynamique, Nantes (France)*, pages 371–84, 1993.

[121] B. Child. *On the configuration of arrays of floating wave energy converters*. PhD thesis, The University of Edinburgh, 2011.

[122] B. Child, J. Cruz, and M. Livingstone. The development of a tool for optimising of arrays of wave energy converters. In *Proceedings of the 9th European Wave and Tidal Energy Conference (EWTEC)*, Southampton, UK, 2011.

[123] B. Child and P. Laporte Weywada. Verification and validation of a wave farm planning tool. In *Proceedings of the 10th European Wave and Tidal Energy Conference (EWTEC)*, Aalborg, Denmark, 2013.

[124] B. F. M. Child and V. Venugopal. Optimal configurations of wave energy device arrays. *Ocean Engineering*, 37(16):1402–1417, 2010.

[125] Y.-C. Chow, Y.-C. Chang, D.-W. Chen, C.-C. Lin, and S.-Y. Tzang. Parametric design methodology for maximizing energy capture of a bottom-hinged flap-type WEC with medium wave resources. *Renewable Energy*, 126:605–616, 2018.

[126] S. D. Chowdhury, J.-R. Nader, A. Fleming, B. Winship, S. Illesinghe, A. Toffoli, A. Babanin, I. Penesis, and R. Manasseh. A review of hydrodynamical interactions into arrays of ocean wave energy converters. Preprint arXiv:1508.00866, 2015.

[127] J. F. Chozas, J. P. Kofoed, and H. C. Sørensen. Predictability and variability of wave and wind: Wave and wind forecasting and diversified energy systems in the Danish North Sea. Report 156, Aalborg University, Wave Energy Research Group, 2016.

[128] J. Chunn. Ocean energy: The next wave. *Ecogeneration*, 95:34–38, 2016.

[129] C. E. Clark and B. DuPont. Reliability-based design optimisation in offshore renewable energy systems. *Renewable and Sustainable Energy Reviews*, 97:390–400, 2018.

[130] C. E. Clark, A. Garcia-Teruel, B. DuPont, and D. Forehand. Towards reliability-based geometry optimisation of a point-absorber with PTO reliability objectives. In *Proceedings of the European Wave and Tidal Energy Conference, Naples, Italy*, pages 1–6, 2019.

[131] E. Clark. `https://newatlas.com/rubber-tube-could-provide-affordable-wave-power/9578/` Anaconda aims for affordable wave power, 2008.

[132] A. Clément, P. McCullen, A. Falcão, A. Fiorentino, F. Gardner, K. Hammarlund, G. Lemonis, T. Lewis, K. Nielsen, S. Petroncini, et al. Wave energy in Europe: current status and perspectives. *Renewable and Sustainable Energy Reviews*, 6(5):405–431, 2002.

[133] D. Clemente, P. Rosa-Santos, and F. Taveira-Pinto. On the potential synergies and applications of wave energy converters: A review. *Renewable and Sustainable Energy Reviews*, 135:110162, 2021.

[134] R. G. Coe, G. Bacelli, D. G. Wilson, O. Abdelkhalik, U. A. Korde, and R. D. Robinett III. A comparison of control strategies for wave energy converters. *International Journal of Marine Energy*, 20:45–63, 2017.

[135] V. D. Colli, P. Cancelliere, F. Marignetti, R. D. Stefano, and M. Scarano. A tubular-generator drive for wave energy conversion. *IEEE Transactions on Industrial Electronics*, 53(4):1152–1159, 2006.

[136] I. Collins, M. Hossain, and I. Masters. A review of flexible membrane structures for wave energy converters. In *Proceedings of the 13th European Wave Tidal and Energy Conference*, Naples, Italy, 2019.

[137] P. Contestabile, G. Crispino, E. Di Lauro, V. Ferrante, C. Gisonni, and D. Vicinanza. Overtopping breakwater for wave energy conversion: Review of state of art, recent advancements and what lies ahead. *Renewable Energy*, 147:705–718, 2020.

[138] P. Contestabile, E. Di Lauro, M. Buccino, and D. Vicinanza. Economic assessment of overtopping breakwater for energy conversion (OBREC): a case study in western australia. *Sustainability*, 9(1):51, 2017.

[139] J. Cordonnier, F. Gorintin, A. De Cagny, A. H. Clément, and A. Babarit. SEAREV: Case study of the development of a wave energy converter. *Renewable Energy*, 80:40–52, 2015.

[140] A. Cornett. A global wave energy resource assessment. In *Proceedings of the 18th International Offshore and Polar Engineering Conference*, Vancouver, Canada, 2008. OnePetro.

[141] CorpowerOcean. https://www.corpowerocean.com/ Corpower Ocean, 08 2021. (Accessed on 08/08/2021).

[142] F. Correia da Fonseca, R. Gomes, J. Henriques, L. Gato, and A. Falcão. Model testing of an oscillating water column spar-buoy wave energy converter isolated and in array: Motions and mooring forces. *Energy*, 112:1207–1218, 2016.

[143] R. Cossu, C. Heatherington, I. Penesis, R. Beecroft, and S. Hunter. Seafloor site characterization for a remote island OWC device near King Island, Tasmania, Australia. *Journal of Marine Science and Engineering*, 8(3):194, 2020.

[144] R. Costello and A. Pecher. Economics of WECs. In A. Pecher and J. P. Kofoed, editors, *Handbook of Ocean Wave Energy*, pages 101–137. Springer International Publishing, Cham, 2017.

[145] S. H. Crowley, R. Porter, and D. V. Evans. A submerged cylinder wave energy converter with internal sloshing power take off. *European Journal of Mechanics - B/Fluids*, 47:108–123, 2014.

[146] W. Cummins. The impulse response function and ship motions. Technical report, David Taylor Model Basin Washington DC, 1962.

[147] L. S. P. da Silva. Nonlinear stochastic analysis of wave energy converters via statistical linearization. Master's thesis, University of São Paulo, Brazil, 2019.

[148] L. S. P. da Silva, B. Cazzolato, N. Sergiienko, and B. Ding. Nonlinear dynamics of a floating offshore wind turbine platform via statistical quadratization - mooring, wave and current interaction. *Ocean Engineering*, 236:109471, 2021.

[149] L. S. P. da Silva, B. S. Cazzolato, N. Y. Sergiienko, B. Ding, H. M. Morishita, and C. P. Pesce. Statistical linearization of the Morison's equation applied to wave energy converters. *Journal of Ocean Engineering and Marine Energy*, pages 1–13, 2020b.

[150] L. S. P. da Silva, H. M. Morishita, C. P. Pesce, and R. T. Gonçalves. Nonlinear analysis of a heaving point absorber in frequency domain via statistical linearization. In *Proceedings of the ASME 38th International Conference on Ocean, Offshore and Arctic Engineering*. American Society of Mechanical Engineers, 2019.

[151] L. S. P. da Silva, C. P. Pesce, H. M. Morishita, and R. T. Gonçalves. Nonlinear analysis of an oscillating water column wave energy device in frequency domain via statistical linearization. In *Proceedings of the ASME 38th International Conference on Ocean, Offshore and Arctic Engineering*. American Society of Mechanical Engineers, 2019.

[152] L. S. P. da Silva, C. Sergiienko, NY nd Pesce, B. Ding, B. Cazzolato, and H. Morishita. Stochastic analysis of nonlinear wave energy converters via statistical linearization. *Applied Ocean Research*, 95, 2020.

[153] L. S. P. da Silva, N. Y. Sergiienko, B. S. Cazzolato, B. Ding, C. P. Pesce, and H. M. Morishita. Nonlinear analysis of an oscillating wave surge converter in frequency domain via statistical linearization. In *Proceedings of the International Conference on Offshore Mechanics and Arctic Engineering*, volume 84416, page V009T09A019. American Society of Mechanical Engineers, 2020.

[154] Y. S. Dai and W. Y. Duan. *Potential flow theory of ship motions in waves*. National Defense Industry Publication, 2008.

[155] G. Dalton, R. Alcorn, and T. Lewis. Case study feasibility analysis of the Pelamis wave energy convertor in Ireland, Portugal and North America. *Renewable Energy*, 35(2):443–455, 2010.

[156] G. J. Dalton, R. Alcorn, and T. Lewis. Case study feasibility analysis of the Pelamis wave energy convertor in Ireland, Portugal and North America. *Renewable Energy*, 35(2):443–455, 2010.

[157] M. DAndrea, M. G. Gonzalez, and R. McKenna. Synergies in offshore energy: a roadmap for the danish sector. *arXiv preprint arXiv:2102.13581*, 2021.

[158] J. Davidson, S. Giorgi, and J. V. Ringwood. Linear parametric hydrodynamic models for ocean wave energy converters identified from numerical wave tank experiments. *Ocean Engineering*, 103:31–39, 2015.

[159] J. Davidson, S. Giorgi, and J. V. Ringwood. Identification of wave energy device models from numerical wave tank data–Part 1: Numerical wave tank identification tests. *IEEE Transactions on Sustainable Energy*, 7(3):1012–1019, 2016.

[160] J. Davidson and J. Ringwood. Mathematical modelling of mooring systems for wave energy converters—a review. *Energies*, 10(5):666, 2017.

[161] A. Day, A. Babarit, A. Fontaine, Y.-P. He, M. Kraskowski, M. Murai, I. Penesis, F. Salvatore, and H.-K. Shin. Hydrodynamic modelling of marine renewable energy devices: A state of the art review. *Ocean Engineering*, 108:46–69, 2015.

[162] A. De Andrés, R. Guanche, L. Meneses, C. Vidal, and I. Losada. Factors that influence array layout on wave energy farms. *Ocean Engineering*, 82:32–41, 2014.

[163] A. De Andres, J. Maillet, J. Hals Todalshaug, P. Möller, D. Bould, and H. Jeffrey. Techno-economic related metrics for a wave energy converters feasibility assessment. *Sustainability*, 8(11):1109, 2016.

[164] S. De Chowdhury and R. Manasseh. Behaviour of eigenmodes of an array of oscillating water column devices. *Wave Motion*, 74:56–72, 2017.

[165] A. de O Falcão. Wave energy utilization: A review of the technologies. *Renewable and Sustainable Energy Reviews*, 14(3):899–918, 2010.

[166] A. de O. Falcão. Wave-power absorption by a periodic linear array of oscillating water columns. *Ocean Engineering*, 29(10):1163–1186, 2002.

[167] M. de Sousa Prado, F. Gardner, M. Damen, and H. Polinder. Modelling and test results of the Archimedes Wave Swing. *Proceedings of the Institution of Mechanical Engineers, Part A: Journal of Power and Energy*, 220(8):855–868, 2006.

[168] Y. Delauré and A. Lewis. 3D hydrodynamic modelling of fixed oscillating water column wave power plant by a boundary element methods. *Ocean Engineering*, 30(3):309–330, 2003.

[169] V. DelliColli, P. Cancelliere, F. Marignetti, R. DiStefano, and M. Scarano. A tubular-generator drive for wave energy conversion. *IEEE Transactions on Industrial Electronics*, 53(4):1152–1159, 2006.

[170] N. Delmonte, D. Barater, F. Giuliani, P. Cova, and G. Buticchi. Oscillating water column power conversion: A technology review. In *Proceedings of the IEEE Energy Conversion Congress and Exposition (ECCE)*, pages 1852–1859. IEEE, 2014.

[171] V. Denoël. Multiple timescale spectral analysis. *Probabilistic Engineering Mechanics*, 39:69–86, 2015.

[172] S. S. Deshpande, L. Anumolu, and M. F. Trujillo. Evaluating the performance of the two-phase flow solver interfoam. *Computational Science & Discovery*, 5(1):014016, 2012.

[173] B. Devolder, P. Rauwoens, and P. Troch. Towards the numerical simulation of 5 floating point absorber wave energy converters installed in a line array using OpenFOAM. In *Proceedings of the 12th European Wave and Tidal Energy Conference (EWTEC)*, Cork, Ireland, 2017.

[174] B. Devolder, V. Stratigaki, P. Troch, and P. Rauwoens. CFD simulations of floating point absorber wave energy converter arrays subjected to regular waves. *Energies*, 11(3):641, 2018.

[175] E. Dialyna and T. Tsoutsos. Wave energy in the mediterranean sea: Resource assessment, deployed WECs and prospects. *Energies*, 14(16):4764, 2021.

[176] M. Dianat, M. Skarysz, and A. Garmory. A coupled level set and volume of fluid method for automotive exterior water management applications. *International Journal of Multiphase Flow*, 91:19–38, 2017.

[177] B. Ding, B. Cazzolato, M. Arjomandi, P. Hardy, and B. Mills. Sea-state based maximum power point tracking damping control of a fully submerged oscillating buoy. *Ocean Engineering*, 126:229–312, 2016.

[178] B. Ding, P.-Y. Wuillaume, F. Meng, A. Babarit, B. Schubert, N. Sergiienko, and B. Cazzolato. Comparison of wave-body interaction modelling methods for the study of reactively controlled point absorber wave energy converter. In *Proceedings of 32th International Workshop on Water Waves and Floating Bodies*, Newcastle, Australia, 2019.

[179] N. Dizadji and S. E. Sajadian. Modeling and optimisation of the chamber of OWC system. *Energy*, 36(5):2360–2366, 2011.

[180] G. DNV. Sesam user Manual-Wadam. *DNV GL Software*, 2017.

[181] J. Dong, J. Gao, L. Tao, and P. Zheng. Research status of wave energy conversion (WEC) device of raft structure. *AIP Conference Proceedings*, 1890(1):030005, 2017.

[182] M. G. Donley and P. Spanos. *Dynamic analysis of non-linear structures by the method of statistical quadratization*, volume 57. Springer Science & Business Media, 2012.

[183] S. Doyle and G. Aggidis. Advancement of oscillating water column wave energy technologies through integrated applications and alternative systems. *International Journal of Energy and Power Engineering*, 14(12):401–412, 2020.

[184] B. Drew, A. Plummer, and M. Sahinkaya. A review of wave energy converter technology. *Proceedings of The Institution of Mechanical Engineers Part A-Journal of Power and Energy*, 223(8):887–902, 2009.

[185] B. Drew, A. R. Plummer, and M. N. Sahinkaya. A review of wave energy converter technology. *Proceedings of the Institution of Mechanical Engineers, Part A: Journal of Power and Energy*, 223(8):887–902, 2009.

[186] G. Duclos, A. Babarit, and A. H. Clement. Optimizing the power take off of a wave energy converter with regard to the wave climate. *Journal of Offshore Mechanics and Arctic Engineering*, 128(1):56–64, 2005.

[187] M. Durand, A. Babarit, B. Pettinotti, O. Quillard, J. Toularastel, and A. Clément. Experimental validation of the performances of the SEAREV wave energy converter with real time latching control. In *Proceedings of the 7th European Wave and Tidal Energy Conference (EWTEC), Porto, Portugal*, volume 1113, 2007.

[188] R. Eatock Taylor and F. Chau. Wave diffraction theory—some developments in linear and nonlinear theory. *Journal of Offshore Mechanics and Arctic Engineering*, 114(3):185–194, 1992.

[189] O. Edenhofer, R. Pichs-Madruga, Y. Sokona, K. Seyboth, S. Kadner, T. Zwickel, P. Eickemeier, G. Hansen, S. Schlömer, C. von Stechow, et al. *Renewable energy sources and climate change mitigation: Special report of the intergovernmental panel on climate change*. Cambridge University Press, 2011.

[190] K. Edwards and M. Mekhiche. Ocean testing of a wave-capturing PowerBuoy. In *Proceedings of the Marine Energy Technology Symposium (METS)*, Washington, DC, 2013.

[191] K. Edwards, M. Mekhiche, et al. Ocean power technologies Powerbuoy: system-level design, development and validation methodology. In *Proceedings of the 2nd Marine Energy Technology Symposium (METS)*, Seattle, WA, USA, 2014.

[192] H. Eidsmoen. Simulation of a slack-moored heaving-buoy wave-energy converter with phase control. Technical report, Division of Physics, NTNU, Trondheim, Norway, 1996.

[193] R. Ekström, B. Ekergård, and M. Leijon. Electrical damping of linear generators for wave energy converters—a review. *Renewable and Sustainable Energy Reviews*, 42:116–128, 2015.

[194] R. Ekström and M. Leijon. Grid connection of wave power farm using an N-level cascaded H-bridge multilevel inverter. *Journal of Electrical and Computer Engineering*, 2013:1–9, 2013.

[195] C. Elefant. Overview of global regulatory processes for permits, consents and authorization of marine renewables. *Global Status and Critical Developments in Ocean Energy*, page 70, 2009.

[196] A. Elhanafi, G. Macfarlane, A. Fleming, and Z. Leong. Experimental and numerical investigations on the intact and damage survivability of a floating–moored oscillating water column device. *Applied Ocean Research*, 68:276–292, 2017.

[197] A. Elhanafi, G. Macfarlane, and D. Ning. Hydrodynamic performance of single–chamber and dual–chamber offshore–stationary oscillating water column devices using CFD. *Applied Energy*, 228:82–96, 2018.

[198] D. Elwood. Evaluation of the performance of a taut-moored dual-body direct-drive wave energy converter through numerical modeling and physical testing. Master's thesis, Oregon State University, 2008.

[199] D. Elwood, S. Yim, E. Amon, A. von Jouanne, and T. Brekken. Experimental force characterization and numerical modeling of a taut-moored dual-body wave energy conversion system. *Journal of Offshore Mechanics and Arctic Engineering*, 132(1), 2010.

[200] H. Endo. Shallow-water effect on the motions of three-dimensional bodies in waves. *Journal of Ship Research*, 31(01):34–40, 1987.

[201] Energy Technologies Institute. *Wave Energy – Insights from the Energy Technologies Institute*, 2015.

[202] J. Engström, M. Eriksson, M. Göteman, J. Isberg, and M. Leijon. Performance of large arrays of point absorbing direct-driven wave energy converters. *Journal of Applied Physics*, 114(20):204502, 2013.

[203] J. Engström, M. Göteman, M. Eriksson, M. Bergkvist, E. Nilsson, A. Rutgersson, and E. Strömstedt. Energy absorption from parks of point-absorbing wave energy converters in the Swedish exclusive economic zone. *Energy Science & Engineering*, 2019.

[204] M. Eriksson. *Modelling and experimental verification of direct drive wave energy conversion: Buoy-generator dynamics*. PhD thesis, Uppsala University, Sweden, 2007.

[205] M. Eriksson, R. Waters, O. Svensson, J. Isberg, and M. Leijon. Wave power absorption: Experiments in open sea and simulation. *Journal of Applied Physics*, 102(8):084910, 2007.

[206] European Commission and ECORYS and Fraunhofer. *Study on Lessons for Ocean Energy Development*, 2017.

[207] European Union: European Commission. *Report from the Commission to the European Parliament and the Council on Progress of Clean Energy Competitiveness*, October 2020. https://eur-lex.europa.eu/ legal-content/EN/TXT/?uri=CELEX:52020DC0953, Accessed: 2021-08-12.

[208] D. Evans. A theory for wave-power absorption by oscillating bodies. *Journal of Fluid Mechanics*, 77(1):1–25, 1976.

[209] D. Evans. Some theoretical aspects of three-dimensional wave-energy absorbers. In *Proceedings of the 1st Symposium Wave Energy Utilization*, pages 78–112, 1979.

[210] D. Evans. *Some analytic results for two- and three-dimensional wave energy absorbers*, page 213. Power from Sea Waves. Academic Press, 1980.

[211] D. Evans. Maximum wave-power absorption under motion constraints. *Applied Ocean Research*, 3(4):200–203, 1981.

[212] D. V. Evans, D. C. Jeffrey, S. H. Salter, and J. R. M. Taylor. Submerged cylinder wave energy device: theory and experiment. *Applied Ocean Research*, 1(1):3–12, 1979.

[213] N. Faedo, S. Olaya, and J. Ringwood. Optimal control, MPC and MPC-like algorithms for wave energy systems: An overview. *IFAC Journal of Systems and Control*, 1:37–56, 2017.

[214] N. Faedo, Y. Peña-Sanchez, and J. V. Ringwood. Finite-order hydrodynamic model determination for wave energy applications using moment-matching. *Ocean Engineering*, 163:251–263, 2018.

[215] A. d. O. Falcão. First-generation wave power plants: current status and R&D requirements. In *Proceedings of the International Conference on Offshore Mechanics and Arctic Engineering*, volume 36835, pages 723–731, 2003.

[216] A. Falcão. Developments in oscillating water column wave energy converters and air turbines. *Renewable Energies Offshore. London: Taylor & Francis Group*, pages 3–11, 2015.

[217] A. Falcão and J. Henriques. Effect of non-ideal power take-off efficiency on performance of single-and two-body reactively controlled wave energy converters. *Journal of Ocean Engineering and Marine Energy*, 1(3):273–286, 2015.

[218] A. d. O. Falcão. The shoreline OWC wave power plant at the azores. In *Proceedings of 4th European Wave Energy Conference*, pages 42–47, 2000.

[219] A. F. Falcão and J. C. Henriques. Oscillating-water-column wave energy converters and air turbines: A review. *Renewable Energy*, 85:1391–1424, 2016.

[220] A. F. Falcão and J. C. Henriques. The spring-like air compressibility effect in oscillating-water-column wave energy converters: Review and analyses. *Renewable and Sustainable Energy Reviews*, 112:483–498, 2019.

[221] A. F. Falcão, J. C. Henriques, and J. J. Cândido. Dynamics and optimisation of the OWC spar buoy wave energy converter. *Renewable Energy*, 48:369–381, 2012.

[222] A. F. Falcão, A. J. Sarmento, L. M. Gato, and A. Brito-Melo. The Pico OWC wave power plant: Its lifetime from conception to closure 1986–2018. *Applied Ocean Research*, 98:102104, 2020.

[223] A. F. Falcão and J. C. Henriques. Oscillating-water-column wave energy converters and air turbines: A review. *Renewable Energy*, 85:1391–1424, 2016.

[224] J. Falnes. Radiation impedance matrix and optimum power absorption for interacting oscillators in surface waves. *Applied Ocean Research*, 2(2):75–80, 1980.

[225] J. Falnes. On non-causal impulse response functions related to propagating water waves. *Applied Ocean Research*, 17(6):379–389, 1995.

[226] J. Falnes. *Ocean waves and oscillating systems*. Cambridge University Press, Cambridge, UK, 2002.

[227] J. Falnes. *Ocean waves and oscillating systems: linear interactions including wave-energy extraction*. Cambridge, United Kingdom, 2002.

[228] J. Falnes. A review of wave-energy extraction. *Marine Structures*, 20(4):185–201, 2007.

[229] J. Falnes and J. Hals. Heaving buoys, point absorbers and arrays. *Philosophical Transactions of the Royal Society A: Mathematical, Physical and Engineering Sciences*, 370(1959):246–277, 2012.

[230] J. Falnes and J. Løvseth. Ocean wave energy. *Energy Policy*, 19(8):768–775, 1991.

[231] H.-W. Fang, Y.-Z. Feng, and G.-P. Li. Optimisation of wave energy converter arrays by an improved differential evolution algorithm. *Energies*, 11(12):3522, 2018.

[232] E. Faraggiana, I. Masters, and J. Chapman. Design of an optimisation scheme for the WaveSub array. In G. Soares, editor, *Advances in Renewable Energies Offshore: In Proceedings of the 3rd International Conference on Renewable Energies Offshore (RENEW)*. Taylor & Francis Group, 2019.

[233] F. J. M. Farley and R. C. T. Rainey. Anaconda: The bulge wave sea energy converter. Report, Maritime Energy Development Ltd., 2006.

[234] N. Farrell, C. O'Donoghue, and K. Morrissey. Quantifying the uncertainty of wave energy conversion device cost for policy appraisal: An Irish case study. *Energy Policy*, 78:62–77, 2015.

[235] F.-X. Faÿ, J. Kelly, J. Henriques, A. Pujana, M. Abusara, M. Mueller, I. Touzon, and P. Ruiz-Minguela. Numerical simulation of control strategies at Mutriku wave power plant. In *Proceedings of the International Conference on Offshore Mechanics and Arctic Engineering*, volume 51319, page V010T09A029. American Society of Mechanical Engineers, 2018.

[236] F.-X. Faÿ, E. Robles, M. Marcos, E. Aldaiturriaga, and E. F. Camacho. Sea trial results of a predictive algorithm at the Mutriku wave power plant and controllers assessment based on a detailed plant model. *Renewable Energy*, 146:1725–1745, 2020.

[237] Y. Feng, X. Liang, J. An, T. Jiang, and Z. L. Wang. Soft-contact cylindrical triboelectric-electromagnetic hybrid nanogenerator based on swing structure for ultra-low frequency water wave energy harvesting. *Nano Energy*, 81:105625, 2021.

[238] Y. Feng, G. Zhang, Y. Shen, Y. You, and J. Sun. Design of Duck wave power generation device based on wave energy. In *Proceedings of the IOP Conference Series: Earth and Environmental Science*, number 1 in 696, page 012039. IOP Publishing, 2021.

[239] J. Fenton. Wave forces on vertical bodies of revolution. *Journal of Fluid Mechanics*, 85(2):241–255, 1978.

[240] F. Ferri. Computationally efficient optimisation algorithms for WECs arrays. In *Proceedings of the 12th European Wave and Tidal Energy Conference (EWTEC)*, Cork, Ireland, 2017.

[241] F. Ferri, S. Ambühl, B. Fischer, and J. P. Kofoed. Balancing power output and structural fatigue of wave energy converters by means of control strategies. *Energies*, 7(4):2246–2273, 2014.

[242] R. A. Fine and F. J. Millero. Compressibility of water as a function of temperature and pressure. *The Journal of Chemical Physics*, 59(10):5529–5536, 1973.

[243] C. Fitzgerald and G. Thomas. A preliminary study on the optimal formation of an array of wave power devices. In *Proceedings of the 7th European Wave and Tidal Energy Conference, Porto, Portugal*, pages 11–14, 2007.

[244] F. Flavià and A. Clément. Extension of Haskind's relations to cylindrical wave fields in the context of an interaction theory. *Applied Ocean Research*, 66:1–12, 2017.

[245] F. F. Flavià, C. McNatt, F. Rongère, A. Babarit, and A. H. Clément. A numerical tool for the frequency domain simulation of large arrays of identical floating bodies in waves. *Ocean Engineering*, 148:299–311, 2018.

[246] F. F. Flavià and M. H. Meylan. An extension of general identities for 3D water-wave diffraction with application to the diffraction transfer matrix. *Applied Ocean Research*, 84:279–290, 2019.

[247] M. Folley. Spectral-domain models. In M. Folley, editor, *Numerical modelling of wave energy converters: state-of-the-art techniques for single devices and arrays*, pages 67–80. Academic Press, 2016.

[248] M. Folley. The wave energy resource. In A. Pecher and J. P. Kofoed, editors, *Handbook of ocean wave energy*, pages 43–79. Springer International Publishing, Cham, 2017.

[249] M. Folley, A. Babarit, B. Child, D. Forehand, L. O'Boyle, K. Silverthorne, J. Spinneken, V. Stratigaki, and P. Troch. A review of numerical modelling of wave energy converter arrays. In *Proceedings of the ASME 31st International Conference on Ocean, Offshore and Arctic Engineering*, 2013.

[250] M. Folley and T. Whittaker. Spectral modelling of wave energy converters. *Coastal Engineering*, 57(10):892–897, 2010.

[251] M. Folley and T. Whittaker. Preliminary cross-validation of wave energy converter array interactions. In *Proceedings of the 32nd International Conference on Ocean, Offshore and Arctic Engineering*, pages V008T09A055–V008T09A055. American Society of Mechanical Engineers, 2013.

[252] M. Folley and T. Whittaker. Validating a spectral-domain model of an OWC using physical model data. *International Journal of Marine Energy*, 2:1–11, 2013.

[253] M. Folley, T. Whittaker, and J. Van't Hoff. The design of small seabed-mounted bottom-hinged wave energy converters. In *Proceedings of the 7th European Wave and Tidal Energy Conference (EWTEC)*, volume 455, page 312. Citeseer, 2007.

[254] M. Folley and T. J. Whittaker. Identification of non-linear flow characteristics of the LIMPET shoreline OWC. In *Proceedings of the Twelfth International Offshore and Polar Engineering Conference*. OnePetro, 2002.

[255] D. Forehand, A. Kiprakis, A. Nambiar, and A. Wallace. A fully coupled wave-to-wire model of an array of wave energy converters. *IEEE Transactions on Sustainable Energy*, 7(1):118–128, 2015.

[256] Foynes Engineering LTD. `https://foynesengineering.wearetherift.com/our_project/sea-power-wave-energy-converter/` Sea Power wave energy converter. 2015.

[257] V. Franzitta, A. Colucci, D. Curto, V. Di Dio, and M. Trapanese. A linear generator for a WaveRoller power device. In *Proceedings of the OCEANS 2017-Aberdeen*, pages 1–5. IEEE, 2017.

[258] A. Fredriksen. Tapered channel wave power plants. In *Energy for Rural and Island Communities*, pages 179–182. Elsevier, 1986.

[259] G. Freebury and W. Musial. Determining equivalent damage loading for full-scale wind turbine blade fatigue tests. In *Proceedings of the ASME Wind Energy Symposium*, page 50, 2000.

[260] Z.-F. Fu and J. He. *Modal analysis*. Butterworth-Heinemann, 2001.

[261] F. Fusco and J. V. Ringwood. Short-term wave forecasting for real-time control of wave energy converters. *IEEE Transactions on Sustainable Energy*, 1(2):99–106, 2010.

[262] F. Fusco and J. V. Ringwood. Short-term wave forecasting for real-time control of wave energy converters. *IEEE Transactions on Sustainable Energy*, 1(2):99–106, 2010.

[263] F. Fusco and J. V. Ringwood. Short-term wave forecasting with AR models in real-time optimal control of wave energy converters. In *Proceedings of the IEEE International Symposium on Industrial Electronics (ISIE)*, pages 2475–2480. IEEE, 2010.

[264] F. Fusco and J. V. Ringwood. A study of the prediction requirements in real-time control of wave energy converters. *IEEE Transactions on Sustainable Energy*, 3(1):176–184, 2012.

[265] F. Fusco and J. V. Ringwood. A simple and effective real-time controller for wave energy converters. *IEEE Transactions on Sustainable Energy*, 4(1):21–30, 2013.

[266] D. Galván-Pozos and F. Ocampo-Torres. Dynamic analysis of a six-degree of freedom wave energy converter based on the concept of the Stewart-Gough platform. *Renewable Energy*, 146:1051–1061, 2020.

[267] D. E. Galvan-Pozos and F. J. Ocampo-Torres. Kinematic and dynamic analysis of the motion of a wave energy converter with more than one degree of freedom. In *Proceedings of the 12th European Wave and Tidal Energy Conference (EWTEC)*, pages 636-1–636-10, Cork, Ireland, 2017.

[268] H. Gao and Y. Yu. The dynamics and power absorption of cone-cylinder wave energy converters with three degree of freedom in irregular waves. *Energy*, 143:833–845, 2018.

[269] Z. Gao and T. Moan. Mooring system analysis of multiple wave energy converters in a farm configuration. In *Proceedings the 8th European Wave and Tidal Energy Conference (EWTEC)*, Uppsala, Sweden, 2009.

[270] P. Garcia-Rosa, G. Bacelli, and J. Ringwood. Control-informed optimal array layout for wave farms. *IEEE Transactions on Sustainable Energy*, 6(2):575–582, 2015.

[271] P. B. Garcia-Rosa, G. Kulia, J. V. Ringwood, and M. Molinas. Real-time passive control of wave energy converters using the Hilbert-Huang transform. *IFAC-PapersOnLine*, 50(1):14705–14710, 2017.

[272] A. Garcia-Teruel, B. DuPont, and D. I. Forehand. Hull geometry optimisation of wave energy converters: On the choice of the objective functions and the optimisation formulation. *Applied Energy*, 298:117153, 2021.

[273] A. Garcia-Teruel and D. Forehand. A review of geometry optimisation of wave energy converters. *Renewable and Sustainable Energy Reviews*, 139:110593, 2021.

[274] D. García-Violini, Y. Peña-Sanchez, N. Faedo, and J. V. Ringwood. An energy-maximising linear time invariant controller (LiTe-Con) for wave energy devices. *IEEE Transactions on Sustainable Energy*, 11(4):2713–2721, 2020.

[275] C. Garrett. Wave forces on a circular dock. *Journal of Fluid Mechanics*, 46(01):129–139, 1971.

[276] T. Garrison. *Oceanography: An invitation to marine science*. Thomson Brooks/Cole, Belmont, USA, 7th edition, 2010.

[277] L. Gato and A. d. O. Falcão. On the theory of the Wells turbine. *Journal of Engineering for Gas Turbines and Power*, 106(3):628–633, 1984.

[278] R. Genest and J. V. Ringwood. A critical comparison of model-predictive and pseudospectral control for wave energy devices. *Journal of Ocean Engineering and Marine Energy*, pages 1–15, 2016.

[279] M. Giassi, V. Castellucci, J. Engström, and M. Göteman. An economical cost function for the optimisation of wave energy converter arrays. In *Proceedings of the 29th International Ocean and Polar Engineering Conference (ISOPE)*, Honolulu, USA, 2019.

[280] M. Giassi, V. Castellucci, and M. Göteman. Economical layout optimisation of wave energy parks clustered in electrical subsystems. *Applied Ocean Research*, 101:102274, 2020.

[281] M. Giassi, J. Engström, J. Isberg, and M. Göteman. Comparison of wave energy park layouts by experimental and numerical methods. *Journal of Marine Science and Engineering*, 8(10):750, 2020.

[282] M. Giassi and M. Göteman. Parameter optimisation in wave energy design by a genetic algorithm. In *Proceedings of 32nd International Workshop on Water Waves and Floating Bodies (IWWWFB)*, Dalian, China, 2017.

[283] M. Giassi and M. Göteman. Layout design of wave energy parks by a genetic algorithm. *Ocean Engineering*, 154:252–261, 2018.

[284] M. Giassi, M. Göteman, S. Thomas, J. Engström, M. Eriksson, and J. Isberg. Multi-parameter optimisation of hybrid arrays of point absorber wave energy converters. In *Proceedings of 12th European Wave and Tidal Energy Conference (EWTEC)*, Cork, Ireland, 2017.

[285] M. Giassi, S. Thomas, Z. Shahroozi, J. Engström, J. Isberg, T. Tosdevin, M. Hann, and M. Göteman. Preliminary results from a scaled test of arrays of point-absorbers with 6 DOF. In *Proceedings of the 13th European Wave and Tidal Conference (EWTEC)*, Naples, Italy, 2019.

[286] M. Giassi, S. Thomas, T. Tosdevin, J. Engström, M. Hann, J. Isberg, and M. Göteman. Capturing the experimental behaviour of a point-absorber WEC

by simplified numerical models. *Journal of Fluids and Structures*, 99:103143, 2020.

[287] G. Giorgi and J. V. Ringwood. Nonlinear Froude-Krylov and viscous drag representations for wave energy converters in the computation/fidelity continuum. *Ocean Engineering*, 141:164–175, 2017.

[288] G. Giorgi and J. V. Ringwood. Analytical representation of nonlinear Froude-Krylov forces for 3-DoF point absorbing wave energy devices. *Ocean Engineering*, 164:749–759, 2018.

[289] G. Giorgi and J. V. Ringwood. Articulating parametric resonance for an OWC spar buoy in regular and irregular waves. *Journal of Ocean Engineering and Marine Energy*, 4(4):311–322, 2018.

[290] S. Giorgi, J. Davidson, and J. Ringwood. Identification of nonlinear excitation force kernels using numerical wave tank experiments. In *Proceedings of the European Wave and Tidal Energy Conference*, pages 1–10, Nantes, France, 2015. European Wave and Tidal Energy Conference 2015.

[291] S. Giorgi, J. Davidson, and J. V. Ringwood. Identification of wave energy device models from numerical wave tank data—Part 2: Data-based model determination. *IEEE Transactions on Sustainable Energy*, 7(3):1020–1027, 2016.

[292] B. Godderidge. *A phenomenological rapid sloshing model for use as an operator guidance system on liquid natural gas carriers*. PhD thesis, University of Southampton, 2009.

[293] D. Golbaz, R. Asadi, E. Amini, H. Mehdipour, M. Nasiri, M. M. Nezhad, S. T. O. Naeeni, and M. Neshat. Ocean wave energy converters optimisation: A comprehensive review on research directions. Preprint arXiv:2105.07180, 2021.

[294] R. Gonzalez-Santamaria, Q.-P. Zou, and S. Pan. Impacts of a wave farm on waves, currents and coastal morphology in South West England. *Estuaries and Coasts*, 38(1):159–172, 2015.

[295] J.-S. Goo and K. Yoshida. A numerical method for huge semisubmersible responses in waves. *Transactions-Society of Naval Architects and Marine Engineers*, 98:365–387, 1990.

[296] M. Góralczyk. Life-cycle assessment in the renewable energy sector. *Applied Energy*, 75(3-4):205–211, 2003.

[297] E. Gorr-Pozzi, H. García-Nava, M. Jaramillo-Torres, M. Larrañaga-Fu, and F. Ocampo-Torres. Evaluation of wave energy converter arrays within a sheltered bay. In *Proceedings of the 13th European Wave and Tidal Conference (EWTEC)*, Naples, Italy, 2019.

[298] M. Göteman. Wave energy parks with point-absorbers of different dimensions. *Journal of Fluids and Structures*, 74:142–157, 2017.

[299] M. Göteman. Iterative multiple cluster scattering. In *Proceedings of 36th International Workshop on Water Waves and Floating Bodies (IWWWFB)*, Seoul, Korea, 2021.

[300] M. Göteman, J. Engström, M. Eriksson, and J. Isberg. Fast modeling of large wave energy farms using interaction distance cut-off. *Energies*, 8(12):13741–13757, 2015.

[301] M. Göteman, J. Engström, M. Eriksson, and J. Isberg. Optimizing wave energy parks with over 1000 interacting point-absorbers using an approximate analytical method. *International Journal of Marine Energy*, 10:113–126, 2015.

[302] M. Göteman, J. Engström, M. Eriksson, J. Isberg, and M. Leijon. Methods of reducing power fluctuations in wave energy parks. *Journal of Renewable and Sustainable Energy*, 6(4):043103, 2014.

[303] M. Göteman, M. Giassi, J. Engström, and J. Isberg. Advances and challenges in wave energy park optimisation–a review. *Frontiers in Energy Research*, 8:26, 2020.

[304] M. Göteman, C. McNatt, M. Giassi, J. Engström, and J. Isberg. Arrays of point-absorbing wave energy converters in short-crested irregular waves. *Energies*, 11(4):964, 2018.

[305] A. Gray, D. Findlay, and L. Johanning. Operations and maintenance planning for community-scale, off-grid wave energy devices. In *Advances in Renewable Energies Offshore: Proceedings of the 2nd International Conference on Renewable Energies Offshore (RENEW)*. CRC Press, Taylor & Francis Group, 2016.

[306] D. Greaves and G. Iglesias. *Wave and tidal energy*. John Wiley & Sons, 2018.

[307] M. Greenhow, T. Vinje, P. Brevig, and J. Taylor. A theoretical and experimental study of the capsize of Salter's Duck in extreme waves. *Journal of Fluid Mechanics*, 118:221–239, 1982.

[308] Y. Gu, B. Ding, N. Y. Sergiienko, and B. S. Cazzolato. Power maximising control of a heaving point absorber wave energy converter. *IET Renewable Power Generation*, 2021.

[309] O. T. Gudmestad and G. Moe. Hydrodynamic coefficients for calculation of hydrodynamic loads on offshore truss structures. *Marine Structures*, 9(8):745–758, 1996.

[310] B. Guo. *Study of scale modelling, verification and control of a heaving point absorber wave energy converter*. PhD thesis, University of Hull, 2017.

[311] B. Guo, R. Patton, M. Abdelrahman, and J. Lan. A continuous control approach to point absorber wave energy conversion. In *Proceedings of the UKACC 11th International Conference on Control (CONTROL)*, pages 1–6. IEEE, 2016.

[312] B. Guo, R. Patton, and S. Jin. Identification and validation of excitation force for a heaving point absorber wave energy convertor. In *Proceedings of the European Wave and Tidal Energy Conference*, pages 1–9, Cork, Ireland, 2017.

[313] B. Guo, R. Patton, S. Jin, J. Gilbert, and D. Parsons. Nonlinear modeling and verification of a heaving point absorber for wave energy conversion. *IEEE Transactions on Sustainable Energy*, 9(1):453–461, 2018.

[314] B. Guo and R. J. Patton. Non-linear viscous and friction effects on a heaving point absorber dynamics and latching control performance. *IFAC-PapersOnLine*, 50(1):15657–15662, 2017.

[315] B. Guo, R. J. Patton, S. Jin, and J. Lan. Numerical and experimental studies of excitation force approximation for wave energy conversion. *Renewable Energy*, 125:877–889, 2018.

[316] B. Guo and J. V. Ringwood. Modelling of a vibro-impact power take-off mechanism for wave energy conversion. In *Proceedings of the European Control Conference (ECC)*, pages 1348–1353. IEEE, 2020.

[317] B. Guo and J. V. Ringwood. Non-linear modeling of a vibro-impact wave energy converter. *IEEE Transactions on Sustainable Energy*, 12(1):492–500, 2020.

[318] B. Guo and J. V. Ringwood. Parametric study of a vibro-impact wave energy converter. *IFAC-PapersOnLine*, 53(2):12283–12288, 2020.

[319] B. Guo and J. V. Ringwood. Geometric optimisation of wave energy conversion devices: A survey. *Applied Energy*, 297:117100, 2021.

[320] B. Guo, R. Wang, D. Ning, L. Chen, and W. Sulisz. Hydrodynamic performance of a novel WEC-breakwater integrated system consisting of triple dual-freedom pontoons. *Energy*, 209:118463, 2020.

[321] Y. Guo, Y.-H. Yu, J. A. van Rij, and N. M. Tom. Inclusion of structural flexibility in design load analysis for wave energy converters. Technical report, National Renewable Energy Lab.(NREL), Golden, CO (United States), 2017.

[322] K. T. Gürsel, D. Ünsalan, G. NEŞER, M. Taner, E. Altunsaray, and M. Önal. a technological assessment of the wave energy converter. *Scientific Bulletin of Naval Academy*, 19(1):408–417, 2016.

[323] G. Hagerman. Wave energy resource and economic assessment for the state of hawaii. Technical report, SEASUN Power Systems, 1992.

[324] G. M. Hagerman and T. Heller. Wave energy: a survey of twelve near-term technologies. In *Proceedings of the International Renewable Energy Conference*, pages 98–110, Honolulu, Hawaii, September 1988.

[325] M. C. Haller, A. Porter, P. Lenee-Bluhm, K. Rhinefrank, E. Hammagren, H. Özkan-Haller, and D. Newborn. Laboratory observation of waves in the vicinity of WEC-arrays. In *Proceedings of the 9th European Wave and Tidal Energy Conference (EWTEC)*, Southampton, UK, 2011.

[326] J. Hals, J. Falnes, and T. Moan. A comparison of selected strategies for adaptive control of wave energy converters. *Journal of Offshore Mechanics and Arctic Engineering*, 133(3):031101, 2011.

[327] J. Hals, J. Falnes, and T. Moan. Constrained optimal control of a heaving buoy wave-energy converter. *Journal of Offshore Mechanics and Arctic Engineering*, 133(1):011401, 2011.

[328] J. Hals Todalshaug. Practical limits to the power that can be captured from ocean waves by oscillating bodies. *International Journal of Marine Energy*, 3:e70–e81, 2013.

[329] J. Hals Todalshaug. Hydrodynamics of WECs. In A. Pecher and J. P. Kofoed, editors, *Handbook of ocean wave energy*, pages 139–158. Springer International Publishing, Cham, 2017.

[330] J. Hals Todalshaug, G. S. Ásgeirsson, E. Hjálmarsson, J. Maillet, P. Möller, P. Pires, M. Guérinel, and M. Lopes. Tank testing of an inherently phase-controlled wave energy converter. *International Journal of Marine Energy*, 15:68–84, 2016.

[331] M. Hann. *A numerical and experimental study of a multi-cell fabric distensible wave energy converter*. PhD thesis, University of Southampton, 2013.

[332] M. Hann, J. Chaplin, and F. Farley. Assessment of a multi-cell fabric structure as an attenuating wave energy converter. In *Proceedings of the World Renewable Energy Congress-Sweden*, 057, pages 2143–2150, Linköping; Sweden, May 2011.

[333] J. E. Hanssen, L. Margheritini, K. O'Sullivan, P. Mayorga, I. Martinez, A. Arriaga, I. Agos, J. Steynor, D. Ingram, R. Hezari, and J. H. Todalshaug. Design and performance validation of a hybrid offshore renewable energy platform. In *10th International Conference on Ecological Vehicles and Renewable Energies (EVER)*, pages 1–8, 2015.

[334] P. Hardy, B. Cazzolato, B. Ding, and Z. Prime. A maximum capture width tracking controller for ocean wave energy converters in irregular waves. *Ocean Engineering*, 121:516–529, 2016.

[335] P. Haren and C. C. Mei. Wave power extraction by a train of rafts: hydrodynamic theory and optimum design. *Applied Ocean Research*, 1(3):147–157, 1979.

[336] R. E. Harris, L. Johanning, and J. Wolfram. Mooring systems for wave energy converters: A review of design issues and choices. https://

www.semanticscholar.org/paper/Mooring-systems-for-wave-energy-converters%3A-A-of-Harris-Johanning/0fcddf19543ff03e757e8241a07 febf75a29e43c *Marec 2004*, 2004.

[337] J. Harrison. Floating offshore wind: Installation, operation & maintenance challenges. Technical Report BF008-001-RE, Blackfish Engineering Design and Product Development, 2020.

[338] M. Haskind. The exciting forces and wetting of ships in waves, Izvestia Akademii Nauk SSSR, Otdelenie Tekhnicheskikh Nauk, 7, 65-79. *David Taylor Model Basin Translation*, 307, 1962.

[339] K. Hasselmann, T. Barnett, E. Bouws, H. Carlson, D. Cartwright, K. Enke, J. Ewing, H. Gienapp, D. Hasselmann, P. Kruseman, et al. Measurements of wind-wave growth and swell decay during the Joint North Sea Wave Project (JONSWAP). Technical report, Deutches Hydrographisches Institut, 1973.

[340] K. Hatalis and S. Kishore. On enabling autonomous operation and control of ocean wave energy farms. In *Proceedings of the 3rd Marine Energy Technology Symposium (METS)*, 2015.

[341] F. He, Z. Huang, and A. W.-K. Law. Hydrodynamic performance of a rectangular floating breakwater with and without pneumatic chambers: An experimental study. *Ocean Engineering*, 51:16–27, 2012.

[342] F. He, Z. Huang, and A. W.-K. Law. An experimental study of a floating breakwater with asymmetric pneumatic chambers for wave energy extraction. *Applied Energy*, 106:222–231, 2013.

[343] F. He, J. Leng, and X. Zhao. An experimental investigation into the wave power extraction of a floating box-type breakwater with dual pneumatic chambers. *Applied Ocean Research*, 67:21–30, 2017.

[344] T. Heath. A review of oscillating water columns. *Philosophical Transactions of the Royal Society A: Mathematical, Physical and Engineering Sciences*, 370(1959):235–245, 2012.

[345] V. Heller. Technology readiness level approach for the development of WECs, 2012. Presentation presented at the 4th CoastLab Teaching School, Wave and Tidal Energy, Porto.

[346] V. Heller, J. Chaplin, F. Farley, M. Hann, G. Hearn, et al. Physical model tests of the Anaconda wave energy converter. In *Proceedings of the 1st IAHR European Congress*, 2000.

[347] M. Hemer and D. Griffin. The wave energy resource along Australia's southern margin. *Journal of Renewable and Sustainable Energy*, 2(4):043108, 2010.

[348] M. Hemer, T. Pitman, K. McInnes, and U. Rosebrock. The Australian wave energy Atlas project overview and final report. Report, CSIRO, 2018.

[349] R. Henderson. Design, simulation, and testing of a novel hydraulic power take-off system for the Pelamis wave energy converter. *Renewable Energy*, 31(2):271–283, 2006.

[350] U. Henfridsson, V. Neimane, K. Strand, R. Kapper, H. Bernhoff, O. Danielsson, M. Leijon, J. Sundberg, K. Thorburn, E. Ericsson, and K. Bergman. Wave energy potential in the Baltic Sea and the Danish part of the North Sea, with reflections on the Skagerrak. *Renewable Energy*, 32(12):2069–2084, 2007.

[351] J. L. Hess and A. O. Smith. Calculation of potential flow about arbitrary bodies. *Progress in Aerospace Sciences*, 8:1–138, 1967.

[352] A. Hillis, A. Brask, and C. Whitlam. Real-time wave excitation force estimation for an experimental multi-DOF WEC. *Ocean Engineering*, 213:107788, 2020.

[353] A. Hillis, C. Whitlam, A. Brask, J. Chapman, and A. Plummer. Active control for multi-degree-of-freedom wave energy converters with load limiting. *Renewable Energy*, 159:1177–1187, 2020.

[354] A. Hillis, C. Whitlam, A. Brask, J. Chapman, and A. R. Plummer. Power capture gains for the WaveSub submerged WEC using active control. In *Proceedings of the Thirteenth European Wave and Tidal Energy Conference*, 2019.

[355] A. Hillis, J. Yardley, A. Plummer, and A. Brask. Model predictive control of a multi-degree-of-freedom wave energy converter with model mismatch and prediction errors. *Ocean Engineering*, 212:107724, 2020.

[356] C. Hirt and B. Nichols. Volume of fluid (vof) method for the dynamics of free boundaries. *Journal of Computational Physics*, 39(1):201–225, 1981.

[357] J. Hodges, J. Henderson, L. Ruedy, M. Soede, J. Weber, P. Ruiz-Minguela, H. Jeffrey, E. Bannon, M. Holland, R. Maciver, D. Hume, J. Villate, and T. Ramsey. An international evaluation and guidance framework for ocean energy technology. Technical report, International Energy Agency (IEA) and Ocean Energy Systems (OES) and Wave Energy Scotland, 2021.

[358] Y. Hong, V. Castellucci, M. Eriksson, C. Boström, and R. Waters. Linear generator-based wave energy converter model with experimental verification and three loading strategies. *IET Renewable Power Generation*, 10(3):349–359, 2016.

[359] H. Hotta, Y. Washio, H. Yokozawa, and T. Miyazaki. R&d on wave power device "Mighty Whale". *Renewable Energy*, 9(1-4):1223–1226, 1996.

[360] D. Howe, J.-R. Nader, and G. Macfarlane. Performance analysis of a floating breakwater integrated with multiple oscillating water column wave energy converters in regular and irregular seas. *Applied Ocean Research*, 99:102147, 2020.

[361] J. Hu, B. Zhou, C. Vogel, P. Liu, R. Willden, K. Sun, J. Zang, J. Geng, P. Jin, L. Cui, et al. Optimal design and performance analysis of a hybrid system combing a floating wind platform and wave energy converters. *Applied Energy*, 269:114998, 2020.

[362] A. Hulme. A ring-source/integral-equation method for the calculation of hydrodynamic forces exerted on floating bodies of revolution. *Journal of Fluid Mechanics*, 128:387–412, 1983.

[363] G. Ibarra-Berastegi, J. Sáenz, A. Ulazia, P. Serras, G. Esnaola, and C. Garcia-Soto. Electricity production, capacity factor, and plant efficiency index at the Mutriku wave farm (2014–2016). *Ocean Engineering*, 147:20–29, 2018.

[364] International Energy Agency (IEA). *Deploying renewables - Best and future police practice*, 2011.

[365] International Energy Agency (IEA), Paris, France. *Ocean Energy Systems: International levelised cost of energy for ocean energy technologies*, 2015.

[366] International Energy Agency (IEA), Paris, France. *Electricity Information: Overview*, 2020.

[367] International Energy Agency (IEA). *Global Energy Review 2019*, 2020.

[368] International Renewable Energy Agency (IRENA), Abu Dhabi. *Innovation outlook: Ocean energy technologies*, 2020.

[369] International Renewable Energy Agency (IRENA). *Renewable Power Generation Costs in 2020*, 2021.

[370] I. R. E. A. (IRENA). *Ocean power: Technologies, patents, deployment status and outlook*, 2014.

[371] A. Iturrioz, R. Guanche, J. Armesto, M. Alves, C. Vidal, and I. Losada. Time-domain modeling of a fixed detached oscillating water column towards a floating multi-chamber device. *Ocean Engineering*, 76:65–74, 2014.

[372] C. Iuppa, L. Cavallaro, R. E. Musumeci, D. Vicinanza, and E. Foti. Empirical overtopping volume statistics at an obrec. *Coastal Engineering*, 152:103524, 2019.

[373] J. Izquierdo-Pérez, B. M. Brentan, J. Izquierdo, N.-E. Clausen, A. Pegalajar-Jurado, and N. Ebsen. Layout optimisation process to minimize the cost of energy of an offshore floating hybrid wind–wave farm. *Processes*, 8(2):139, 2020.

[374] J.-L. Jacob. Financing renewable energy innovations in Europe: Investment criteria of early stage investors and guidelines for the business plan development in the context of wave energy conversion. Master's thesis, Beuth University of Applied Sciences Berlin and MBA Renewables, 2016.

[375] P. Jean, A. Wattez, G. Ardoise, C. Melis, R. Van Kessel, A. Fourmon, E. Barrabino, J. Heemskerk, and J. Queau. Standing wave tube electro active polymer wave energy converter. In *Proceedings of the Electroactive Polymer Actuators and Devices (EAPAD) 2012*, volume 8340, page 83400C. International Society for Optics and Photonics, 2012.

[376] X. Ji, S. Liu, H. Bingham, and J. Li. Multi-directional random wave interaction with an array of cylinders. *Ocean Engineering*, 110:62–77, 2015.

[377] S. Jin and D. Greaves. Wave energy in the UK: Status review and future perspectives. *Renewable and Sustainable Energy Reviews*, 143:110932, 2021.

[378] S. Jin, R. J. Patton, and B. Guo. Viscosity effect on a point absorber wave energy converter hydrodynamics validated by simulation and experiment. *Renewable Energy*, 129:500–512, 2018.

[379] S. Jin, R. J. Patton, and B. Guo. Enhancement of wave energy absorption efficiency via geometry and power take-off damping tuning. *Energy*, 169:819–832, 2019.

[380] L. Johanning and G. Smith. Improved measurement technologies for floating wave energy converter (WEC) mooring arrangements. *Underwater Technology*, 27(4):175–184, 2008.

[381] L. Johanning, G. H. Smith, and J. Wolfram. Measurements of static and dynamic mooring line damping and their importance for floating WEC devices. *Ocean Engineering*, 34(14-15):1918–1934, 2007.

[382] S. John Ashlin, S. Sannasiraj, and V. Sundar. Hydrodynamic performance of an array of oscillating water column device exposed to oblique waves. In *Proceedings of the 12th International Conference on Hydrodynamics (ICHD)*, Egmond aan Zee, The Netherlands, 2016.

[383] J. Joslin, E. Cotter, P. Murphy, P. Gibbs, R. Cavagnaro, C. Crisp, A. R. Stewart, B. Polagye, P. S. Cross, E. Hjetland, et al. The wave-powered adaptable monitoring package: hardware design, installation, and deployment. In *Proceedings of the European Wave and Tidal Energy Conference*, pages 1–6, 2019.

[384] C. Josset, A. Babarit, and A. Clément. A wave-to-wire model of the SEAREV wave energy converter. *Proceedings of the institution of mechanical engineers, Part M: Journal of Engineering for the Maritime Environment*, 221(2):81–93, 2007.

[385] J. Joubert and J. Van Niekerk. Recent developments in wave energy along the coast of Southern Africa. In *Proceedings of the 8th European Wave Tidal Energy Conference*, Uppsala, Sweden, 2009.

[386] J. Joubert and J. L. Van Niekerk. Designing the ShoreSWEC as a breakwater and wave energy converter. In *Proceedings of the CRSES Annual Student Symposium*, 2011.

[387] M. Jusoh, M. Ibrahim, M. Daud, A. Albani, and Z. Mohd Yusop. Hydraulic power take-off concepts for wave energy conversion system: a review. *Energies*, 12(23):4510, 2019.

[388] H. Kagemoto, M. Murai, M. Saito, B. Molin, and S. Malenica. Experimental and theoretical analysis of the wave decay along a long array of vertical cylinders. *Journal of Fluid Mechanics*, 456:113, 2002.

[389] H. Kagemoto and D. Yue. Interactions among multiple three-dimensional bodies in water waves: an exact algebraic method. *Journal of Fluid Mechanics*, 166:189–209, 1986.

[390] S. Kalogirou. Chapter 1 - introduction. In S. A, editor, *Solar Energy Engineering (Second Edition)*, pages 1–49. Academic Press, Boston, second edition edition, 2014.

[391] M. Kamarlouei, J. Gaspar, M. Calvário, T. Hallak, M. J. Mendes, F. Thiebaut, and C. G. Soares. Experimental analysis of wave energy converters concentrically attached on a floating offshore platform. *Renewable Energy*, 152:1171–1185, 2020.

[392] F. Kara. Time domain prediction of power absorption from ocean waves with wave energy converter arrays. *Renewable Energy*, 92:30–46, 2016.

[393] M. Kashiwagi. Hydrodynamic interactions among a great number of columns supporting a very large flexible structure. *Journal of Fluids and Structures*, 14(7):1013–1034, 2000.

[394] M. Kashiwagi. Wave-induced local steady forces on a column-supported very large floating structure. In *Proceeding of the 11th International Offshore and Polar Engineering Conference, Stavanger, Norway*. International Society of Offshore and Polar Engineers, 2001.

[395] M. Kashiwagi and T. Kohjo. A calculation method for hydrodynamic interactions of multiple bodies supporting a huge floating body. In *Proceeding of the 13th Symposium of Ocean Engineering, Society of Naval Architects of Japan*, pages 247–254, 1995. (in Japanese).

[396] G. M. Katsaounis, S. Polyzos, and S. A. Mavrakos. An experimental study of the hydrodynamic behavior of a TLP platform for a 5MW wind turbine with OWC devices. In *MARINE VII: Proceedings of the VII International Conference on Computational Methods in Marine Engineering*, pages 722–731. CIMNE, 2017.

[397] T. Kelly, J. Campbell, T. Dooley, and J. Ringwood. Modelling and results for an array of 32 oscillating water columns. In *Proceedings of the 10th European Wave and Tidal Energy Conference (EWTEC)*, Aalborg, Denmark, 2013.

[398] C. Khairallah, E. Eid, P. Rahme, and C. B. Mosleh. Analysis of a Wave Roller energy-harvesting device. In *Proceedings of the 3rd International Conference on Advances in Computational Tools for Engineering Applications (ACTEA)*, pages 32–36. IEEE, 2016.

[399] D. Kim, S. Poguluri, H. Ko, H. Lee, and Y. Bae. Numerical and experimental study on linear behavior of Salter's Duck wave energy converter. *Journal of Ocean Engineering and Technology*, 33(2):116–122, 2019.

[400] J. Kim, H.-M. Kweon, W.-M. Jeong, I.-H. Cho, and H.-Y. Cho. Design of the dual-buoy wave energy converter based on actual wave data of East Sea. *International Journal of Naval Architecture and Ocean Engineering*, 7(4):739–749, 2015.

[401] J.-S. Kim, B. W. Nam, K.-H. Kim, S. Park, S. H. Shin, and K. Hong. A numerical study on hydrodynamic performance of an inclined OWC wave energy converter with nonlinear turbine–chamber interaction based on 3D potential flow. *Journal of Marine Science and Engineering*, 8(3):176, 2020.

[402] K.-H. Kim, B. Nam, S. Park, J.-S. Kim, G. Kim, C.-H. Lim, and K. Hong. Initial design of OWC WEC applicable to breakwater in remote island. In *Proceedings of the 4th Asian Wave Tidal Energy Conference, Taipei, Taiwan*, pages 9–13, 2018.

[403] D. Kisacik, V. Stratigaki, M. Wu, L. Cappietti, I. Simonetti, P. Troch, A. Crespo, C. Altomare, J. Domínguez, M. Hall, et al. Efficiency and survivability of a floating oscillating water column wave energy converter moored to the seabed: An overview of the EsflOWC MaRINET2 database. *Water*, 12(4):992, 2020.

[404] J. P. Kofoed. *Wave Overtopping of Marine Structures: utilization of wave energy*. PhD thesis, Aalborg University, Denmark, 2002.

[405] J. P. Kofoed. Model testing of the wave energy converter Seawave Slot-Cone Generator. Technical report, Department of Civil Engineering, Aalborg University, 2005.

[406] J. P. Kofoed. The wave energy sector. In *Handbook of Ocean Wave Energy*, pages 17–42. Springer, Cham, 2017.

[407] J. P. Kofoed and M. P. Antonishen. Hydraulic evaluation of the Crest Wing wave energy converter. Technical Report 42, Department of Civil Engineering, Aalborg University, 2008.

[408] J. P. Kofoed and M. P. Antonishen. The Crest Wing wave energy device: 2nd phase testing. Technical Report 59, Department of Civil Engineering, Aalborg University, 2009.

[409] J. P. Kofoed, P. Frigaard, E. Friis-Madsen, and H. C. Sørensen. Prototype testing of the wave energy converter Wave Dragon. *Renewable Energy*, 31(2):181–189, 2006.

[410] J. P. Kofoed, D. Vicinanza, and E. Osaland. Estimation of design wave loads on the SSG WEC pilot plant based on 3-D model tests. In *Proceedings of the 16th International Offshore and Polar Engineering Conference*. OnePetro, 2006.

[411] J. Kolbusz, D. Harrowfield, M. Cocho, and D. W. Velasco. Integration of wave measurement device into a wave energy array. In *Proceedings of the Asian Wave and Tidal Energy Conference (AWTEC)*, Singapore, 2016.

[412] S. Koley and K. Trivedi. Mathematical modeling of Oyster wave energy converter device. In *Proceedings of the AIP Conference Proceedings*, volume 2277, page 130014. AIP Publishing LLC, 2020.

[413] D. Konispoliatis and S. Mavrakos. Hydrodynamics and power absorption characteristics of free floating and moored arrays of OWC's devices. In *Proceedings of the ASME 2014 33rd International Conference on Ocean, Offshore and Arctic Engineering*, San Francisco, USA, 2014.

[414] D. Konispoliatis and S. Mavrakos. Hydrodynamic analysis of an array of interacting free-floating oscillating water column (OWC's) devices. *Ocean Engineering*, 111:179–197, 2016.

[415] W. Koo and M.-H. Kim. Nonlinear time-domain simulation of a land-based oscillating water column. *Journal of Waterway, Port, Coastal, and Ocean Engineering*, 136(5):276–285, 2010.

[416] U. A. Korde, J. Lyu, R. D. Robinett III, D. G. Wilson, G. Bacelli, and O. O. Abdelkhalik. Constrained near-optimal control of a wave energy converter in three oscillation modes. *Applied Ocean Research*, 69:126–137, 2017.

[417] T. Kovaltchouk, S. Armstrong, A. Blavette, H. Ahmed, and B. Multon. Wave farm flicker severity: Comparative analysis and solutions. *Renewable Energy*, 91:32–39, 2016.

[418] D. R. B. Kraemer. *The motions of hinged-barge systems in regular seas*. PhD thesis, The Johns Hopkins University, 2001.

[419] M. Kramer, J. Andersen, S. Thomas, F. B. Bendixen, and et al. Highly accurate experimental heave decay tests with a floating sphere: a public benchmark dataset for model validation of fluid-structure interaction. *Energies*, 14:269, 2021.

[420] M. Kramer, L. Marquis, P. Frigaard, Performance evaluation of the WaveStar prototype. In *Proceedings of the 9th European Wave and Tidal Energy Conference (EWTEC)*, pages 5–9, 2011.

[421] K. Krishnanand and D. Ghose. Glowworm swarm optimisation for simultaneous capture of multiple local optima of multimodal functions. *Swarm Intelligence*, 3(2):87–124, 2009.

[422] E. Kristiansen, Å. Hjulstad, and O. Egeland. State-space representation of radiation forces in time-domain vessel models. *Ocean Engineering*, 32(17):2195–2216, 2005.

[423] V. Krivtsov and B. Linfoot. Basin testing of wave energy converters in trondheim: investigation of mooring loads and implications for wider research. *Journal of Marine Science and Engineering*, 2(2):326–335, 2014.

[424] S. Krüner and C. Hackl. Experimental identification of the optimal current vectors for a permanent-magnet synchronous machine in wave energy converters. *Energies*, 12(5):862, 2019.

[425] K. Kudo. Optimal design of Kaimei-type wave power absorber. *Journal of the Society of Naval Architects of Japan*, 1984(156):245–254, 1984.

[426] S.-Y. Kung. A new identification and model reduction algorithm via singular value decomposition. In *Proceedings of the 12th Asilomar conference on circuits, systems and computers, Pacific Grove*, pages 705–714, 1978.

[427] A. Kurniawan, M. Grassow, and F. Ferri. Numerical modelling and wave tank testing of a self-reacting two-body wave energy device. *Ships and Offshore Structures*, pages 1–13, 2019.

[428] S. Lattanzio and J. Scruggs. Maximum power generation of a wave energy converter in a stochastic environment. In *Proceedings of the IEEE international conference on control applications (CCA)*, pages 1125–1130, Denver, CO, 2011. IEEE.

[429] S. M. Lau and G. E. Hearn. Suppression of irregular frequency effects in fluid–structure interaction problems using a combined boundary integral equation method. *International Journal for Numerical Methods in Fluids*, 9(7):763–782, 1989.

[430] G. Lavidas and H. Polinder. Wave energy in the Netherlands: Past, present and future perspectives. In *Proceedings of the 13th European Wave and Tidal Energy Conference*, page 1226. EWTEC, 2019.

[431] C.-H. Lee. *WAMIT theory manual*. Massachusetts Institute of Technology, Department of Ocean Engineering, 1995.

[432] C.-H. Lee, J. N. Newman, and X. Zhu. An extended boundary integral equation method for the removal of irregular frequency effects. *International Journal for Numerical Methods in Fluids*, 23(7):637–660, 1996.

[433] J. Lee and F. Zhao. GWEC global wind report 2019. Technical report, Wind Global Energy Council, 2020.

[434] M. J. Legaz, D. Coronil, P. Mayorga, and J. Fernández. Study of a hybrid renewable energy platform: W2Power. In *Proceedings of the International Conference on Offshore Mechanics and Arctic Engineering*, volume 51326, page V11AT12A040. American Society of Mechanical Engineers, 2018.

[435] M. Leijon, C. Boström, O. Danielsson, S. Gustafsson, K. Haikonen, O. Langhamer, E. Strömstedt, M. Stålberg, J. Sundberg, O. Svensson, S. Tyrberg, and R. Waters. Wave energy from the north sea: Experiences from the Lysekil research site. *Surveys in Geophysics*, 29(3):221–240, 2008.

[436] M. Leijon, O. Danielsson, M. Eriksson, K. Thorburn, H. Bernhoff, J. Isberg, J. Sundberg, I. Ivanova, E. Sjöstedt, O. Ågren, et al. An electrical approach to wave energy conversion. *Renewable Energy*, 31(9):1309–1319, 2006.

[437] E. Lejerskog, C. Boström, L. Hai, R. Waters, and M. Leijon. Experimental results on power absorption from a wave energy converter at the Lysekil wave energy research site. *Renewable Energy*, 77:9–14, 2015.

[438] E. Lemos, N. Haraczy, and A. Shah. Estimating the generating efficiency of the AquaBuOY 2.0. *Mathematical Methods in Engineering Penn State Harrisburg, Middletown, PA*, 17057, 2018.

[439] A. R. Lesemann and E. J. Hammagren. Searay autonomous offshore power system AOPS: Results of sea trials and payload support demonstration. In *Proceedings of the Offshore Technology Conference*. OnePetro, 2021.

[440] B. Li, D. E. Macpherson, and J. K. H. Shek. Direct drive wave energy converter control in irregular waves. In *Proceedings on the IET Conference on Renewable Power Generation (RPG 2011)*, pages 1–6, Edinburgh, UK, 2011.

[441] G. Li and M. Belmont. Model predictive control of sea wave energy converters–Part II: The case of an array of devices. *Renewable Energy*, 68:540–549, 2014.

[442] L. Li, Y. Gao, D. Ning, and Z. Yuan. Development of a constraint non-causal wave energy control algorithm based on artificial intelligence. *Renewable and Sustainable Energy Reviews*, page 110519, 2020.

[443] X. Li, C. Chen, Q. Li, L. Xu, C. Liang, K. Ngo, R. G. Parker, and L. Zuo. A compact mechanical power take-off for wave energy converters: Design, analysis, and test verification. *Applied Energy*, 278:115459, 2020.

[444] X.-M. Li, S.-T. Quek, and C.-G. Koh. Stochastic response of offshore platforms by statistical cubicization. *Journal of Engineering Mechanics*, 121(10):1056–1068, 1995.

[445] Y. Li and C. Mei. Bragg scattering by a line array of small cylinders in a waveguide. Part 1. Linear aspects. *Journal of Fluid Mechanics*, 583:161–187, 2007.

[446] H. Liang, C. Ouled Housseine, X. B. Chen, and Y. Shao. Efficient methods free of irregular frequencies in wave and solid/porous structure interactions. *Journal of Fluids and Structures*, 98:103130, 2020.

[447] H. Liang, H. Wu, and F. Noblesse. Validation of a global approximation for wave diffraction-radiation in deep water. *Applied Ocean Research*, 74:80–86, 2018.

[448] C. Linton. Rapidly convergent representations for Green functions for Laplace equation. *Proceedings of the Royal Society of London. Series A: Mathematical, Physical and Engineering Sciences*, 455(1985):1767–1797, 1999.

[449] C. Linton and D. Evans. The interaction of waves with a row of circular cylinders. *Journal of Fluid Mechanics*, 251:687–708, 1993.

[450] C. M. Linton and P. McIver. *Handbook of mathematical techniques for wave/structure interactions*. CRC Press, 2001.

[451] C. Liu. Current research status and challenge for direct-drive wave energy conversions. *IETE Journal of Research*, pages 1–13, 2021.

[452] C. Liu, M. Hu, Z. Zhao, Y. Zeng, W. Gao, J. Chen, H. Yan, J. Zhang, Q. Yang, G. Bao, S. Chen, D. Wei, and S. Min. Latching control of a raft-type wave energy converter with a hydraulic power take-off system. *Ocean Engineering*, 236:109512, 2021.

[453] J. Liu, A. Guo, Q. Fang, H. Li, H. Hu, and P. Liu. Wave action by arrays of vertical cylinders with arbitrary smooth cross-section. *Journal of Hydrodynamics*, pages 1–11, 2019.

[454] Y. Liu. HAMS: a frequency-domain preprocessor for wave-structure interactions—theory, development, and application. *Journal of Marine Science and Engineering*, 7(3):81, 2019.

[455] Y. Liu, P. Cong, Y. Gou, S. Yoshida, and M. Kashiwagi. Enhanced Endo's approach for evaluating free-surface Green's function with application to wave-structure interactions. *Ocean Engineering*, 207:107377, 2020.

[456] Y. Liu, H. Iwashita, and C. Hu. A calculation method for finite depth free-surface Green function. *International Journal of Naval Architecture and Ocean Engineering*, 7(2):375–389, 2015.

[457] Y. Liu, H. Liang, M. Kashiwagi, and P. Cong. Alternative approaches of evaluating diffraction transfer matrix and radiation characteristics using the hybrid source-dipole formulation. *Applied Ocean Research*, 114:102769, 2021.

[458] Y. Liu, S. Yoshida, C. Hu, M. Sueyoshi, L. Sun, J. Gao, P. Cong, and G. He. A reliable open-source package for performance evaluation of floating renewable energy systems in coastal and offshore regions. *Energy Conversion and Management*, 174:516–536, 2018.

[459] Z. Liu, B.-S. Hyun, J. Jin, and K. Hong. Practical evaluation method on the performance of pilot OWC system in Korea. In *Proceedings of the Twenty-first International Offshore and Polar Engineering Conference*. OnePetro, 2011.

[460] Z. Liu, Y. Wang, and X. Hua. Proposal of a novel analytical wake model and array optimisation of oscillating wave surge converter using differential evolution algorithm. *Ocean Engineering*, 219:108380, 2021.

[461] Z. Liu, R. Zhang, H. Xiao, and X. Wang. Survey of the mechanisms of power take-off (PTO) devices of wave energy converters. *Acta Mechanica Sinica*, 36:644–658, 2020.

[462] T. T. Loh, D. Greaves, T. Maeki, M. Vuorinen, D. Simmonds, and A. Kyte. Numerical modelling of the WaveRoller device using OpenFOAM. In *Proceedings of the 3rd Asian Wave & Tidal Energy Conference*, 2016.

[463] I. López, J. Andreu, S. Ceballos, I. M. de Alegría, and I. Kortabarria. Review of wave energy technologies and the necessary power-equipment. *Renewable and Sustainable Energy Reviews*, 27:413–434, 2013.

[464] I. López, R. Carballo, D. M. Fouz, and G. Iglesias. Design selection and geometry in OWC wave energy converters for performance. *Energies*, 14(6):1707, 2021.

[465] I. López, R. Carballo, and G. Iglesias. Site-specific wave energy conversion performance of an oscillating water column device. *Energy Conversion and Management*, 195:457–465, 2019.

[466] A. López-Ruiz, R. Bergillos, J. Raffo-Caballero, and M. Ortega-Sánchez. Towards an optimum design of wave energy converter arrays through an integrated approach of life cycle performance and operational capacity. *Applied Energy*, 209:20–32, 2018.

[467] B. Lotfi and L. Huang. A novel wave energy converter using the Stewart platform. *Journal of Green Engineering*, 4(1):33–48, 2013.

[468] J. Lucas, M. Livingstone, M. Vuorinen, and J. Cruz. Development of a wave energy converter (WEC) design tool–application to the WaveRoller WEC including validation of numerical estimates. In *Proceedings of the 4th International Conference on Ocean Energy*, volume 17, 2012.

[469] T. Lykke Andersen, M. Kramer, and P. Frigaard. AquaBuOY. model tests at Aalborg University. Technical report, Aalborg Univiversity, Denmark, 2003.

[470] J. Lyu, O. Abdelkhalik, and L. Gauchia. Optimisation of dimensions and layout of an array of wave energy converters. *Ocean Engineering*, 192:106543, 2019.

[471] I. López, J. Andreu, S. Ceballos, I. Martínez de Alegría, and I. Kortabarria. Review of wave energy technologies and the necessary power-equipment. *Renewable and Sustainable Energy Reviews*, 27:413–434, 2013.

[472] L.-M. Macadré, F. McAuliffe, O. Keysan, M. Donovan, S. Armstrong, J. Murphy, and K. Lynch. Optimal power aggregation methods for marine renewable energy converters; a combined economic and reliability approach. In *Proceedings of the 11th European Wave and Tidal Energy Conference (EWTEC)*, Nantes, France, 2015.

[473] E. Mackay. Consistent expressions for the free-surface Green function in finite water depth. *Applied Ocean Research*, 93:101965, 2019.

[474] E. Mackay, J. Cruz, M. Livingstone, and P. Arnold. Validation of a time-domain modelling tool for wave energy converter arrays. *Proceedings of the European Wave and Tidal Energy Conference (EWTEC)*, 2013.

[475] E. Mackay, J. Cruz, C. Retzler, P. Arnold, E. Bannon, and R. Pascal. Validation of a new wave energy converter design tool with large scale single machine experiments. In *Proceedings of the 1st Asian wave and tidal conference series*, 2012.

[476] A. Madrigal. *Powering the dream: The history and promise of green technology.* Da Capo Press, 2011.

[477] D. Magagna. Ocean energy technology development report 2018, EUR 29907, EN. Technical report, Technical report Publications Office of the European Union, 2019.

[478] D. Magagna. Ocean energy - technology development report 2020 - JRC123159. Technical report, Publications Office of the European Union, Luxembourg, 2020. EUR 30509 EN.

[479] D. Magagna, D. Carr, D. Stagonas, A. Mcnabola, L. Gill, and G. Muller. Experimental evaluation of the performances of an array of multiple oscillating water columns. In *Proceedings of the 9th European Wave and Tidal Energy Conference (EWTEC)*, Southampton, UK, 2011.

[480] D. Magagna, L. Margheritini, A. Alessi, E. Bannon, E. Boelman, D. Bould, V. Coy, E. De Marchi, P. Frigaard, C. Soares, et al. Workshop on identification of future emerging technologies in the ocean energy sector: JRC conference and workshop reports. In *Proceedings of the JRC Conference and Workshop: Workshop on identification of future emerging technologies in the ocean energy sector*. European Commission* Office for Official Publications of the European Union, 2018.

[481] D. Magagna, L. Margheritini, A. Moro, and P. Schild. Considerations on future emerging technologies in the ocean energy sector. In *Advances in Renewable Energies Offshore: Proceedings of the 3rd International Conference on Renewable Energies Offshore (RENEW)*, page 3, Lisbon, Portugal, October 2018. CRC Press.

[482] D. Magagna, R. Monfardini, and A. Uihlein. JRC ocean energy status report 2016 edition. *Publications Office of the European Union: Luxembourg*, 2016.

[483] D. Magagna, D. Stagonas, and G. Muller. Physical investigation into an array of onshore OWCPs designed for water delivery. In *Proceedings of the World Renewable Energy Congress-Sweden; 8-13 May; 2011; Linköping; Sweden*, 057, pages 2206–2213. Linköping University Electronic Press, 2011.

[484] D. Magagna and A. Uihlein. Ocean energy development in Europe: Current status and future perspectives. *International Journal of Marine Energy*, 11:84–104, 2015.

[485] F. Mahnamfar and A. Altunkaynak. Comparison of numerical and experimental analyses for optimizing the geometry of OWC systems. *Ocean Engineering*, 130:10–24, 2017.

[486] C. Maisondieu. WEC survivability threshold and extractable wave power. In *Proceedings of the 11th European Wave Tidal Energy Conference (EWTEC)*, Nantes, France, 2015.

[487] K. Mala, J. Jayaraj, J. Ninoy, V. Jayashankar, Techno-commercial study of twin unidirectional turbine based wave energy plants. In *Proceedings of the Eighth ISOPE Ocean Mining Symposium.* International Society of Offshore and Polar Engineers, 2009.

[488] P. Malali and K. Marchand. Assessment of currently available ocean wave energy conversion systems using technology readiness levels. *International Journal of Renewable Energy Technology*, 11(2):126–146, 2020.

[489] O. Malmo and A. Reitan. Development of the Kvaerner multiresonant OWC. In *Hydrodynamics of Ocean Wave-Energy Utilization*, pages 57–67. Springer, 1986.

[490] R. Manasseh, S. Sannasiraj, K. McInnes, V. Sundar, and P. Jalihal. Integration of wave energy and other marine renewable energy sources with the needs of coastal societies. *The International Journal of Ocean and Climate Systems*, 8(1):19–36, 2017.

[491] H. Mankle, Y. Yu, and B. DuPont. WEC-Sim array simulation development and experimental comparison study. In *Proceedings of the 13th European Wave and Tidal Conference (EWTEC)*, Naples, Italy, 2019.

[492] H. Mankle, Y.-H. Yu, and B. DuPont. WEC-Sim array development and experimental comparison study. Technical report, National Renewable Energy Lab.(NREL), Golden, CO (United States), 2019.

[493] L. D. Mann. Application of ocean observations & analysis: The CETO wave energy project. In *Operational Oceanography in the 21st Century*, pages 721–729. Springer, 2011.

[494] P. Maragos, J. F. Kaiser, and T. F. Quatieri. Energy separation in signal modulations with application to speech analysis. *IEEE Transactions on Signal Processing*, 41(10):3024–3051, 1993.

[495] L. Margheritini, P. Frigaard, and G. Iglesias. Technological and commercial comparison of OWC and SSG wave energy converters built into breakwaters. In *Developments in Renewable Energies Offshore*, pages 167–178. CRC Press, 2020.

[496] A. Maria-Arenas, A. J. Garrido, E. Rusu, and I. Garrido. Control strategies applied to wave energy converters: state of the art. *Energies*, 12(16):3115, 2019.

[497] Marine Power Systems. WaveSub: Class leading performance, 2021. Accessed on June 2021.

[498] Marine Renewables Infrastructure Network. *REWEC3-Electrical PTO system Optimisation*, 2014.

[499] L. Marquis, M. Kramer, and P. Frigaard. First power production figures from the Wave Star Roshage wave energy converter. In *Proceedings of the 3rd International Conference on Ocean Energy (ICOE-2010), Bilbao, Spain*, volume 68, 2010.

[500] G. Marsh. Unbridled power. *Renewable Energy Focus*, 15(1):32–35, 2014.

[501] L. Martinelli, P. Ruol, E. Fassina, F. Giuliani, and N. Delmonte. A wave-2-wire experimental investigation of the new" Seabreath wave energy converter: The hydraulic response. *Coastal Engineering Proceedings*, 34:29–29, 2014.

[502] L. Martinelli and B. Zanuttigh. Effects of mooring compliancy on the mooring forces, power production, and dynamics of a floating wave activated body energy converter. *Energies*, 11(12), 2018.

[503] Y. Masuda. An experience of wave power generator through tests and improvement. In *Hydrodynamics of ocean wave-energy utilization*, pages 445–452. Springer, 1986.

[504] Y. Masuda and T. Miyazaki. Wave power electric generation study in Japan. In *Proceedings of the International Symposium on Wave and Tidal Energy*, volume 1, pages B6_85–B6_92, 1978.

[505] Y. Masuda and T. Miyazaki. The sea trial of the wave power generator'Kaimei'. *Energy Developments in Japan*, 3:165–179, 1980.

[506] Y. Masuda, T. Yamazaki, Y. Outa, and M. McCormick. Study of Backward Bent Duct Buoy. In *Proceedings of the OCEANS'87*, pages 384–389. IEEE, 1987.

[507] S. Mavrakos. Hydrodynamic coefficients for groups of interacting vertical axisymmetric bodies. *Ocean Engineering*, 18(5):485–515, 1991.

[508] S. Mavrakos and A. Kalofonos. Power absorption by arrays of interacting vertical axisymmetric wave-energy devices. *Journal of Offshore Mechanics and Arctic Engineering*, 119(4):244–251, 1997.

[509] S. Mavrakos and P. Koumoutsakos. Hydrodynamic interaction among vertical axisymmetric bodies restrained in waves. *Applied Ocean Research*, 9(3):128–140, 1987.

[510] S. Mavrakos and P. McIver. Comparison of methods for computing hydrodynamic characteristics of arrays of wave power devices. *Applied Ocean Research*, 19(5-6):283–291, 1997.

[511] R. Mayon. *Investigation of wave impacts on porous structures for coastal defences*. PhD thesis, University of Southampton, 2017.

[512] R. Mayon, D. Ning, C. Zhang, L. Chen, and R. Wang. Wave energy capture by an omnidirectional point sink oscillating water column system. *Applied Energy*, 304:117795, 2021.

[513] R. Mayon and D.-Z. Ning. A nonlinear dual phase numerical model to investigate the efficiency response of varied OWC chamber geometries. In *Proceedings of the International Conference on Offshore Mechanics and Arctic Engineering*, volume 84416, page V009T09A017. American Society of Mechanical Engineers, 2020.

[514] P. Mayorga, J. E. Hanssen, R. Hezari, T. Davey, J. Steynor, and D. Ingram. Experimental validation of the w2power hybrid floating platform. In *13th Deep Sea Offshore Wind R&D Conference*, Trondheim, Norway, January 2016.

[515] T. Mazarakos, D. Konispoliatis, G. Katsaounis, S. Polyzos, D. Manolas, S. Voutsinas, T. Soukissian, and S. A. Mavrakos. Numerical and experimental studies of a multi-purpose floating TLP structure for combined wind and wave energy exploitation. *Mediterranean Marine Science*, 20(4):745–763, 2019.

[516] F. D. McAuliffe and G. Dalton. Report of existing installation, operations, maintenance and logistics model available best suited for wave and tidal applications. Technical Report 8.3, MaREI, University College Cork, 2020.

[517] M. McCormick. *Ocean Wave Energy Conversion*. Dover Civil and Mechanical Engineering Series. Dover Publications, 2007.

[518] M. McCormick, J. Murthagh, and P. McCab. Large-scale experimental study of a hinged-barge wave energy conversion system. In *Proceedings of the 3rd European Wave Energy Conference Patras*, pages 215–222, Greece, 1998.

[519] J. McGuinness and G. Thomas. Optimal arrangements of elementary arrays of wave-power devices. In *Proceedings of the 11th European Wave and Tidal Energy Conference (EWTEC)*, Nantes, France, 2015.

[520] J. McGuinness and G. Thomas. Hydrodynamic optimisation of small arrays of heaving point absorbers. *Journal of Ocean Engineering and Marine Energy*, 2(4):439–457, 2016.

[521] J. McGuinness and G. Thomas. The constrained optimisation of small linear arrays of heaving point absorbers. part i: The influence of spacing. *International Journal of Marine Energy*, 20:33–44, 2017.

[522] J. McGuinness and G. Thomas. Optimisation of wave-power arrays without prescribed geometry over incident wave angle. *International Marine Energy Journal*, 4(1):1–10, 2021.

[523] P. McIver. Wave forces on arrays of floating bodies. *Journal of Engineering Mathematics*, 18(4):273–285, 1984.

[524] P. McIver. Some hydrodynamic aspects of arrays of wave-energy devices. *Applied Ocean Research*, 16(2):61–69, 1994.

[525] P. McIver and D. Evans. Approximation of wave forces on cylinder arrays. *Applied Ocean Research*, 6(2):101–107, 1984.

[526] P. McIver and D. Evans. An approximate theory for the performance of a number of wave-energy devices set into a reflecting wall. *Applied Ocean Research*, 10:58, 1988.

[527] P. McIver and D. V. Evans. The occurrence of negative added mass in free-surface problems involving submerged oscillating bodies. *Journal of Engineering Mathematics*, 18(1):7–22, 1984.

[528] J. C. McNatt, A. Porter, and K. Ruehl. Comparison of numerical methods for modeling the wave field effects generated by individual wave energy converters and multiple converter wave farms. *Journal of Marine Science and Engineering*, 8(3):168, 2020.

[529] J. C. McNatt, V. Venugopal, and D. Forehand. The cylindrical wave field of wave energy converters. *International Journal of Marine Energy*, 3:e26–e39, 2013.

[530] J. C. McNatt, V. Venugopal, and D. Forehand. A novel method for deriving the diffraction transfer matrix and its application to multi-body interactions in water waves. *Ocean Engineering*, 94:173–185, 2015.

[531] E. Medina-Lopez, W. Allsop, A. Dimakopoulos, and T. Bruce. Conjectures on the failure of the OWC breakwater at Mutriku. In *Coastal structures and solutions to coastal disasters 2015: Resilient coastal communities*, pages 592–603. American Society of Civil Engineers Reston, VA, 2017.

[532] E. Medina-Lopez, W. Allsop, A. Dimakopoulos, and T. Bruce. Damage to the Mutriku OWC breakwater–some lessons from further analysis. In *Coasts,*

Marine Structures and Breakwaters 2017: Realising the Potential, pages 957–967. ICE Publishing, 2018.

[533] E. Mehlum. Tapchan. In *Hydrodynamics of Ocean Wave-Energy Utilization*, pages 51–55. Springer, 1986.

[534] J. A. Melby and W. Appleton. Evaluation of wave transmission characteristics of OSPREY wave power plant for Noyo Bay, California. Technical report, Army Engineer Waterways Experiment Station Vicksburg MS Coastal Hydraulics Lab, 1997.

[535] A. B. Melo, E. Sweeney, and J. L. Villate. Global review of recent ocean energy activities. *Marine Technology Society Journal*, 47(5):97–103, 2013.

[536] E. Mendoza, R. Silva, B. Zanuttigh, E. Angelelli, T. L. Andersen, L. Martinelli, J. Q. H. Nørgaard, and P. Ruol. Beach response to wave energy converter farms acting as coastal defence. *Coastal Engineering*, 87:97–111, 2014.

[537] P. Mercadé Ruiz, F. Ferri, and J. Kofoed. Experimental validation of a wave energy converter array hydrodynamics tool. *Sustainability*, 9(1):115, 2017.

[538] A. Mérigaud. *A harmonic balance framework for the numerical simulation of non-linear wave energy converter models in random seas*. PhD thesis, National University of Ireland Maynooth, 2018.

[539] A. Merigaud, J.-C. Gilloteaux, and J. V. Ringwood. A nonlinear extension for linear boundary element methods in wave energy device modelling. In *Proceedings of the International Conference on Offshore Mechanics and Arctic Engineering*, pages 1–7, Rio de janeiro, Brazil, 2012. American Society of Mechanical Engineers.

[540] A. Mérigaud and J. V. Ringwood. A nonlinear frequency-domain approach for numerical simulation of wave energy converters. *IEEE Transactions on Sustainable Energy*, 9(1):86–94, 2017.

[541] P.-E. Meunier, A. Clément, J.-C. Gilloteaux, and K. Sofien. Development of a methodology for collaborative control within a WEC array. *International Marine Energy Journal*, 1(1):51–59, 2018.

[542] C. Michailides, Z. Gao, and T. Moan. Experimental and numerical study of the response of the offshore combined wind/wave energy concept SFC in extreme environmental conditions. *Marine Structures*, 50:35–54, 2016.

[543] B. Mills and B. Ding. Improving wave-absorber power. In *Proceedings of the 12th European Wave and Tidal Energy Conference (EWTEC)*, Cork, Ireland, 2017.

[544] S. K. Mishra, D. K. Mohanta, B. Appasani, and E. Kabalci. *OWC-Based Ocean Wave Energy Plants*. Springer, 2021.

[545] T. Miyazaki and Y. Masuda. Tests on the wave power generator "Kaimei". In *Proceedings of the Offshore Technology Conference*, Houston, Texas, 1980.

[546] M. Moarefdoost, L. Snyder, and B. Alnajjab. Layouts for ocean wave energy farms: Models, properties, and optimisation. *Omega*, 66:185–194, 2017.

[547] Mocean Energy. https://www.mocean.energy/#green-energy Green energy out of the blue. 2019.

[548] L. Mofor, J. Goldsmith, and F. Jones. Ocean energy: Technology readiness, patents, deployment status and outlook. Technical report, International Renewable Energy Agency (IRENA) Report, Paris, 2014.

[549] K. Monk. *Forecasting for control and environmental impacts of wave energy converters*. PhD thesis, Plymouth University, 2016.

[550] G. Moretti, M. S. Herran, D. Forehand, M. Alves, H. Jeffrey, R. Vertechy, and M. Fontana. Advances in the development of dielectric elastomer generators for wave energy conversion. *Renewable and Sustainable Energy Reviews*, 117:109430, 2020.

[551] J. Morison, J. Johnson, and S. Schaaf. The force exerted by surface waves on piles. *Journal of Petroleum Technology*, 2(05):149–154, 1950.

[552] M. T. Morris-Thomas, R. J. Irvin, and K. P. Thiagarajan. An Investigation Into the Hydrodynamic Efficiency of an Oscillating Water Column. *Journal of Offshore Mechanics and Arctic Engineering*, 129(4):273–278, 07 2006.

[553] F. Mouwen. Presentation on Wavebob to engineers Ireland. *December, 9th,* 2008.

[554] G. Muller. The Californian wave power craze of the late 19th and early 20th century. In *Proceedings of the 12th European Wave and Tidal Energy Conference (EWTEC)*, Cork, Ireland, 2017.

[555] T. Mundon, B. Nair, and J. Vining. Wave energy converter, 2017. Oscilla Power, Inc., US 20170009732 A1, 12 January.

[556] P.-H. Musiedlak, E. J. Ransley, D. Greaves, M. Hann, G. Iglesias, and B. Child. Investigation of model validity for numerical survivability testing of WECs. In *Proceedings of the 12th European Wave and Tidal Energy Conference (EWTEC)*, Cork, Ireland, 2017.

[557] M. A. Mustapa, O. Yaakob, Y. M. Ahmed, C.-K. Rheem, K. Koh, and F. A. Adnan. Wave energy device and breakwater integration: A review. *Renewable and Sustainable Energy Reviews*, 77:43–58, 2017.

[558] J.-R. Nader, A. Fleming, G. Macfarlane, I. Penesis, and R. Manasseh. Novel experimental modelling of the hydrodynamic interactions of arrays of wave energy converters. *International Journal of marine energy*, 20:109–124, 2017.

[559] J.-R. Nader, S.-P. Zhu, P. Cooper, and B. Stappenbelt. A finite-element study of the efficiency of arrays of oscillating water column wave energy converters. *Ocean Engineering*, 43:72–81, 2012.

[560] S. Nagata, K. Toyota, Y. Yasutaka, T. Setoguchi, Y. Kyozuka, and Y. Masuda. Experimental research on primary conversion of a floating OWC "Backward Bent Duct Buoy". In *Proceedings of the Seventeenth International Offshore and Polar Engineering Conference*. OnePetro, 2007.

[561] A. J. Nambiar, D. I. Forehand, M. M. Kramer, R. H. Hansen, and D. M. Ingram. Effects of hydrodynamic interactions and control within a point absorber array on electrical output. *International Journal of Marine Energy*, 9:20–40, 2015.

[562] NEMOS GmbH. The NEMOS wave energy converter, 2017. Accessed 31 October 2017.

[563] M. Neshat, E. Abbasnejad, Q. Shi, B. Alexander, and M. Wagner. Adaptive neuro-surrogate-based optimisation method for wave energy converters placement optimisation. In *Proceedings of the International Conference on Neural Information Processing*, pages 353–366. Springer, 2019.

[564] M. Neshat, B. Alexander, N. Sergiienko, and M. Wagner. A hybrid evolutionary algorithm framework for optimising power take off and placements of wave energy converters. In *Proceedings of the Genetic and Evolutionary Computation Conference (GECCO'19)*, Prague, Czech Republic, 2019. Preprint arXiv:1904.07043.

[565] M. Neshat, B. Alexander, N. Y. Sergiienko, and M. Wagner. New insights into position optimisation of wave energy converters using hybrid local search. *Swarm and Evolutionary Computation*, 59:100744, 2020.

[566] M. Neshat, B. Alexander, and M. Wagner. A hybrid cooperative co-evolution algorithm framework for optimising power take off and placements of wave energy converters. *Information Sciences*, 534:218–244, 2020.

[567] M. Neshat, B. Alexander, M. Wagner, and Y. Xia. A detailed comparison of meta-heuristic methods for optimising wave energy converter placements. In *Proceedings of the Genetic and Evolutionary Computation Conference (GECCO'18)*, Kyoto, Japan, 2018.

[568] M. Neshat, N. Y. Sergiienko, E. Amini, M. Majidi Nezhad, D. Astiaso Garcia, B. Alexander, and M. Wagner. A new bi-level optimisation framework for optimising a multi-mode wave energy converter design: A case study for the Marettimo Island, Mediterranean Sea. *Energies*, 13(20):5498, 2020.

[569] M. Neshat, N. Y. Sergiienko, S. Mirjalili, M. Majidi Nezhad, G. Piras, and D. Astiaso Garcia. Multi-mode wave energy converter design optimisation using an improved moth flame optimisation algorithm. *Energies*, 14(13):3737, 2021.

[570] J. N. Newman. The exciting forces on fixed bodies in waves. *Journal of Ship Research*, 6(4):10–17, 1962.

[571] J. N. Newman. The interaction of stationary vessels with regular waves. In *Proceedings of the 11th Symposium on Naval Hydrodynamics*, London, 1976.

[572] J. N. Newman. Algorithms for the free-surface Green function. *Journal of Engineering Mathematics*, 19(1):57–67, 1985.

[573] J. N. Newman. Wave effects on deformable bodies. *Applied Ocean Research*, 16(1):47–59, 1994.

[574] H. Nguyen, C. Wang, Z. Tay, and V. Luong. Wave energy converter and large floating platform integration: A review. *Ocean Engineering*, 213:107768, 2020.

[575] H. N. Nguyen and P. Tona. Wave excitation force estimation for wave energy converters of the point-absorber type. *IEEE Transactions on Control Systems Technology*, 26(6):2173–2181, 2017.

[576] H. P. Nguyen and B. Eng. Extracting wave energy while reducing hydroelastic response of very large floating structures. PhD presentation, The University of Queensland, Australia, 2020.

[577] S. R. Nielsen, Q. Zhou, B. Basu, M. T. Sichani, and M. M. Kramer. Optimal control of an array of non-linear wave energy point converters. *Ocean Engineering*, 88:242–254, 2014.

[578] D. Ning, R. Wang, and C. Zhang. Numerical simulation of a dual-chamber oscillating water column wave energy converter. *Sustainability*, 9(9):1599, 2017.

[579] D. Ning, Y. Zhou, and C. Zhang. Hydrodynamic modeling of a novel dual-chamber OWC wave energy converter. *Applied Ocean Research*, 78:180–191, 2018.

[580] D.-Z. Ning, S. Ke, R. Mayon, and C. Zhang. Numerical investigation on hydrodynamic performance of an OWC wave energy device in the stepped bottom. *Frontiers in Energy Research*, 7:152, 2019.

[581] D.-Z. Ning, J. Shi, Q.-P. Zou, and B. Teng. Investigation of hydrodynamic performance of an OWC (oscillating water column) wave energy device using a fully nonlinear HOBEM (higher-order boundary element method). *Energy*, 83:177–188, 2015.

[582] D.-z. Ning, R.-q. Wang, L.-f. Chen, and K. Sun. Experimental investigation of a land-based dual-chamber OWC wave energy converter. *Renewable and Sustainable Energy Reviews*, 105:48–60, 2019.

[583] D.-Z. Ning, R.-Q. Wang, Y. Gou, M. Zhao, and B. Teng. Numerical and experimental investigation of wave dynamics on a land-fixed OWC device. *Energy*, 115:326–337, 2016.

[584] D.-Z. Ning, R.-Q. Wang, Q.-P. Zou, and B. Teng. An experimental investigation of hydrodynamics of a fixed OWC wave energy converter. *Applied Energy*, 168:636–648, 2016.

[585] D.-z. Ning, Y. Zhou, R. Mayon, and L. Johanning. Experimental investigation on the hydrodynamic performance of a cylindrical dual-chamber oscillating water column device. *Applied Energy*, 260:114252, 2020.

[586] I. Noad and R. Porter. Optimisation of arrays of flap-type oscillating wave surge converters. *Applied Ocean Research*, 50:237–253, 2015.

[587] M. K. Ochi. *Ocean waves: the stochastic approach*. 6. Cambridge University Press, 2005.

[588] A. O'Dea, M. C. Haller, and H. T. Özkan-Haller. The impact of wave energy converter arrays on wave-induced forcing in the surf zone. *Ocean Engineering*, 161:322–336, 2018.

[589] T. F. Ogilvie. First-and second-order forces on a cylinder submerged under a free surface. *Journal of Fluid Mechanics*, 16(3):451–472, 1963.

[590] T. F. Ogilvie. Recent progress toward the understanding and prediction of ship motions. In *Proceedings of the 5th Symposium on Naval Hydrodynamics*, volume 1, pages 2–5. Bergen, Norway, 1964.

[591] M. Ohkusu. Hydrodynamic forces on multiple cylinders in waves. In *Proceedings of the International Symposium on the Dynamics of Marine Vehicles and Structures in Waves*. Institute of Mechanical Engineers, 1974.

[592] S. Ohmatsu. A new simple method to eliminate the irregular frequencies in the theory water wave radiation problems. Technical Report 70, Papers of Ship Research Institute, 1983.

[593] H. Ohneda, S. Igarashi, O. Shinbo, S. Sekihara, K. Suzuki, H. Kubota, H. Ogino, and H. Morita. Construction procedure of a wave power extracting caisson breakwater. In *Proceedings of 3rd Symposium on Ocean Energy Utilization*, pages 171–179, 1991.

[594] M. Ohno. Interim report on the second stage of field experiments on a wave power extracting caisson in Sakata Port. In *Proceedings of the International Symposium on Ocean Energy Development*, pages 173–182, 1993.

[595] S. Oliveira-Pinto, P. Rosa-Santos, and F. Taveira-Pinto. Electricity supply to offshore oil and gas platforms from renewable ocean wave energy: Overview and case study analysis. *Energy Conversion and Management*, 186:556–569, 2019.

[596] S. Oliveira-Pinto, P. Rosa-Santos, and F. Taveira-Pinto. Assessment of the potential of combining wave and solar energy resources to power supply worldwide offshore oil and gas platforms. *Energy Conversion and Management*, 223:113299, 2020.

[597] OPT. Ocean power technologies, 08 2021. (Accessed on 08/08/2021).

[598] S. Ornes. Turning water into watts. *Physics World*, 33(3):31, 2020.

[599] J. Orszaghova, H. Wolgamot, S. Draper, R. Eatock Taylor, P. Taylor, and A. Rafiee. Transverse motion instability of a submerged moored buoy. *Proceedings of the Royal Society A*, 475(2221):20180459, 2019.

[600] J. Orszaghova, H. Wolgamot, S. Draper, P. H. Taylor, and A. Rafiee. Onset and limiting amplitude of yaw instability of a submerged three-tethered buoy. *Proceedings of the Royal Society A*, 476(2235):20190762, 2020.

[601] H. Osawa, Y. Washio, T. Ogata, Y. Tsuritani, and Y. Nagata. The offshore floating type wave power device" Mighty Whale" open sea tests performance of the prototype–. In *Proceedings of the Twelfth International Offshore and Polar Engineering Conference*. OnePetro, 2002.

[602] H. T. Özkan-Haller, M. C. Haller, J. C. McNatt, A. Porter, and P. Lenee-Bluhm. Analyses of wave scattering and absorption produced by WEC arrays: physical/numerical experiments and model assessment. In *Marine Renewable Energy*, pages 71–97. Springer, 2017.

[603] L. O'Boyle, K. Doherty, J. van't Hoff, and J. Skelton. The value of full scale prototype data-testing Oyster 800 at EMEC, Orkney. In *Proceedings of the 11th European Wave and Tidal Energy Conference (EWTEC), Nantes, France*, pages 6–11, 2015.

[604] L. O'Boyle, B. Elsäßer, and T. Whittaker. Experimental measurement of wave field variations around wave energy converter arrays. *Sustainability*, 9(1):70, 2017.

[605] A. Palha, L. Mendes, C. J. Fortes, A. Brito-Melo, and A. Sarmento. The impact of wave energy farms in the shoreline wave climate: Portuguese pilot zone case study using Pelamis energy wave devices. *Renewable Energy*, 35(1):62–77, 2010.

[606] J. Palm, C. Eskilsson, G. Paredes, and L. Bergdahl. Coupled mooring analysis for floating wave energy converters using CFD: Formulation and validation. *International Journal of Marine Energy*, 16:83–99, 2016.

[607] G. Palma, P. Contestabile, B. Zanuttigh, S. M. Formentin, and D. Vicinanza. Integrated assessment of the hydraulic and structural performance of the OBREC device in the Gulf of Naples, Italy. *Applied Ocean Research*, 101:102217, 2020.

[608] S. Park, K.-H. Kim, B.-W. Nam, J.-S. Kim, and K. Hong. Experimental and numerical analysis of performance of oscillating water column wave energy converter applicable to breakwaters. In *Proceedings of the International Conference on Offshore Mechanics and Arctic Engineering*, volume 58899, page V010T09A044. American Society of Mechanical Engineers, 2019.

[609] K. Parsa, M. Mekhiche, J. Sarokhan, and D. Stewart. Performance of OPT's commercial PB3 PowerBuoy during 2016 ocean deployment and comparison to projected model results. In *Proceedings of the International Conference on Offshore Mechanics and Arctic Engineering*, volume 57786, page V010T09A021. American Society of Mechanical Engineers, 2017.

[610] A. Parwal, M. Fregelius, I. Temiz, M. Göteman, J. de Oliveira, C. Boström, and M. Leijon. Energy management for a grid-connected wave energy park through a hybrid energy storage system. *Applied Energy*, 231:399–411, 2018.

[611] B. Pattanaik, Y. N. Rao, D. Leo, and P. Jalihal. Experimental studies on development of power take off system for wave powered navigational buoy. In *Proceedings of the IEEE 13th International Conference on Industrial and Information Systems (ICIIS)*, pages 367–370. IEEE, 2018.

[612] B. Pattanaik, Y. N. Rao, P. K. Murthy, A. Viswanath, and P. Jalihal. Wave powered navigational buoy electrical power assessment during open sea trial. In *Proceedings of the International Conference on Power Electronics & IoT Applications in Renewable Energy and its Control (PARC)*, pages 428–431. IEEE, 2020.

[613] J. S. Pawlowski and D. W. Bass. A theoretical and numerical model of ship motions in heavy seas. *SNAME Transactions*, 99:319, 1991.

[614] A. Pecher, J. P. Kofoed, and E. Angelelli. Experimental study on the WavePiston wave energy converter. Technical Report 73, Department of Civil Engineering, Aalborg University, 2010.

[615] J. Peckolt, T. Runkel, S. Baumann, J. Putz, N. Schneider, J. Wegner, J. A. Lucas, and B. Friedhoff. Infrastructure Access Report: 1:5 scale tests of the NEMOS wave energy converter in natural waves at Nissum Bredning Test Site. Report, Marine Renewable Infrastructure Network, 2015.

[616] R. Pelc and R. M. Fujita. Renewable energy from the ocean. *Marine Policy*, 26(6):471–479, 2002.

[617] R. Pena, J. Clare, and G. Asher. Doubly fed induction generator using back-to-back PWM converters and its application to variable-speed wind-energy generation. *IEE Proceedings-Electric Power Applications*, 143(3):231–241, 1996.

[618] Y. Pena-Sanchez, M. Garcia-Abril, F. Paparella, and J. V. Ringwood. Estimation and forecasting of excitation force for arrays of wave energy devices. *IEEE Transactions on Sustainable Energy*, 9(4):1672–1680, 2018.

[619] Y. Peña-Sanchez, C. Windt, J. Davidson, and J. V. Ringwood. A critical comparison of excitation force estimators for wave-energy devices. *IEEE Transactions on Control Systems Technology*, 28(6):2263–2275, 2019.

[620] M. Penalba, G. Giorgi, and J. V. Ringwood. Mathematical modelling of wave energy converters: a review of nonlinear approaches. *Renewable and Sustainable Energy Reviews*, 78:1188–1207, 2017.

[621] M. Penalba, T. Kelly, and J. Ringwood. Using NEMOH for modelling wave energy converters: A comparative study with WAMIT. In *Proceedings of the 12th European Wave and Tidal Energy Conference (EWTEC)*, Cork, Ireland, 2017.

[622] M. Penalba and J. Ringwood. A review of wave-to-wire models for wave energy converters. *Energies*, 9(7):506, 2016.

[623] M. Penalba and J. V. Ringwood. A high-fidelity wave-to-wire model for wave energy converters. *Renewable Energy*, 134:367–378, 2019.

[624] M. Penalba and J. V. Ringwood. Systematic complexity reduction of wave-to-wire models for wave energy system design. *Ocean Engineering*, 217:107651, 2020.

[625] M. Penalba, I. Touzón, J. Lopez-Mendia, and V. Nava. A numerical study on the hydrodynamic impact of device slenderness and array size in wave energy farms in realistic wave climates. *Ocean Engineering*, 142:224–232, 2017.

[626] W. Peng, Y. Zhang, Q. Zou, X. Yang, Y. Liu, and J. Zhang. Experimental investigation of a triple pontoon wave energy converter and breakwater hybrid system. *IET Renewable Power Generation*, 2021.

[627] N. Pereira, D. de Oliveira Valério, and P. Beirão. ISWEC devices on a wave farm handled by a multi-agent system. *Applied Ocean Research*, 111:102659, 2021.

[628] T. Pérez and T. I. Fossen. Time-vs. frequency-domain identification of parametric radiation force models for marine structures at zero speed. *Modeling, Identification and Control*, 29(1):1–19, 2008.

[629] T. Perez and T. I. Fossen. A Matlab toolbox for parametric identification of radiation-force models of ships and offshore structures. *Modeling, Identification and Control*, 30(1):1, 2009.

[630] T. Perez and T. I. Fossen. Practical aspects of frequency-domain identification of dynamic models of marine structures from hydrodynamic data. *Ocean Engineering*, 38(2):426–435, 2011.

[631] C. Pérez-Collazo, D. Greaves, and G. Iglesias. A review of combined wave and offshore wind energy. *Renewable and Sustainable Energy Reviews*, 42:141–153, 2015.

[632] M. A. Peter and M. H. Meylan. Infinite-depth interaction theory for arbitrary floating bodies applied to wave forcing of ice floes. *Journal of Fluid Mechanics*, 500:145–167, 2004.

[633] A. Pichard, C. Wale, and A. Rafiee. Techno-economical tools for WEC scale optimisation. In *Proceedings of the 13th European Wave and Tidal Energy Conference (EWTEC)*, pages 1729–1–1729–8, 2019.

[634] M. Pidcock. The calculation of Green's functions in three dimensional hydrodynamic gravity wave problems. *International Journal for Numerical Methods in Fluids*, 5(10):891–909, 1985.

[635] W. J. Pierson and L. Moskowitz. A proposed spectral form for fully developed wind seas based on the similarity theory of SA Kitaigorodskii. *Journal of Geophysical Research*, 69(24):5181–5190, 1964.

[636] V. Piscopo, G. Benassai, R. Della Morte, and A. Scamardella. Cost-based design and selection of point absorber devices for the Mediterranean Sea. *Energies*, 11(4):946, 2018.

[637] D. Pizer. Maximum wave-power absorption of point absorbers under motion constraints. *Applied Ocean Research*, 15(4):227–234, 1993.

[638] B. Poppelaars. Mooring and installation of wave energy converter Wavebob. Master's thesis, Delft University of Technology, The Netherlands, 2009.

[639] G. Poteras, G. Deak, A.-G. Baraitaru, M. V. Olteanu, D. S. C. Halin, et al. Bioengineering technologies used for the development and equipment of complex installations to obtain energy from three renewable sources. complex installations for coastal areas. In *Proceedings of the IOP Conference Series: Earth and Environmental Science*, volume 1 of *616*, page 012028. IOP Publishing, 2020.

[640] M. Prado and H. Polinder. 9 - case study of the Archimedes Wave Swing (AWS) direct drive wave energy pilot plant. In M. Mueller and H. Polinder, editors, *Electrical Drives for Direct Drive Renewable Energy Systems*, Woodhead Publishing Series in Energy, pages 195–218. Woodhead Publishing, 2013.

[641] M. Previsic. Cost breakdown structure for WEC. Technical report, Sandia National Laboratory, 2012. [online:] `https://energy.sandia.gov/download/23667/`.

[642] T. Price. James Blyth—Britain's first modern wind power pioneer. *Wind Engineering*, 29(3):191–200, 2005.

[643] C. Pérez-Collazo, D. Greaves, and G. Iglesias. A review of combined wave and offshore wind energy. *Renewable and Sustainable Energy Reviews*, 42:141–153, 2015.

[644] D. Qiao, R. Haider, J. Yan, D. Ning, and B. Li. Review of wave energy converter and design of mooring system. *Sustainability*, 12(19):8251, 2020.

[645] S. Quek, X. Li, and C. Koh. Stochastic response of jack-up platform by the method of statistical quadratization. *Applied Ocean Research*, 16(2):113–122, 1994.

[646] A. Rafiee and J. Fiévez. Numerical prediction of extreme loads on the CETO wave energy converter. In *Proceedings of the 11th European Wave and Tidal Energy Conference*, Nantes, France, 2015.

[647] A. Rafiee, H. Wolgamot, S. Draper, J. Orszaghova, J. Fiévez, and T. Sawyer. Identifying the design wave group for the extreme response of a point absorber wave energy converter. In *Proceedings of the Asian Wave and Tidal Energy Conference (AWTEC), Singapore*, volume 2428, 2016.

[648] S. Raghunathan, C. Tan, and N. Wells. Theory and performance of a Wells turbine. *Journal of Energy*, 6(2):157–160, 1982.

[649] M. Rahm. *Ocean wave energy: underwater substation system for wave energy converters*. PhD thesis, Uppsala University, Sweden, 2010.

[650] M. Rahm, O. Svensson, C. Boström, R. Waters, and M. Leijon. Experimental results from the operation of aggregated wave energy converters. *IET Renewable Power Generation*, 6(3):149–160, 2012.

[651] E. Ransley, D. Greaves, A. Raby, D. Simmonds, and M. Hann. Survivability of wave energy converters using CFD. *Renewable Energy*, 109:235–247, 2017.

[652] E. J. Ransley. *Survivability of wave energy converter and mooring coupled system using CFD*. PhD thesis, Plymouth University, 2015.

[653] E. J. Ransley and et al. Focused wave interactions with floating structures: a blind comparative study. *Proceedings of the Institution of Civil Engineers: Engineering and Computational Mechanics*, 174(1):46–61, 2021.

[654] R. Read and H. Bingham. Time-and frequency-domain comparisons of the Wavepiston wave energy converter. In *Proceedings of 33rd International Workshop on Water Waves and Floating Bodies (IWWWFB)*, Guidel-Plages, France, 2018.

[655] B. G. Reguero, I. J. Losada, and F. J. Méndez. A global wave power resource and its seasonal, interannual and long-term variability. *Applied Energy*, 148:366–380, 2015.

[656] G. Reikard, B. Robertson, and J.-R. Bidlot. Combining wave energy with wind and solar: Short-term forecasting. *Renewable Energy*, 81:442–456, 2015.

[657] J. Ren, P. Jin, Y. Liu, and J. Zang. Wave attenuation and focusing by a parabolic arc pontoon breakwater. *Energy*, 217:119405, 2021.

[658] Rendel Palmer & Tritton and Kennedy & Donkin. United Kingdom wave energy program - consultant's 1981 assessment. Report, UK: Department of Energy, 1982.

[659] E. Renzi and F. Dias. Resonant behaviour of an oscillating wave energy converter in a channel. *Journal of Fluid Mechanics*, 701:482–510, 2012.

[660] E. Renzi and F. Dias. Relations for a periodic array of flap-type wave energy converters. *Applied Ocean Research*, 39:31–39, 2013.

[661] E. Renzi, K. Doherty, A. Henry, and F. Dias. How does Oyster work? the simple interpretation of Oyster mathematics. *European Journal of Mechanics-B/Fluids*, 47:124–131, 2014.

[662] K. Rezanejad, J. Bhattacharjee, and C. Guedes Soares. Analytical and numerical study of dual-chamber oscillating water columns on stepped bottom. *Renewable Energy*, 75:272–282, 2015.

[663] K. Rezanejad, J. Bhattacharjee, and C. G. Soares. Analytical and numerical study of dual-chamber oscillating water columns on stepped bottom. *Renewable Energy*, 75:272–282, 2015.

[664] K. Rezanejad, J. Gadelho, and C. G. Soares. Hydrodynamic analysis of an oscillating water column wave energy converter in the stepped bottom condition using CFD. *Renewable Energy*, 135:1241–1259, 2019.

[665] K. Rezanejad, J. F. Gadelho, I. López, R. Carballo, and C. Guedes Soares. Improving the hydrodynamic performance of OWC wave energy converter by attaching a step. In *Proceedings of the International Conference on Offshore Mechanics and Arctic Engineering*, volume 58899, page V010T09A043. American Society of Mechanical Engineers, 2019.

[666] K. Rezanejad, A. Souto-Iglesias, and C. Guedes Soares. Experimental investigation on the hydrodynamic performance of an L-shaped duct oscillating water column wave energy converter. *Ocean Engineering*, 173:388–398, 2019.

[667] P. Ricci. Time-domain models. In *Numerical Modelling of Wave Energy Converters*, pages 31–66. Elsevier, 2016.

[668] D. P. Rijnsdorp, J. E. Hansen, and R. J. Lowe. Understanding coastal impacts by nearshore wave farms using a phase-resolving wave model. *Renewable Energy*, 150:637–648, 2020.

[669] G. Rinaldi, P. Thies, R. Walker, and L. Johanning. On the analysis of a wave energy farm with focus on maintenance operations. *Journal of Marine Science and Engineering*, 4(3):51, 2016.

[670] J. V. Ringwood. Wave energy control: status and perspectives 2020. *IFAC-PapersOnLine*, 53(2):12271–12282, 2020.

[671] J. V. Ringwood. Wave energy control: Status and perspectives 2020. In *Proceedings of the IFAC World Congress*, pages 1–12, Berlin, Germany, 2020.

[672] J. V. Ringwood, G. Bacelli, and F. Fusco. Energy-maximizing control of wave-energy converters: the development of control system technology to optimize their operation. *IEEE Control Systems*, 34(5):30–55, 2014.

[673] J. V. Ringwood, A. Mérigaud, N. Faedo, and F. Fusco. An analytical and numerical sensitivity and robustness analysis of wave energy control systems. *IEEE Transactions on Control Systems Technology*, 28(4):1337–1348, 2019.

[674] J. Roberts and P. Spanos. *Random vibration and statistical linearization.* Courier Corporation, 2003.

[675] A. Roessling and J. Ringwood. Finite order approximations to radiation forces for wave energy applications. *Renewable Energies Offshore*, page 359, 2015.

[676] A. Roessling and J. V. Ringwood. Finite order approximations to radiation forces for wave energy applications. In *Proceedings of the International Conference on Renewable Energies Offshore*, pages 359–366, Lison, Portugal, 2014.

[677] P. Ropero-Giralda, A. J. Crespo, B. Tagliafierro, C. Altomare, J. M. Domínguez, M. Gómez-Gesteira, and G. Viccione. Efficiency and survivability analysis of a point-absorber wave energy converter using DualSPHysics. *Renewable Energy*, 162:1763–1776, 2020.

[678] B. J. Rosenberg and T. Mundon. Numerical and physical modeling of a flexibly-connected two-body wave energy converter. Report, Oscilla Power, 2016.

[679] K. M. Ruehl, J. D. Roberts, A. Porter, G. Chang, C. C. Chartrand, and H. Smtih. Development verification and application of the SNL-SWAN open source wave farm code. Technical report, Sandia National Lab.(SNL-NM), Albuquerque, NM (United States), 2015.

[680] P. Ruiz, V. Nava, M. Topper, P. Minguela, F. Ferri, and J. Kofoed. Layout optimisation of wave energy converter arrays. *Energies*, 10(9):1262, 2017.

[681] P. Ruol, L. Martinelli, and P. Pezzutto. Multi-chamber OWC devices to reduce and convert wave energy in harbour entrance and inner channels. In *Proceedings of the Twenty-first International Offshore and Polar Engineering Conference*. OnePetro, 2011.

[682] E. Rusu. Evaluation of the wave energy conversion efficiency in various coastal environments. *Energies*, 7(6):4002–4018, 2014.

[683] E. Rusu and C. G. Soares. Coastal impact induced by a Pelamis wave farm operating in the Portuguese nearshore. *Renewable Energy*, 58:34–49, 2013.

[684] Z. Sabeur, W. Roberts, and A. Cooper. Development and use of an advanced numerical model using the volume of fluid method for the design of coastal structures. *Numerical Methods for Fluid Dynamics*, 5:565–573, 1995.

[685] M. Safonov and R. Chiang. A Schur method for balanced model reduction. In *Proceedings of the American Control Conference*, pages 1036–1040. IEEE, 1988.

[686] S. H. Salter. Wave power. *Nature*, 249(5459):720–724, 1974.

[687] S. H. Salter. Progress on Edinburgh Ducks. In *Hydrodynamics of Ocean Wave-Energy Utilization*, pages 35–50. Springer, 1986.

[688] Y. Sang, H. B. Karayaka, Y. Yan, J. Z. Zhang, D. Bogucki, and Y.-H. Yu. A rule-based phase control methodology for a slider-crank wave energy converter power take-off system. *International Journal of Marine Energy*, 19:124–144, 2017.

[689] D. Sarkar, E. Contal, N. Vayatis, and F. Dias. Prediction and optimisation of wave energy converter arrays using a machine learning approach. *Renewable Energy*, 97:504–517, 2016.

[690] D. Sarkar, E. Renzi, and F. Dias. Wave power extraction by an oscillating wave surge converter in random seas. In *Proceedings of the International Conference on Offshore Mechanics and Arctic Engineering*, volume 55423, page V008T09A008. American Society of Mechanical Engineers, 2013.

[691] A. Sarmento. Model-test optimisation of an OWC wave power plant. *International Journal of Offshore and Polar Engineering*, 3(01), 1993.

[692] T. Sarpkaya. Force on a circular cylinder in viscous oscillatory flow at low keulegan—carpenter numbers. *Journal of Fluid Mechanics*, 165:61–71, 1986.

[693] SBM Offshore. `https://www.sbmoffshore.com/newsroom/news-events/sea-trial-principality-monaco-announced-sbm-offshores-innovative-s3r-wave` Sea trial in the Principality of Monaco announced for SBM Offshore's innovative S3 Wave Energy Converter. 2019.

[694] C. Schauder and H. Mehta. Vector analysis and control of advanced static var compensators. In *Proceedings of the IEE C (Generation, Transmission and Distribution)*, number 4 in 140, pages 299–306. IET, 1993.

[695] S. Schlömer, T. Bruckner, L. Fulton, E. Hertwich, A. McKinnon, D. Perczyk, J. Roy, R. Schaeffer, R. Sims, P. Smith, and R. Wiser. Annex III: Technology-specific cost and performance parameters. Report, 2014.

[696] J. Scruggs, S. Lattanzio, A. Taflanidis, and I. Cassidy. Optimal causal control of a wave energy converter in a random sea. *Applied Ocean Research*, 42:1–15, 2013.

[697] Seapower.ie. `http://www.seapower.ie/our-technology/Seapower` Technology. Accessed 20 April 2022.

[698] N. Sergiienko. *Three-tether wave energy converter: hydrodynamic modelling, performance assessment and control.* Doctoral thesis, University of Adelaide, Australia, 2018.

[699] N. Sergiienko, B. Cazzolato, M. Arjomandi, B. Ding, and L. S. P. da Silva. Considerations on the control design for a three-tether wave energy converter. *Ocean Engineering*, 183:469–477, 2019.

[700] N. Sergiienko, B. Cazzolato, B. Ding, P. Hardy, and M. Arjomandi. Performance comparison of the floating and fully submerged quasi-point absorber wave energy converters. *Renewable Energy*, 108:425–437, 2017.

[701] N. Sergiienko, A. Rafiee, B. Cazzolato, B. Ding, and M. Arjomandi. Feasibility study of the three-tether axisymmetric wave energy converter. *Ocean Engineering*, 150:221–233, 2018.

[702] N. Y. Sergiienko, B. S. Cazzolato, B. Ding, and M. Arjomandi. An optimal arrangement of mooring lines for the three-tether submerged point-absorbing wave energy converter. *Renewable Energy*, 93:27–37, 2016.

[703] N. Y. Sergiienko, B. S. Cazzolato, B. Ding, and M. Arjomandi. Three-tether axisymmetric wave energy converter: Estimation of energy delivery. In *Proceedings of the 3rd Asian Wave and Tidal Energy Conference*, volume 1, pages 163–171, Singapore, 2016.

[704] N. Y. Sergiienko, B. S. Cazzolato, P. Hardy, M. Arjomandi, and B. Ding. Internal-model-based velocity tracking control of a submerged three-tether wave energy converter. In *Proceedings of the 12th European Wave and Tidal Energy Conference*, pages 1126-1–1126-8, Cork, Ireland, 2017.

[705] N. Y. Sergiienko, M. Cocho, B. S. Cazzolato, and A. Pichard. Effect of a model predictive control on the design of a power take-off system for wave energy converters. *Applied Ocean Research*, 115:102836, 2021.

[706] N. Y. Sergiienko, L. S. P. da Silva, B. Ding, and B. S. Cazzolato. Importance of drivetrain optimisation to maximise electrical power from wave energy converters. *IET Renewable Power Generation*, 2021.

[707] N. Y. Sergiienko, M. Neshat, L. S. P. da Silva, B. Alexander, and M. Wagner. Design optimisation of a multi-mode wave energy converter. In *Proceedings of the International Conference on Offshore Mechanics and Arctic Engineering*, volume 84416, page V009T09A039. American Society of Mechanical Engineers, 2020.

[708] F. Sharkey, E. Bannon, M. Conlon, and K. Gaughan. Dynamic electrical ratings and the economics of capacity factor for wave energy converter arrays. In *Proceedings of the 9th European Wave and Tidal Energy Conference (EWTEC)*, Southampton, UK, 2011.

[709] F. Sharkey, M. Conlon, and K. Gaughan. Impacts on the electrical system economics from critical design factors of wave energy converters and arrays. In *Proceedings of the 10th European Wave and Tidal Energy Conference (EWTEC)*, Aalborg, Denmark, 2013.

[710] C. Sharp. *Wave Energy Converter Array Optimisation: Algorithm Development and Investigation of Layout Design Influences*. PhD thesis, Oregon State University, 2018.

[711] C. Sharp and B. DuPont. Wave energy converter array optimisation: A genetic algorithm approach and minimum separation distance study. *Ocean Engineering*, 163:148–156, 2018.

[712] C. Sharp, B. DuPont, B. Bosma, P. Lomonaco, and B. Batten. Array optimisation of fixed oscillating water columns for active device control. In *Proceedings of the 12th European Wave and Tidal Energy Conference (EWTEC)*, Cork, Ireland, 2017.

[713] S. Sheng, K. Wang, H. Lin, Y. Zhang, Y. You, Z. Wang, A. Chen, J. Jiang, W. Wang, and Y. Ye. Model research and open sea tests of 100 kW wave energy converter Sharp Eagle Wanshan. *Renewable Energy*, 113:587–595, 2017.

[714] W. Sheng, B. Flannery, A. Lewis, and R. Alcorn. Experimental studies of a floating cylindrical OWC WEC. In *Proceedings of the International Conference on Offshore Mechanics and Arctic Engineering*, volume 44946, pages 169–178. American Society of Mechanical Engineers, 2012.

[715] H. Shi, S. Huang, and F. Cao. Hydrodynamic performance and power absorption of a multi-freedom buoy wave energy device. *Ocean Engineering*, 172:541–549, 2019.

[716] E. H. Shoeib, E. Hamin Infield, and H. Renski. Measuring the impacts of wind energy projects on US rural counties' community services and cost of living. *Energy Policy*, 153:112279, 2021.

[717] P. Siddorn and R. Eatock Taylor. Diffraction and independent radiation by an array of floating cylinders. *Ocean Engineering*, 35(13):1289–1303, 2008.

[718] M. J. Simon. Multiple scattering in arrays of axisymmetric wave-energy devices. Part 1. A matrix method using a plane-wave approximation. *Journal of Fluid Mechanics*, 120:1–25, 1982.

[719] A. Sinha, D. Karmakar, and C. Soares. Performance of optimally tuned arrays of heaving point absorbers. *Renewable Energy*, 92:517–531, 2016.

[720] S. A. Sirigu, L. Foglietta, G. Giorgi, M. Bonfanti, G. Cervelli, G. Bracco, and G. Mattiazzo. Techno-economic optimisation for a wave energy converter via genetic algorithm. *Journal of Marine Science and Engineering*, 8(7):482, 2020.

[721] L. Sjökvist and M. Göteman. Peak forces on a point absorbing wave energy converter impacted by tsunami waves. *Renewable Energy*, 133:1024–1033, 2019.

[722] J. Sjolte, C. M. Sandvik, E. Tedeschi, and M. Molinas. Exploring the potential for increased production from the wave energy converterLifesaver by reactive control. *Energies*, 6(8):3706–3733, 2013.

[723] J. Sjolte, G. Tjensvoll, and M. Molinas. Power collection from wave energy farms. *Applied Sciences*, 3(2):420–436, 2013.

[724] J. Sjolte, G. Tjensvoll, and M. Molinas. Self-sustained all-electric wave energy converter system. *COMPEL: The International Journal for Computation and Mathematics in Electrical and Electronic Engineering*, 2014.

[725] S. Skogestad and I. Postlethwaite. *Multivariable feedback control: Analysis and design*, volume 2. Wiley New York, 2007.

[726] Z. Smith and K. Taylor. *Renewable and alternative energy resources: a reference handbook*. ABC-CLIO, 2008.

[727] L. Snyder and M. Moarefdoost. Optimizing wave farm layouts under uncertainty. In *Proceedings of the 3rd Marine Energy Technology Symposium (METS)*, Washington DC, USA, 2015.

[728] L. Socha. Linearization in analysis of nonlinear stochastic systems: recent results—part I: theory. *Applied Mechanics Reviews*, 58(3):178–205, 2005.

[729] L. Socha. Linearization in analysis of nonlinear stochastic systems, recent results—part II: applications. *Applied Mechanics Reviews*, 58(5):303–315, 2005.

[730] L. Socha. *Linearization methods for stochastic dynamic systems*. Springer, Berlin, 2008.

[731] H. Soerensen, E. Friis-Madsen, W. Panhauser, D. Dunce, J. Nedkvintne, P. B. Frigaard, J. P. Kofoed, W. Knapp, S. Riemann, E. Holmén, et al. Development of Wave Dragon from scale 1: 50 to prototype. In *Proceedings from the Fifth European Wave Energy Conference: Cork, Ireland, 2003*, 2003.

[732] E. Solomin, E. Sirotkin, E. Cuce, S. P. Selvanathan, and S. Kumarasamy. Hybrid floating solar plant designs: A review. *Energies*, 14(10), 2021.

[733] D. Son and R. W. Yeung. Optimizing ocean-wave energy extraction of a dual coaxial-cylinder WEC using nonlinear model predictive control. *Applied Energy*, 187:746–757, 2017.

[734] H. Sorensen, E. Friis-Madsen, L. Christensen, J. P. Kofoed, P. B. Frigaard, and W. Knapp. The results of two years testing in real sea of Wave Dragon. In *Proceedings of 6th European Wave and Tidal Energy Conference: EWTEC 2005: 6th Reflectors to Focus Wave Energy*, pages 481–487. Department of Mechanical Engineering, University of Strathclyde, 2005.

[735] P. Spanos, M. Di Paola, and G. Failla. A galerkin approach for power spectrum determination of nonlinear oscillators. *Meccanica*, 37(1):51–65, 2002.

[736] P. Spanos and M. Donley. Equivalent statistical quadratization for nonlinear systems. *Journal of Engineering Mechanics*, 117(6):1289–1310, 1991.

[737] P. Spanos, F. Strati, G. Malara, and F. Arena. Stochastic dynamic analysis of U-OWC wave energy converters. In *Proceedings of the 36th International Conference on Ocean, Offshore and Arctic Engineering*, pages 1–8. American Society of Mechanical Engineers, 2017.

[738] P. D. Spanos, F. Arena, A. Richichi, and G. Malara. Efficient dynamic analysis of a nonlinear wave energy harvester model. *Journal of Offshore Mechanics and Arctic Engineering*, 138(4):041901, 2016.

[739] D. V. Spitzley and G. A. Keoleian. Life cycle environmental and economic assessment of willow biomass electricity: A comparision with other renewable and non-renewable sources. Report, University of Michigan, Centre for Sustainable Systems, 2005.

[740] B. H. Spring and P. L. Monkmeyer. Interaction of plane waves with vertical cylinders. In *Proceeding of the 14th International Conference on Coastal Engineering*, pages 1828–1847, Copenhagen, Denmark, 1974.

[741] M. A. Srokosz. The submerged sphere as an absorber of wave power. *Journal of Fluid Mechanics*, 95(4):717–741, 1979.

[742] P. Stansby and E. Carpintero Moreno. Study of snap loads for idealized mooring configurations with a buoy, inextensible and elastic cable combinations for the multi-float M4 wave energy converter. *Water*, 12(10), 2020.

[743] P. Stansby, E. Carpintero Moreno, and T. Stallard. Capture width of the three-float multi-mode multi-resonance broadband wave energy line absorber M4 from laboratory studies with irregular waves of different spectral shape and directional spread. *Journal of Ocean Engineering and Marine Energy*, 1:287–298, 2015.

[744] P. Stansby, E. Carpintero Moreno, and T. Stallard. Large capacity multi-float configurations for the wave energy converter M4 using a time-domain linear diffraction model. *Applied Ocean Research*, 68:53–64, 2017.

[745] P. Stansby, E. Carpintero Moreno, T. Stallard, and A. Maggi. Three-float broad-band resonant line absorber with surge for wave energy conversion. *Renewable Energy*, 78:132–140, 2015.

[746] C. Stokes and D. C. Conley. Modelling offshore wave farms for coastal process impact assessment: Waves, beach morphology, and water users. *Energies*, 11(10):2517, 2018.

[747] V. Stratigaki and L. Mouffe. Ocean energy in belgium-2019. Technical report, Ocean Energy Systems (OES), 2019.

[748] V. Stratigaki, P. Troch, T. Stallard, D. Forehand, J. Kofoed, M. Folley, M. Benoit, A. Babarit, and J. Kirkegaard. Wave basin experiments with large wave energy converter arrays to study interactions between the converters and effects on other users in the sea and the coastal area. *Energies*, 7(2):701–734, 2014.

[749] J. Straub. In search of technology readiness level (TRL) 10. *Aerospace Science and Technology*, 46:312–320, 2015.

[750] L. Sun, R. E. Taylor, and Y. S. Choo. Responses of interconnected floating bodies. *The IES Journal Part A: Civil & Structural Engineering*, 4(3):143–156, 2011.

[751] M. Suzuki, C. Arakawa, and S. Takahashi. Performance of wave power generating system installed in breakwater at Sakata port in Japan. In *Proceedings of the Fourteenth International Offshore and Polar Engineering Conference*. OnePetro, 2004.

[752] R. Taghipour, T. Perez, and T. Moan. Hybrid frequency–time domain models for dynamic response analysis of marine structures. *Ocean Engineering*, 35(7):685–705, 2008.

[753] K. Tarrant and C. Meskell. Investigation on parametrically excited motions of point absorbers in regular waves. *Ocean Engineering*, 111:67–81, 2016.

[754] K. R. Tarrant. *Numerical modelling of parametric resonance of a heaving point absorber wave energy converter*. PhD thesis, Trinity College Dublin, 2015.

[755] Z. Tay and V. Venugopal. Hydrodynamic interactions of oscillating wave surge converters in an array under random sea state. *Ocean Engineering*, 145:382–394, 2017.

[756] J. Tedd and P. Frigaard. Short term wave forecasting, using digital filters, for improved control of wave energy converters. In *Proceedings of the 17th International Offshore and Polar Engineering Conference*, Lisbon, Portugal, 2007. International Society of Offshore and Polar Engineers.

[757] J. Tedd and J. P. Kofoed. Measurements of overtopping flow time series on the Wave Dragon, wave energy converter. *Renewable Energy*, 34(3):711–717, 2009.

[758] E. Tedeschi and M. Santos-Mugica. Modeling and control of a wave energy farm including energy storage for power quality enhancement: the bimep case study. *IEEE Transactions on Power Systems*, 29(3):1489–1497, 2013.

[759] B. Teillant, R. Costello, J. Weber, and J. Ringwood. Productivity and economic assessment of wave energy projects through operational simulations. *Renewable Energy*, 48:220–230, 2012.

[760] P. R. Teixeira, D. P. Davyt, E. Didier, and R. Ramalhais. Numerical simulation of an oscillating water column device using a code based on Navier–Stokes equations. *Energy*, 61:513–530, 2013.

[761] J. Telste and F. Noblesse. Numerical evaluation of the Green function of water-wave radiation and diffraction. *Journal of Ship Research*, 30(02):69–84, 1986.

[762] I. Temiz, J. Leijon, B. Ekergård, and C. Boström. Economic aspects of latching control for a wave energy converter with a direct drive linear generator power take-off. *Renewable Energy*, 128:57–67, 2018.

[763] B. Teng and R. E. Taylor. New higher-order boundary element methods for wave diffraction/radiation. *Applied Ocean Research*, 17(2):71–77, 1995.

[764] Tethys. Wello Penguin at `https://tethys.pnnl.gov/project-sites/wello-penguin-emec` EMEC, 08 2021 (Accessed on 08/08/2021).

[765] A. Tetu. Power take-off systems for wecs. In A. Pecher and J. P. Kofoed, editors, *Handbook of ocean wave energy*, pages 203–220. Springer International Publishing, Cham, 2017.

[766] A. Têtu, F. Ferri, M. B. Kramer, and J. Hals Todalshaug. Physical and mathematical modeling of a wave energy converter equipped with a negative spring mechanism for phase control. *Energies*, 11(9):2362, 2018.

[767] A. Tetu, J. P. Kofoed, F. Schlütter, and P. Hammer. Forecast model for current, wave and wind climate at the Danish test site for wave energy, DanWEC. In *Proceedings of the International Conference on Time Series and Forecasting*, pages 1328–1339. Godle Impresiones Digitales SL, 2018.

[768] The State Council Information Office of the People's Republic of China. *Energy in China's New Era*, December 2020.

[769] P. R. Thies, L. Johanning, and T. Gordelier. Component reliability testing for wave energy converters: Rationale and implementation. In *Proceedings of the European Wave and Tidal Energy Conference (EWTEC)*, Aalborg, Denmark, 2013.

[770] S. Thomas, M. Eriksson, M. Göteman, M. Hann, J. Isberg, and J. Engström. Experimental and numerical collaborative latching control of wave energy converter arrays. *Energies*, 11(11):3036, 2018.

[771] S. Thomas, M. Giassi, M. Eriksson, M. Göteman, J. Isberg, E. Ransley, M. Hann, and J. Engström. A model free control based on machine learning for energy converters in an array. *Big Data and Cognitive Computing*, 2(4):36, 2018.

[772] S. Thomas, S. Weller, and T. Stallard. Float response within an array: Numerical and experimental comparison. In *Proceedings of the 2nd International Conference on Ocean Energy (ICOE)*, volume 1517, Brest, France, 2008.

[773] T. Thorpe. An overview of wave energy technologies: status, performance and costs. In *Wave Power: Moving Towards Commercial Viability*. Wiley, 1999.

[774] T. W. Thorpe. A brief review of wave energy. Technical Report ETSU-R120, The UK Department of Trade and Industry, 1999.

[775] J. Tissandier, A. Babarit, A. Clément, et al. Study of the smoothing effect on the power production in an array of SEAREV wave energy converters. In *Proceedings of the 18th International Offshore and Polar Engineering Conference*, 2008.

[776] M. A. Tognarelli, J. Zhao, and A. Kareem. Equivalent statistical cubiciza-tion for system and forcing nonlinearities. *Journal of Engineering Mechanics*, 123(8):890–893, 1997.

[777] M. A. Tognarelli, J. Zhao, K. Rao, and A. Kareem. Equivalent statistical quadratization and cubicization for nonlinear systems. *Journal of Engineering Mechanics*, 123(5):512–523, 1997.

[778] N. Tom, M. Lawson, Y. Yu, and A. Wright. Spectral modeling of an oscillating surge wave energy converter with control surfaces. *Applied Ocean Research*, 56:143–156, 2016.

[779] Y. Torre-Enciso, I. Ortubia, L. L. De Aguileta, and J. Marqués. Mutriku wave power plant: from the thinking out to the reality. In *Proceedings of the 8th European Wave and Tidal Energy Conference (EWTEC)*, volume 710, pages 319–329, Uppsala, Sweden, 2009.

[780] K. Toyota, S. Nagata, Y. Imai, and T. Setoguchi. Effects of hull shape on primary conversion characteristics of a floating OWC" Backward Bent Duct Buoy". *Journal of Fluid Science and Technology*, 3(3):458–465, 2008.

[781] N. Tran, N. Sergiienko, B. Cazzolato, B. Ding, M. Ghayesh, and M. Arjomandi. The impact of pitch-surge coupling on the performance of a submerged cylin-drical wave energy converter. *Applied Ocean Research*, 104:102377, 2020.

[782] N. Tran, N. Y. Sergiienko, B. S. Cazzolato, M. H. Ghayesh, and M. Arjomandi. The effect of nonlinear pitch-surge coupling on the performance of multi-dof submerged wecs. In *Proceedings of the 30th International Ocean and Polar Engineering Conference*. OnePetro, 2020.

[783] P. Troch, V. Stratigaki, T. Stallard, D. Forehand, M. Folley, J. Kofoed, M. Benoit, A. Babarit, D. Sánchez, L. Bosscher, et al. Physical modelling of an array of 25 heaving wave energy converters to quantify variation of re-sponse and wave conditions. In *Proceedings of the 10th European Wave and Tidal Energy Conference (EWTEC)*, Aalborg, Denmark, 2013.

[784] A. Tustin. The effects of backlash and of speed-dependent friction on the sta-bility of closed-cycle control systems. *Journal of the Institution of the Electrical Engineers-Part IIA: Automatic Regulators and Servo Mechanisms*, 94(1):143–151, 1947.

[785] J. Twidell and T. Weir. *Renewable energy resources*. Routledge, 2015.

[786] A. Ulazia, M. Penalba, A. Rabanal, G. Ibarra-Berastegi, J. Ringwood, and J. Sáenz. Historical evolution of the wave resource and energy production off the chilean coast over the 20th century. *Energies*, 11(9):2289, 2018.

[787] L. Ulvgård, L. Sjökvist, M. Göteman, and M. Leijon. Line force and damping at full and partial stator overlap in a linear generator for wave power. *Journal of Marine Science and Engineering*, 4(4):81, 2016.

[788] University of Southampton, Energy & Climate Change. 20 April 2022. `https://energy.soton.ac.uk/anaconda-wave-energy-converter-concept/` Anaconda Wave Energy Converter Concept.

[789] F. Ursell. Surface waves on deep water in the presence of a submerged circular cylinder. I. In *Mathematical Proceedings of the Cambridge Philosophical Society*, volume 46, pages 141–152. Cambridge University Press, 1950.

[790] US Department of Energy. Wave Energy Prize Rules (5.26.15 R1), 2016.

[791] US Department of Energy. Wave Energy Prize - 1/20th Testing - Oscilla Power, 2017.

[792] D. Valério, P. Beirão, and J. da Costa. Optimisation of wave energy extraction with the Archimedes Wave Swing. *Ocean Engineering*, 34(17-18):2330–2344, 2007.

[793] D. Valério, M. J. Mendes, P. Beirão, and J. S. da Costa. Identification and control of the AWS using neural network models. *Applied Ocean Research*, 30(3):178–188, 2008.

[794] D. Valério, M. D. Ortigueira, and J. S. da Costa. Identifying a transfer function from a frequency response. *Journal of Computational and Nonlinear Dynamics*, 3(2):021207, 2008.

[795] M. van Grieken and B. Dower. Chapter 23 - wind turbines and landscape. In T. Letcher, editor, *Wind Energy Engineering*, pages 493–515. Academic Press, 2017.

[796] R. van 't Veer and H. J. Tholen. Added resistance of moonpools in calm water. In *Nick Newman Symposium on Marine Hydrodynamics; Yoshida and Maeda Special Symposium on Ocean Space Utilization; Special Symposium on Offshore Renewable Energy*, volume 6 of *Proceedings of the International Conference on Offshore Mechanics and Arctic Engineering*, pages 153–162, 06 2008.

[797] S. Venafra, M. Morelli, and A. Masini. Satellite remote sensing applied to off-shore wind energy. *Proceedings of the EARSeL eProceedings*, 13(1):1, 2014.

[798] G. Verao Fernández, P. Balitsky, V. Stratigaki, and P. Troch. Coupling methodology for studying the far field effects of wave energy converter arrays over a varying bathymetry. *Energies*, 11(11):2899, 2018.

[799] G. Verao Fernández, V. Stratigaki, and P. Troch. Irregular wave validation of a coupling methodology for numerical modelling of near and far field effects of wave energy converter arrays. *Energies*, 12(3):538, 2019.

[800] T. Verbrugghe, V. Stratigaki, P. Troch, R. Rabussier, and A. Kortenhaus. A comparison study of a generic coupling methodology for modeling wake effects of wave energy converter arrays. *Energies*, 10(11):1697, 2017.

[801] M. G. Verduzco-Zapata and F. J. Ocampo Torres. Study of a 6 DOF wave energy converter interacting with regular waves using 3D CFD. In *Proceedings of the 11th European Wave and Tidal Energy Conference*, Nantes, France, 2015.

[802] M. Vicente, M. Alves, and A. Sarmento. Layout optimisation of wave energy point absorbers arrays. In *Proceedings of the 10th European Wave and Tidal Energy Conference (EWTEC)*, Aalborg, Denmark, 2013.

[803] P. Vicente, A. de O Falcão, L. Gato, and P. Justino. Dynamics of arrays of floating point-absorber wave energy converters with inter-body and bottom slack-mooring connections. *Applied Ocean Research*, 31(4):267–281, 2009.

[804] P. Vicente, A. de O Falcão, and P. Justino. A time domain analysis of arrays of floating point-absorber wave energy converters including the effect of nonlinear mooring forces. In *Proceedings of the 3rd International Conference on Ocean Energy (ICOE)*, Bilbao, Spain, 2010.

[805] D. Vicinanza, L. Margheritini, P. Contestabile, J. P. Kofoed, and P. Frigaard. Seawave Slot-cone Generator: an innovative caisson breakwaters for energy production. In *Coastal Engineering 2008: (In 5 Volumes)*, pages 3694–3705. World Scientific, 2009.

[806] D. Vicinanza, J. H. Nørgaard, P. Contestabile, and T. L. Andersen. Wave loadings acting on overtopping breakwater for energy conversion. *Journal of Coastal Research*, 65(10065):1669–1674, 2013.

[807] A. Viviano, R. E. Musumeci, D. Vicinanza, and E. Foti. Pressures induced by regular waves on a large scale OWC. *Coastal Engineering*, 152:103528, 2019.

[808] A. Wacher and K. Nielsen. Mathematical and numerical modeling of the AquaBuOY wave energy converter. *Mathematics-in-Industry Case Studies Journal.*, 2:16–33, 2010.

[809] W. Wan Nik, O. Sulaiman, R. Rosliza, Y. Prawoto, and A. Muzathik. Wave energy resource assessment and review of the technologies. *International Journal of Energy & Environment*, 2(6):1101–1112, 2011.

[810] C. Wang and B. T. Wang. Floating solutions for challenges facing humanity. In *Proceedings of the International Conference on Sustainable Civil Engineering and Architecture ICSCEA 2019*, pages 3–29. Springer, 2020.

[811] J. Wang, S. Yang, C. Jiang, Q. Yan, and P. Lund. A novel 2-stage dish concentrator with improved optical performance for concentrating solar power plants. *Renewable Energy*, 108:92–97, 2017.

[812] L. Wang, J. Engström, M. Leijon, and J. Isberg. Coordinated control of wave energy converters subject to motion constraints. *Energies*, 9(6):475, 2016.

[813] L. Wang, J. Engström, M. Leijon, and J. Isberg. Performance of arrays of direct-driven wave energy converters under optimal power take-off damping. *AIP Advances*, 6(8):085313, 2016.

[814] L. Wang and J. Isberg. Nonlinear passive control of a wave energy converter subject to constraints in irregular waves. *Energies*, 8(7):6528–6542, 2015.

[815] L. Wang, J. Isberg, and E. Tedeschi. Review of control strategies for wave energy conversion systems and their validation: the wave-to-wire approach. *Renewable and Sustainable Energy Reviews*, 81:366–379, 2018.

[816] L. Wang and J. V. Ringwood. Control-informed ballast and geometric optimisation of a three-body hinge-barge wave energy converter using two-layer optimisation. *Renewable Energy*, 171:1159–1170, 2021.

[817] R. Wang, D. Ning, and Q. Zou. Wave loads on a land-based dual-chamber oscillating water column wave energy device. *Coastal Engineering*, 160:103744, 2020.

[818] Y. Washio, H. Osawa, Y. Nagata, F. Fujii, H. Furuyama, and T. Fujita. The offshore floating type wave power device Mighty Whale: open sea tests. In *Proceedings of the Tenth International Offshore and Polar Engineering Conference*. OnePetro, 2000.

[819] R. Waters, M. Stålberg, O. Danielsson, O. Svensson, S. Gustafsson, E. Strömstedt, M. Eriksson, J. Sundberg, and M. Leijon. Experimental results from sea trials of an offshore wave energy system. *Applied Physics Letters*, 90(3):034105, 2007.

[820] G. N. Watson. *A treatise on the theory of Bessel functions*. Cambridge University Press, 1995.

[821] V. H. Wavegen. Siadar wave energy project Siadar scoping report, 2012.

[822] Wavestar. Wavestar, 08 2021. http://wavestarenergy.com/ (Accessed on 08/08/2021).

[823] J. Weber. Representation of non-linear aero-thermodynamic effects during small scale physical modelling of OWC WECs. In *Proceedings of the 7th European Wave and Tidal Energy Conference*, Porto, Portugal, 2007.

[824] J. Weber, R. Costello, F. Mouwen, J. Ringwood, and G. Thomas. Techno-economic WEC system optimisation - Methodology applied to Wavebob system definition. In *Proceedings of the 3rd International Conference on Ocean Energy*, pages 1–5, Bilbao, Spain, 2010.

[825] J. Weber, F. Mouwen, A. Parish, and D. Robertson. Wavebob—research & development network and tools in the context of systems engineering. In *Proceedings of the 8th European Wave and Tidal Energy Conference (EWTEC)*, pages 416–420, Uppsala, Sweden, 2009.

[826] J. V. Wehausen. Causality and the radiation condition. *Journal of Engineering Mathematics*, 26(1):153–158, 1992.

[827] A. Weinstein, G. Fredrikson, M. Parks, and K. Nielsen. AquaBuOY-the offshore wave energy converter numerical modeling and optimisation. In *Proceedings of the Oceans' 04 MTS/IEEE Techno-Ocean'04 (IEEE Cat. No. 04CH37600)*, volume 4, pages 1854–1859. IEEE, 2004.

[828] S. Weller, T. Stallard, and P. Stansby. Interaction factors for a rectangular array of heaving floats in irregular waves. *IET Renewable Power Generation*, 4(6):628–637, 2010.

[829] Wello. The future of wave energy, 08 2021. https://wello.eu/#:~: text=Wello's%20Penguin%20Wave%20Energy%20converter,the%20middle% 20of%20the%20ocean. (Accessed on 08/08/2021).

[830] F. Wendt, Y.-H. Yu, K. Nielsen, K. Ruehl, et al. International energy agency ocean energy sytems task 10 wave energy converter modeling verification and validation. In *Proceedings of the 12th European Wave and Tidal Energy Conference*, Cork, Ireland, 2017.

[831] J. Westphalen, D. Greaves, C. Williams, P. Taylor, D. Causon, C. Mingham, Z. Hu, P. Stansby, B. Rogers, and P. Omidvar. Extreme wave loading on offshore wave energy devices using CFD: a hierarchical team approach. In *Proceedings of the 8th European Wave and Tidal Energy Conference*, pages 500–508, 2009.

[832] M. Wexler. A sociological framing of the NIMBY (not-in-my-backyard) syndrome. *International Review of Modern Sociology*, pages 91–110, 1996.

[833] C. Whitlam, J. Chapman, A. J. Hillis, J. Roesner, G. Foster, G. Stockman, D. Greaves, and H. Martyn. Validation of variable depth operation of a novel wave energy convertor using scale model testing. In *Proceedings of the 12th European Wave and Tidal Energy Conference (EWTEC)*, pages 838-1–838-6, Cork, Ireland, 2017.

[834] T. Whittaker. Performance of the limpet wave power plant-prediction, measurement & potential. In *Proceedings of the 5th European Wave Power Conference*, 2003.

[835] D. Wilson, G. Bacelli, R. G. Coe, D. L. Bull, O. Abdelkhalik, U. A. Korde, and R. D. Robinett III. A comparison of WEC control strategies. Technical Report SAND2016-4293, Sandia National Laboratories, 2016.

[836] M. Witt, E. Sheehan, S. Bearhop, A. Broderick, D. Conley, S. Cotterell, E. Crow, W. Grecian, C. Halsband, and D. Hodgson. Assessing wave energy effects on biodiversity: the Wave Hub experience. *Philosophical Transactions of the Royal Society A*, 370(1959):502–529, 2012.

[837] H. Wolgamot, M. Meylan, and C. Reid. Multiply heaving bodies in the time-domain: Symmetry and complex resonances. *Journal of Fluids and Structures*, 69:232–251, 2017.

[838] H. Wolgamot, P. Taylor, R. Eatock Taylor, T. Van Den Bremer, A. Raby, and C. Whittaker. Experimental observation of a near-motion-trapped mode: free motion in heave with negligible radiation. *Journal of Fluid Mechanics*, 786, 2016.

[839] H. Wolgamot, P. Taylor, and R. E. Taylor. The interaction factor and directionality in wave energy arrays. *Ocean Engineering*, 47:65–73, 2012.

[840] M. Wolley and J. Platts. Energy on the crest of a wave. *New Scientist*, pages 241 – 243, 1975.

[841] B. Wu, M. Li, R. Wu, T. Chen, Y. Zhang, and Y. Ye. BBDB wave energy conversion technology and perspective in China. *Ocean Engineering*, 169:281–291, 2018.

[842] B. Wu, M. Li, R. Wu, Y. Zhang, and W. Peng. Experimental study on primary efficiency of a new pentagonal backward bent duct buoy and assessment of prototypes. *Renewable Energy*, 113:774–783, 2017.

[843] H. Wu, C. Zhang, Y. Zhu, W. Li, D. Wan, and F. Noblesse. A global approximation to the Green function for diffraction radiation of water waves. *European Journal of Mechanics-B/Fluids*, 65:54–64, 2017.

[844] J. Wu, S. Shekh, N. Sergiienko, B. Cazzolato, B. Ding, F. Neumann, and M. Wagner. Fast and effective optimisation of arrays of submerged wave energy converters. In *Proceedings of the Genetic and Evolutionary Computation Conference (GECCO'16)*, Denver, USA, 2016.

[845] J. Wu, Y. Yao, L. Zhou, N. Chen, H. Yu, W. Li, and M. Göteman. Performance analysis of solo Duck wave energy converter arrays under motion constraints. *Energy*, 139:155–169, 2017.

[846] J. Wu, Y. Yao, L. Zhou, N. Chen, H. Yu, W. Li, and M. Göteman. Performance analysis of solo Duck wave energy converter arrays under motion constraints. *Energy*, 139:155–169, 2017.

[847] J. Wu, Y. Yao, L. Zhou, and M. Göteman. Real-time latching control strategies for the solo Duck wave energy converter in irregular waves. *Applied Energy*, 222:717–728, 2018.

[848] S. Xu, S. Wang, and C. G. Soares. Review of mooring design for floating wave energy converters. *Renewable and Sustainable Energy Reviews*, 111:595–621, 2019.

[849] J. Yang, D.-h. Zhang, Y. Chen, H. Liang, M. Tan, W. Li, and X.-d. Ma. Design, optimisation and numerical modelling of a novel floating pendulum wave energy converter with tide adaptation. *China Ocean Engineering*, 31(5):578–588, 2017.

[850] S.-H. Yang, J. Ringberg, and E. Johnson. Analysis of interaction effects between WECs in four types of wave farms. In G. Soares, editor, *Advances in Renewable Energies Offshore: In Proceedings of the 3rd International Conference on Renewable Energies Offshore (RENEW)*. Taylor & Francis Group, 2019.

[851] S.-H. Yang, J. W. Ringsberg, and E. Johnson. Wave energy converters in array configurations—influence of interaction effects on the power performance and fatigue of mooring lines. *Ocean Engineering*, 211:107294, 2020.

[852] Y. Ye, K. Wang, Y. You, and S. Sheng. Research of power take-off system for "Sharp Eagle II" wave energy converter. *China Ocean Engineering*, 33(5):618–627, 2019.

[853] R. Yemm, D. Pizer, C. Retzler, and R. Henderson. Pelamis: experience from concept to connection. *Philosophical Transactions of the Royal Society A: Mathematical, Physical and Engineering Sciences*, 370(1959):365–380, 2012.

[854] O. Yılmaz and A. Incecik. Analytical solutions of the diffraction problem of a group of truncated vertical cylinders. *Ocean Engineering*, 25:6 (SCI), 1998.

[855] Y. You, S. Sheng, B. Wu, and Y. He. Wave energy technology in China. *Philosophical Transactions of the Royal Society A: Mathematical, Physical and Engineering Sciences*, 370(1959):472–480, 2012.

[856] Y. Yu and S. Liu. *Random wave and it applications to engineering (4th Edition) (in Chinese)*. Dalian University of Technology Press, 2011.

[857] Y.-H. Yu and Y. Li. Reynolds-averaged navier–stokes simulation of the heave performance of a two-body floating-point absorber wave energy system. *Computers & Fluids*, 73:104–114, 2013.

[858] Z. Yu and J. Falnes. State-space modelling of a vertical cylinder in heave. *Applied Ocean Research*, 17(5):265–275, 1995.

[859] D. K. Yue, H. S. Chen, and C. C. Mei. A hybrid element method for diffraction of water waves by three-dimensional bodies. *International Journal for Numerical Methods in Engineering*, 12(2):245–266, 1978.

[860] B. Zanuttigh and E. Angelelli. Experimental investigation of floating wave energy converters for coastal protection purpose. *Coastal Engineering*, 80:148–159, 2013.

[861] B. Zanuttigh, E. Angelelli, A. Kortenhaus, K. Koca, Y. Krontira, and P. Koundouri. A methodology for multi-criteria design of multi-use offshore platforms for marine renewable energy harvesting. *Renewable Energy*, 85:1271–1289, 2016.

[862] B. Zanuttigh, L. Martinelli, M. Castagnetti, P. Ruol, J. P. Kofoed, and P. Frigaard. Integration of wave energy converters into coastal protection schemes. In *Proceedings of the 3rd International Conference on Ocean Energy*, 2010.

[863] C. Zhang and D. Ning. Hydrodynamic study of a novel breakwater with parabolic openings for wave energy harvest. *Ocean Engineering*, 182:540–551, 2019.

[864] D. Zhang, W. Li, and Y. Lin. Wave energy in china: Current status and perspectives. *Renewable Energy*, 34(10):2089–2092, 2009.

[865] H. Zhang and G. Aggidis. Nature rules hidden in the biomimetic wave energy converters. *Renewable and Sustainable Energy Reviews*, 97:28–37, 2018.

[866] Y. Zhang, S. Sheng, Y. You, Z. Huang, and W. Wang. Study of hydrodynamic characteristics of a Sharp Eagle wave energy converter. *China Ocean Engineering*, 31(3):364–369, 2017.

[867] X. Zhao, D. Ning, Q. Zou, D. Qiao, and S. Cai. Hybrid floating breakwater-WEC system: A review. *Ocean Engineering*, 186:106126, 2019.

[868] X. Zhao, R. Xue, J. Geng, and M. Göteman. Analytical investigation on the hydrodynamic performance of a multi-pontoon breakwater-WEC system. *Ocean Engineering*, 220:108394, 2021.

[869] X. Zhao, L. Zhang, M. Li, and L. Johanning. Experimental investigation on the hydrodynamic performance of a multi-chamber OWC-breakwater. *Renewable and Sustainable Energy Reviews*, 150:111512, 2021.

[870] S. Zheng. *Study on Hydrodynamic Characteristics of the Raft-type Wave-Powered Desalination Device*, pages 1–183. Springer Singapore, Singapore, 2018.

[871] S. Zheng, A. Antonini, Y. Zhang, D. Greaves, J. Miles, and G. Iglesias. Wave power extraction from multiple oscillating water columns along a straight coast. *Journal of Fluid Mechanics*, 878:445–480, 2019.

[872] S. Zheng and Y. Zhang. Analysis for wave power capture capacity of two interconnected floats in regular waves. *Journal of Fluids and Structures*, 75:158–173, 2017.

[873] S. Zheng, Y. Zhang, and G. Iglesias. Wave-structure interaction in hybrid wave farms. *Journal of Fluids and Structures*, 83:386–412, 2018.

[874] S. Zheng, Y. Zhang, and W. Sheng. Maximum wave energy conversion by two interconnected floaters. *Journal of Energy Resources Technology*, 138(3):032004, 2016.

[875] S. Zheng, Y. Zhang, Y. Zhang, and W. Sheng. Numerical study on the dynamics of a two-raft wave energy conversion device. *Journal of Fluids and Structures*, 58:271–290, 2015.

[876] Q. Zhong and R. Yeung. Model-predictive control strategy for an array of wave-energy converters. *Journal of Marine Science and Application*, 18(1):26–37, 2019.

[877] Q. Zhong and R. W. Yeung. Wave-body interactions among energy absorbers in a wave farm. *Applied Energy*, 233:1051–1064, 2019.

[878] Y. Zhou, D. Ning, W. Shi, L. Johanning, and D. Liang. Hydrodynamic investigation on an OWC wave energy converter integrated into an offshore wind turbine monopile. *Coastal Engineering*, 162:103731, 2020.

[879] Y. Zhou, C. Zhang, and D. Ning. Hydrodynamic investigation of a concentric cylindrical OWC wave energy converter. *Energies*, 11(4):985, 2018.

[880] H. Zhu. *A Seabased wave energy device: An experimental investigation*. PhD thesis, The University of Waikato, New Zealand, 2020.

[881] X. Zhu. *Irregular frequency removal from the boundary integral equation for the wave-body problem*. PhD thesis, Massachusetts Institute of Technology, 1994.

[882] S. Zou and O. Abdelkhalik. On the control of three-degree-of-freedom wave energy converters. In *Proceedings of the 11th International Conference on Energy Sustainability*, pages V001T07A001–V001T07A001. American Society of Mechanical Engineers, 2017.

[883] S. Zou, O. Abdelkhalik, R. Robinett, U. Korde, G. Bacelli, D. Wilson, and R. Coe. Model predictive control of parametric excited pitch-surge modes in wave energy converters. *International Journal of Marine Energy*, 19:32–46, 2017.

Index

For Product Safety Concerns and Information please contact our EU
representative GPSR@taylorandfrancis.com
Taylor & Francis Verlag GmbH, Kaufingerstraße 24, 80331 München, Germany

www.ingramcontent.com/pod-product-compliance
Ingram Content Group UK Ltd.
Pitfield, Milton Keynes, MK11 3LW, UK
UKHW050926180425
457613UK00003B/34